Advanced Nanoindentation in Materials

Special Issue Editors

Ting Tsui

Matt Pharr

MDPI • Basel • Beijing • Wuhan • Barcelona • Belgrade

MDPI

Special Issue Editors

Ting Tsui
University of Waterloo
Canada

Matt Pharr
Texas A&M University
USA

Editorial Office
MDPI AG
St. Alban-Anlage 66
Basel, Switzerland

This edition is a reprint of the Special Issue published online in the open access journal *Materials* (ISSN 1996-1944) in 2017 (available at: http://www.mdpi.com/journal/materials/special_issues/nanoindentation_materials).

For citation purposes, cite each article independently as indicated on the article page online and as indicated below:

Lastname, F.M.; Lastname, F.M. Article title. *Journal Name*. **Year**. *Article number, page range.*

First Edition 2018

ISBN 978-3-03842-749-0 (Pbk)
ISBN 978-3-03842-750-6 (PDF)

Table of Contents

About the Special Issue Editors

Ting Tsui After completing his doctoral degree from Rice University, Houston, Texas, Professor Ting Tsui joined Advanced Micro Devices (AMD) as a Senior Materials Engineer to study the mechanical reliability of copper/low-k technology for the integrated circuits. His interests in processing technology landed him a position at the Texas Instruments Inc. (TI). At TI, Tsui developed low-k thin films for Plasma Enhanced Chemical Vapor Deposited (PECVD) interlayer dielectric and dielectric barriers for 90 nm, 65 nm, and 45 nm technology node. He is also interested in the small-scale mechanical aspect of these materials, such as thin film channel cracking and delamination. In 2007, Tsui began his appointment at the University of Waterloo and research in the areas of nanoindentaion, biological cell morphology control, cell deformation, porous ultra-low-dielectric constant materials, and nano-mechanics.

Matt Pharr, Assistant Professor, Mechanical Engineering, Texas A&M University Matt Pharr received his Ph.D. at Harvard University in 2014 and was postdoctoral researcher at the University of Illinois at Urbana-Champaign from 2014–2016. In 2016, he joined the department of mechanical engineering at Texas A&M University as an assistant professor (with a courtesy appointment in materials science and engineering). His current research interests include materials for energy storage and conversion, deformation and fracture of soft materials, mechanics of stretchable electronics, coupled electro-chemo-mechanics, and mass transport in materials.

materials MDPI

Review

Micro-Mechanical Response of an Al-Mg Hybrid System Synthesized by High-Pressure Torsion

Megumi Kawasaki [1,2] and Jae-il Jang [1,*]

[1] Division of Materials Science and Engineering, Hanyang University, Seoul 04763, Korea; megumi@hanyang.ac.kr

[2] Departments of Aerospace & Mechanical Engineering and Materials Science, University of Southern California, Los Angeles, CA 90089-1453, USA; mkawasak@usc.edu

* Correspondence: jijang@hanyang.ac.kr; Tel.: +82-2-2220-0402

Academic Editor: Daolun Chen

Received: 10 May 2017; Accepted: 27 May 2017; Published: 30 May 2017

Abstract: This paper summarizes recent efforts to evaluate the potential for the formation of a metal matrix nanocomposite (MMNC) by processing two commercial bulk metals of aluminum and magnesium alloy through high-pressure torsion (HPT) at room temperature. After significant evolutions in microstructures, successful fabrication of an Al-Mg hybrid system was demonstrated by observing unique microstructures consisting of a multi-layered structure and MMNC. Moreover, the evolution of small-scale mechanical properties was examined through the novel technique of nanoindentation and the improvement in plasticity was estimated by calculating the strain rate sensitivity of the Al-Mg hybrid system after HPT. The present paper demonstrates that, in addition to conventional tensile testing, the nanoindentation technique is exceptionally promising for ultrafine-grained materials processed by HPT, where the samples may have small overall dimensions and include heterogeneity in the microstructure.

Keywords: intermetallic composite; grain refinement; hardness; high-pressure torsion; nanocomposite; nanoindentation; plasticity

---

1. Introduction

The processing of metals through the application of severe plastic deformation (SPD) provides the potential for achieving exceptional grain refinement in bulk solids. It is now well accepted that ultrafine-grained (UFG) materials are defined as having submicron grains in the range of 100–1000 nm or nanocrystalline grains with sizes in the range of 10–100 nm [1]. Among the reported SPD techniques, one of the most attractive methods refers to the processing by high-pressure torsion (HPT) [2], where this type of processing leads to exceptional grain refinement that is not generally achieved using other procedures [3]. Therefore, HPT has been applied for the consolidation of metallic powders [4–8] and bonding of machining chips [9,10], whereas very limited reports examined the application of HPT for the fabrication of hybrid systems including nanocomposites [11,12].

The fundamental principles of HPT processing were described in detail in an earlier review [2]. Specifically, the sample in a disk shape is applied to receive both very high compressive straining and torsion straining concurrently. Numerous reports are now available demonstrating that bulk materials after HPT having ultrafine and nanometer grain sizes generally show superior mechanical properties, which include high strength dictated through the Hall–Petch relationship at low temperatures and a superplastic forming capability at high temperatures due to possible fast diffusion.

The development of micro-mechanical behavior is observed after significant changes in microstructure through SPD processing and it is of great importance for obtaining practical future applications of these UFG metals. Accordingly, recent developments in characterization techniques lead to a better understanding of the evolution in the mechanical properties of UFG materials processed by SPD. In particular, the novel technique of nanoindentation has become a common tool for the simultaneous measurement of a number of mechanical properties by using a small microstructural scale [13]. Thus, there have been several studies reporting the use of the nanoindentation technique for investigating the plastic deformation properties of SPD-processed metals including Al alloys [14–18], pure Cu [19], pure Cr [20], a ZK60 magnesium alloy [21], pure Nb [22], a Pb-62%Sn eutectic alloy [23], a Zn-22% Al eutectoid alloy [24] and high entropy alloys (HEAs) [25–27], while numerous studies demonstrated conventional mechanical testing in tension and compression. A recent review summarized the available experimental results showing the enhancement in the micro-mechanical response at room temperature (RT) observed by the nanoindentation technique in a range of metals and alloys after several different SPD processing procedures [28].

Accordingly, a new approach of applying conventional HPT processing was demonstrated under quasi-constrained conditions [29,30] for synthesizing a hybrid system and ultimately forming an intermetallic-based metal matrix nanocomposite (MMNC) from separate conventional metals of Al and Mg through diffusion bonding at RT by HPT. It should be emphasized that the synthesis of intermetallic compounds in the present study involves the bulk-state reaction of metals [31], which is different from the procedure of mechanical alloying of powder through the consolidation of powders using HPT [32,33]. The unique microstructure and exceptional hardness was demonstrated in the Al-Mg hybrid system synthesized by HPT through 20 turns. A study of post-deformation annealing (PDA) was applied to determine the microstructural change and influence in mechanical properties in the synthesized alloy system. Moreover, the enhancement in micro-scale deformation behavior was examined through the nanoindentation experiments on the Al-Mg system after HPT and after PDA. Special emphasis is placed on demonstrating the evolution of the micro-mechanical responses in the hybrid system by measuring the strain rate sensitivity.

2. Synthesis of an Al-Mg Hybrid Metal System through Diffusion Bonding

2.1. Microstructural Evolution and Hardness Development

Figure 1 shows optical micrographs demonstrating overviews of the microstructure taken at the cross-sections of the Al-Mg disks after HPT for 1, 5, 10 and 20 turns and 20 HPT turns followed by PDA at 573 K for 1 h from the top [34,35].

A disk after one turn showed a multi-layered structure with fragmented Mg layers with thicknesses of ~200 μm without any segregation at the Al-Mg interfaces throughout the disk diameter. A similar formation of multi-layered microstructure was observed at the central regions at $r < 2.0$ mm after five turns and ~1.0 mm after 10 and 20 turns, where r denotes the radius of the HPT disk. By contrast, the disk peripheries at $r > 2.5$ mm after five turns demonstrated a homogeneous distribution of very fine Mg phases having thicknesses of ~5–10 μm to even true nano-scale sizes of ~100–500 nm within the Al matrix. Furthermore, these fine Mg phases disappeared at the disk edges, and there was no evidence of visible Mg phases at ~$3 < r < 5$ mm after 10 and 20 turns. A similar microstructure of multi-layered formation towards complete mixture of Al-Mg phases along the radial direction was observed after HPT for 20 turns followed by PDA, whereas the outer region was reduced. Accordingly, the synthesized Al-Mg system after HPT and after PDA consists of gradient-type microstructures involving microstructural heterogeneity across the disk diameters.

Detailed microstructural analysis was conducted at the disk edges of $r \approx 4.0$–4.5 mm and the results are shown in Figures 2 and 3 for the Al-Mg system after HPT for 5–10 turns and 20 turns and additional PDA, respectively. After five turns of HPT, true nanostructure was already achieved as shown in Figure 2a, where a layered microstructure is demonstrated at the edge of the Al-Mg disk. The layers have thicknesses of 90–120 nm, and these layers contain numerous dislocations subdividing the layers in a vertical sense. The measurements showed an average grain size of ~190 nm in the Al matrix phase. In this micrograph, a single Mg phase is visible that has a rigid bonding interface with the Al matrix without any visible voids. Moreover, within the Al matrix phase, several thin nano-layers are observed with an average thickness of ~20 nm as indicated by the white arrows.

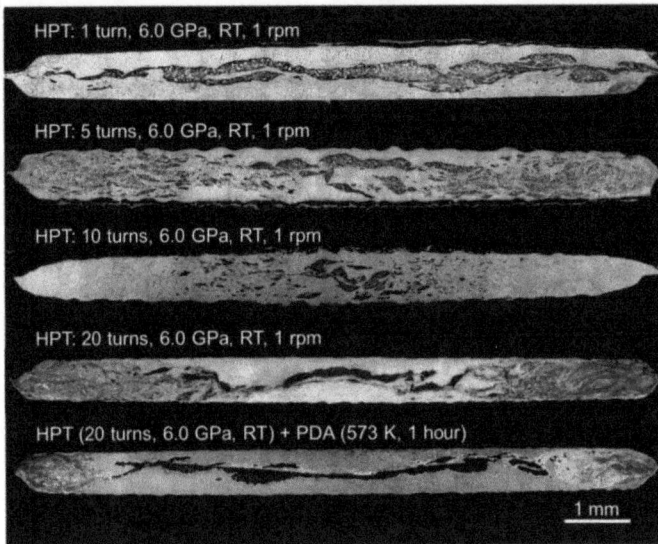

Figure 1. The vertical cross-sections of the Al–Mg system after HPT under a pressure of 6.0 GPa at room temperature for, from the top, 1, 5, 10, 20 turns and for 20 HPT turns followed by PDA [34,35].

Figure 2b shows a finer microstructure after 10 HPT turns at the edge of the Al-Mg disk. There was no evidence of an Mg-rich phase and an average grain size of ~90 nm was observed where, although it is not included in the micrograph, there was a similar type of thin layers as shown in the disk edge after five turns. An earlier study examined these thin layers in the Al matrix closely by element mapping and quantitative chemical analysis [34]. The detailed analysis revealed that the thin layers are composed of an intermetallic compound of β-Al_3Mg_2 that has a low density of ~2.25 g cm^{-3}. Since the thin layers existed randomly in the Al matrix [35], the HPT processing synthesized the intermetallic-based MMNC at the disk edge of the Al-Mg system.

After 10 HPT turns, carefully prepared disk edge was examined by XRD and the profile is shown in Figure 2c [34]. The result was examined by the MAUD software [36] so that the analysis quantified to give 73.4 ± 2.3% of Al, 4.9 ± 0.7% of Mg and 21.7 ± 1.5% of the intermetallic compound γ-$Al_{12}Mg_{17}$. It should be noted that the XRD analysis was not able to detect the presence of β-Al_3Mg_2 due to the small content that is smaller than the detectable limit. Moreover, although the disk edge was carefully prepared, the Mg-rich phase close to the central region was detected, whereas there is no presence of Mg-rich phase at the edge region in the Al-Mg system after HPT for 10 turns. The changes in lattice parameter of Al were calculated through the analysis to provide an estimate of the Mg solubility in the Al solid solution, and it is revealed that the Al matrix in the disk edge includes the supersaturate amount of Mg. Thus, the results showed that the disk edge after 10 turns involves two different types of intermetallic compounds forming a new type of MMNC.

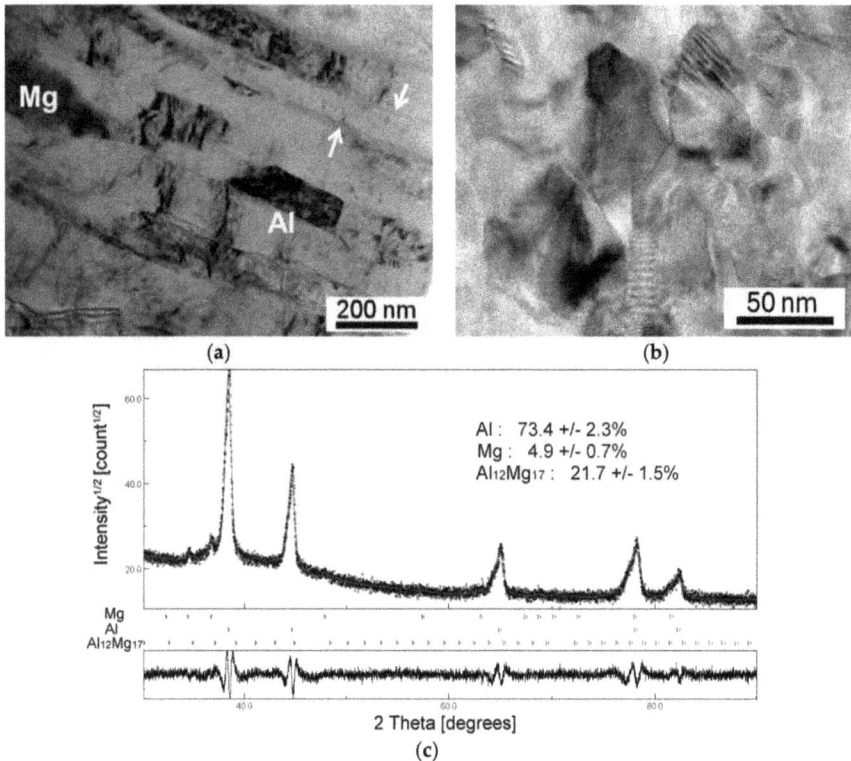

Figure 2. TEM bright-field images taken at the Al-Mg disk edges after HPT for (**a**) five turns and (**b**) 10 turns and (**c**) the XRD profile with the MAUD estimation for the disk edge after 10 HPT turns [34].

The TEM micrographs taken at the peripheral region after HPT for 20 turns are shown in Figure 3a,b, where there is a mixture of microstructures with a nanolayered structure and an equiaxed grain structure, respectively. In practice, the nanolayered microstructure has an average thickness of ~20 nm and these layers contain numerous dislocations vertically. The equiaxed grains showed an average grain size of ~60 nm. By contrast, a PDA treatment demonstrated an apparent microstructural recovery, so that it is apparent from Figure 3c that the Al-Mg system contained a homogeneous equiaxed microstructure with an average grain size of ~380 nm [37].

The X-ray profiles are shown in Figure 3d,e for the carefully prepared Al-Mg disk edges after 20 turns by HPT and after HPT followed by PDA, respectively, where additional compositional analysis based on the X-ray profile through MOUD was displayed as a table in each plot [37]. It should be reminded once more that these disk edges after HPT for 20 turns do not include any Mg-rich phase, whereas the inevitable concentrations of the Mg-rich phase existed close to the mid-radius of the processed samples.

The disk edge immediately after HPT for 20 turns showed that there is evidence of γ-$Al_{12}Mg_{17}$ in the Al matrix, whereas, after PDA, there is an Al-7% Mg solid solution phase with two different intermetallic compounds of β-Al_3Mg_2 and γ-$Al_{12}Mg_{17}$. Thus, these results suggest that processing by HPT for 20 turns and additional PDA produced two different types of deformation-induced MMNCs containing intermetallic compounds at the disk edges of the Al-Mg system, and these MMNCs are different from those observed after five and 10 turns as shown in Figure 2. Moreover, the experimental results anticipate that the formation of these intermetallic compounds provides an excellent potential for reinforcing the Al matrix by improving the hardness and strength at the disk edges of the Al-Mg system.

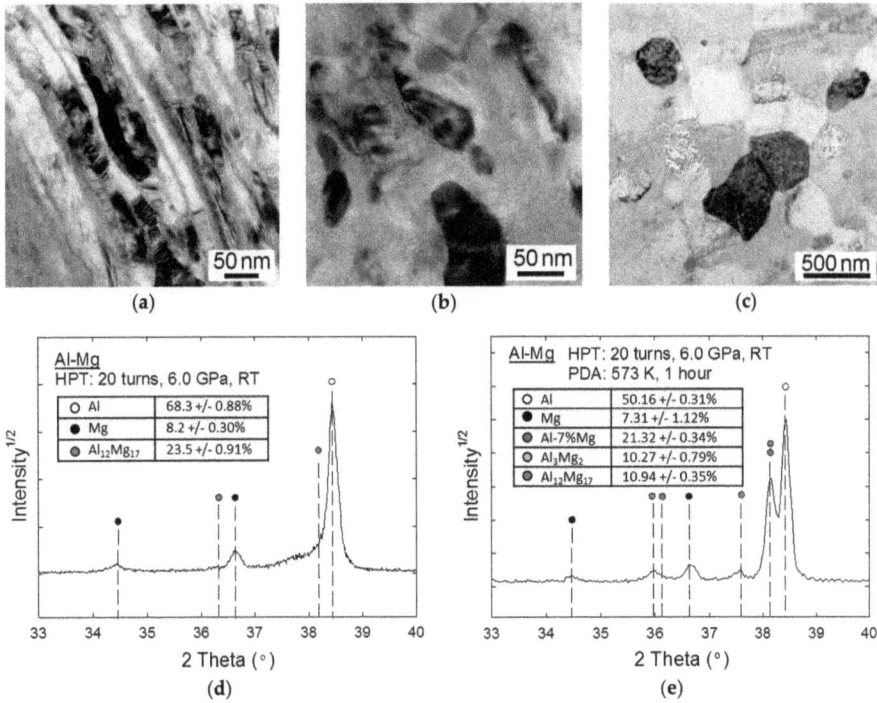

Figure 3. Representative TEM bright-field images taken at the disk edges after (**a,b**) HPT for 20 turns and (**c**) HPT followed by PDA in the Al-Mg system and the X-ray diffraction profiles for the disk edges of the Al-Mg system after (**d**) HPT for 20 turns and (**e**) HPT and PDA [37].

2.2. Hardness Development in the Al-Mg Hybrid System

The distributions of Vickers microhardness were examined over the vertical cross-sections of the processed Al-Mg disks as were shown in Figure 1, and the data set was visualized by constructing color-coded hardness contour maps that are then overlapped with the micrographs. These hardness maps are shown in Figure 4 for the disks after HPT for, from the top, 1, 5, 10 and 20 turns and after HPT followed by PDA, respectively. The hardness values are indicated in the color key on the right. For reference, the Al-1050 alloy and the ZK60 alloy show a saturation hardness of Hv ≈ 65 [38] and ~110 [39] across the disk diameters after HPT for five turns providing sufficient torsional straining.

The overall cross-section of the Al-Mg system after HPT for one turn shows an average microhardness value of ~70, which is similar to the saturated hardness value of ~65 for the base material of the Al alloy processed by HPT. This hardness value remains constant at the centers at $r < 2.5$–3.0 mm of the Al-Mg disks up to 20 turns. On the contrary, processing by HPT for five turns tends to introduce high hardness with a maximum of Hv ≈ 130 at the peripheral region where the fine Mg phase is homogeneously distributed within the Al matrix. Moreover, there is a significant increase in hardness after 10 turns where a maximum hardness of ~270 was recorded at the peripheral region at $r > 3.0$ mm. Moreover, exceptionally high hardness of Hv ≈ 330 was observed at $r > 3.0$ mm, where then the hardness shows the transition to ~150–240 at $r ≈ 2.5$–3.0 mm followed by Hv ≈ 60–70 at the central region at $r < 2.5$ mm. After PDA, the HPT-processed Al-Mg disk for 20 turns showed a slight reduction in hardness to Hv ≈ 30 in the central region at $r \leq 3.0$ mm and to Hv ≈ 220 at the peripheral region at $r \approx 4.0$–5.0 mm.

These high values of hardness at the Al-Mg disk peripheries are much higher than the highest achievable hardness in the base alloys of Al and Mg after HPT, and it is anticipated by the concurrent occurrences of grain refinement, solid solution hardening and precipitation hardening by the intermetallic compounds. It should be emphasized that the rapid diffusivity of Mg atoms into the Al matrix is a key process for the diffusion bonding of Al and Mg and for the formation of intermetallic compounds through HPT [35]. Limited recent studies demonstrated experimental evidence for enhanced atomic diffusion in bulk nanostructured materials processed by equal-channel angular pressing (ECAP) [40] and HPT [25,34]. The fast diffusivity in the SPD-processed materials may be attributed to the processing conditions including severe hydraulic pressure and a limited temperature rise during processing [34] and torsional stress [25] during HPT processing and the high population of lattice defects produced in the nanostructure [40]. A recent review describes the significance of the fast atomic mobility during SPD by recognizing the drastic increase in the vacancy concentration within the processed materials [41].

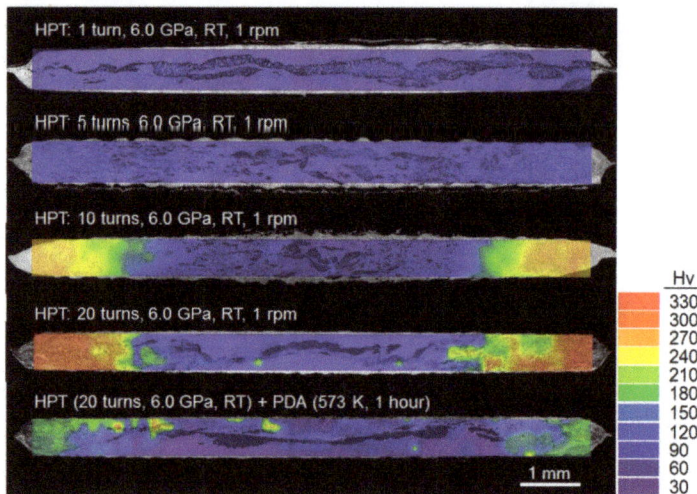

Figure 4. Color-coded hardness contour maps taken at the vertical cross-sections along the disk diameters after HPT for 1, 5, 10 and 20 turns [34] and after HPT for 20 turns followed by PDA.

3. Micro-Mechanical Properties of the Al-Mg Hybrid System

The micro-mechanical response was examined using a nanoindentation technique for the Al-Mg system processed by HPT and after HPT with subsequent PDA. Figure 5 shows the representative curves of load, P, versus displacement, h, measured at an equivalent strain rate of 1.0×10^{-3} s^{-1} at the specific phases of (a) Al and (b) Mg at the disk center and (c) a mixture of Al and Mg phases and an intermetallic compound of β-Al$_3$Mg$_2$ at the disk edge after HPT for five turns [42]. It should be noted that the discontinuity of each curve at the final stage of unloading is due to thermal expansion, and these effects can be omitted from the analysis.

A series of indentation curves for the Al and Mg layers shows less broadening between the fifteen separate measurements as seen in Figure 5a,b, respectively. Therefore, it is demonstrated that there is a reasonably constant mechanical response within the phases in the multi-layered microstructure at the disk center. It is also apparent from these plots that higher ductility is shown in the Al phase with larger displacements under a fixed load in comparison with the Mg phase in the central region after HPT for five turns.

By contrast, Figure 5c shows a wide deviation in micro-mechanical response for all fifteen measurements, thereby indicating the plastic instability at the disk edge forming an MMNC in the Al-Mg system after five turns. It should be noted that, although the instability is demonstrated, all independent measurements showed smaller displacements than the separate Al and Mg phases without any intermetallic compound at the disk center as shown in Figure 5a,b. Thereby, it demonstrates higher hardness at the disk edge reinforced by an intermetallic compound and it is consistent with the hardness results demonstrated in Figure 4. Moreover, this plastic instability was observed at all indentation rates applied in the present measurements, although the slower strain rate shows a lower tendency.

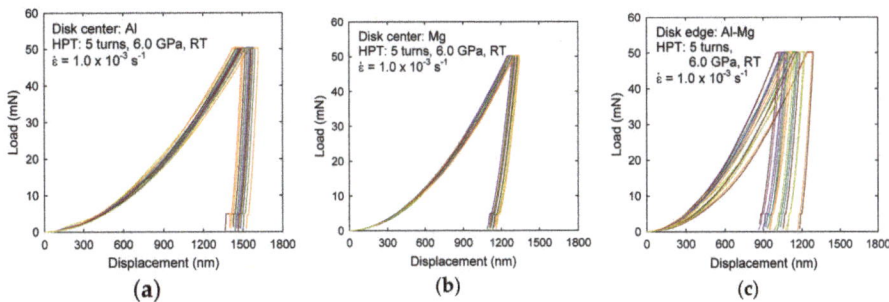

Figure 5. Representative load-displacement curves for (**a**) Al and (**b**) Mg phases at the disk center and for (**c**) the Al-Mg system forming an MMNC at the disk edge when testing at 1.0×10^{-3} s^{-1} for the sample after five turns by HPT [42].

Figure 6 shows representative load-displacement curves for the Al-Mg disk edges after (a) HPT for 20 turns and (b) HPT followed by PDA when measuring at four equivalent strain rates at 1.25×10^{-4}–1.0×10^{-3} s^{-1}. The disk edge consists of an MMNC after 20 HPT turns showed all separate P-h curves at all four strain rates placed in reasonably consistent locations as shown in Figure 6a, thereby indicating no strain rate dependency of plasticity in the strain rate range. By contrast, a different type of MMNC processed by HPT followed by PDA revealed an apparent positive strain rate dependency where there are increasing displacements at slower strain rates of nanoindentation, as shown in Figure 6b. It is also apparent by comparing the load-displacement curves between the two samples that the MMNC immediately after HPT shows much lower displacements than the MMNC after HPT and PDA at all strain rates, thereby demonstrating the high hardness of the MMNC in the Al-Mg disk edge immediately after HPT for 20 turns. This result is fully consistent with the Vickers hardness measurements as shown in Figure 4.

It should be noted that no plastic instability was observed in both MMNCs at the disk edges after HPT for 20 turns and after PDA, which is in contrast with an MMNC in the Al-Mg system after HPT for five turns, as shown in Figure 5c. This apparent dichotomy is probably due to the higher volume fractions and the homogeneous distributions of the hard and fine phases of two intermetallic compounds within the microstructure after higher numbers of HPT turns.

Additional examinations were conducted by nanoindentation to demonstrate the hardness variations along the disk thicknesses at the central regions of the Al-Mg system after HPT for 20 turns and additional PDA. Specifically, the average values of nanoindentation hardness, H, were determined from three separate indentations recorded at uniformly separated points by a distance of 150 μm at the same distance from the mid-section in the disk height.

The optical micrographs showing the multi-layers at the disk centers with indentation marks are shown in Figure 7a,b for the Al-Mg disks after HPT for two turns and after HPT followed by PDA, respectively. It is apparent that the diffusion bonding of the Al and Mg phases were satisfactory so that there was no crack initiation by the severe stress introduced by the nanoindentation process at the interfaces in the multi-layers after 20 HPT turns. On the contrary, although there is consistent strong bonding at the interfaces, there are visible layers with an average thickness of <30 μm at the Al-Mg interfaces as shown in Figure 7b.

The variations of the measured indentation hardness are demonstrated in Figure 7c. The hardness immediately after HPT for 20 turns showed the apparent hardness changes at the Al-Mg interface boundaries. In practice, the hardness at the mid-section of the Al-Mg disk showed the higher hardness measured from the Mg phase, which is directly reduced with distance from the mid-section due to the phase change to Al. However, at the disk center after HPT and PDA, there is an exceptionally high hardness of over 4.0 MPa at the interphase boundaries, and it was determined as the β-Al$_3$Mg$_2$ by the chemical analysis.

Thus, although the hardness at the Al and Mg phases are reduced due to the microstructural recovery by PDA, it is concluded that the additional PDA process introduces further development in the diffusion-bonded interfaces in the Al-Mg system and demonstrating an excellent potential for constructing a unique microstructural formation throughout the disk diameter in such metal systems produced by HPT.

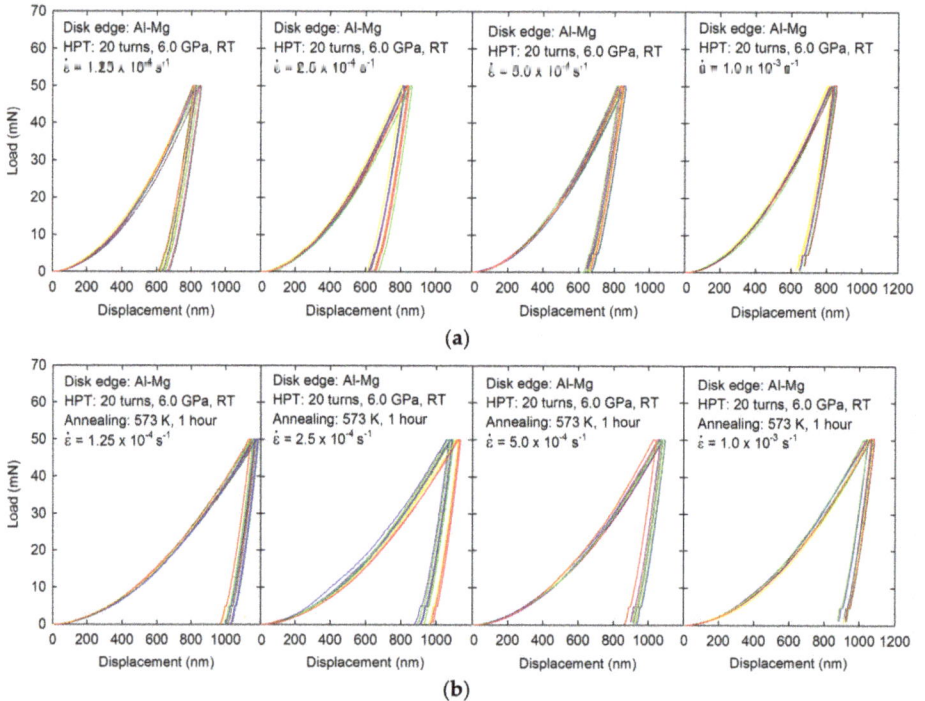

Figure 6. Representative load-displacement curves for the Al-Mg disk edges after (**a**) HPT for 20 turns and (**b**) HPT and PDA when measuring at four strain rates at $1.25 \times 10^{-4} - 1.0 \times 10^{-3}$ s^{-1}.

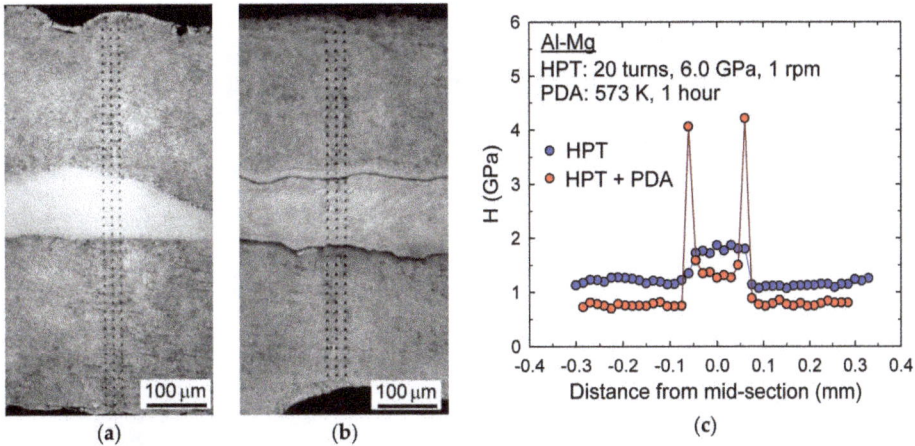

Figure 7. Optical micrographs showing the Al-Mg multi-layers at the disk centers with indentations for the samples after (**a**) HPT for 20 turns and (**b**) HPT and PDA and (**c**) variations of the nanoindentation hardness values for these two sample conditions of the Al-Mg system.

4. Discussion

4.1. Improvement in Micro-Mechanical Response by PDA

The deformation characteristics at RT were evaluated by calculating the strain rate sensitivity, m, from the data set of nanoindentation testing shown in Figures 5 and 6 for the disk edges in the Al-Mg system after HPT for five and 20 turns, respectively. The value of m was determined at a given strain, ε, and absolute temperature, T, by considering Tabor's empirical prediction showing that the flow stress is equivalent to H/3 for fully plastic deformation at a constant strain rate $\dot{\varepsilon}$ [43], where H is the nanoindentation hardness estimated according to the Oliver–Pharr method [44]:

$$m = \left(\frac{\partial \ln \sigma_f}{\partial \ln \dot{\varepsilon}}\right)_{\varepsilon,T} = \left(\frac{\partial \ln(H/3)}{\partial \ln \dot{\varepsilon}}\right)_{\varepsilon,T} \tag{1}$$

Thus, the values of m were calculated from the slopes of the lines in a logarithmic plot of H/3 against $\dot{\varepsilon}$ as shown in Figure 8 for the disk edges of the Al-Mg system after HPT for (a) five turns [42] and (b) 20 turns and after HPT and PDA at 573 k for 1 h [37]. It should be noted that the error bar on each datum point represents the standard deviation of the numbers of measurements, whereas the error ranges are too small to recognize in Figure 8b.

The strain rate sensitivity was estimated as $m \approx 0.01$ at the disk edge after HPT for five turns as shown in Figure 8a. Moreover, with wider error bars due to the plastic instability especially with increasing indentation strain rates, the estimations imply the possibility of a much smaller strain rate sensitivity in the Al-Mg system. However, the analysis estimated the m values of -0.001 and 0.1 for the MMNCs after HPT for 20 turns and HPT followed by PDA, respectively, as shown in Figure 8b. Thus, the strain rate sensitivity was reduced with increasing HPT turns, and, thereafter, the PDA treatment provided a significant enhancement in the m value of the MMNC in the Al-Mg system processed by HPT. A recent review tabulated the available data of the strain rate sensitivity, m, in a series of UFG metals processed by SPD [28].

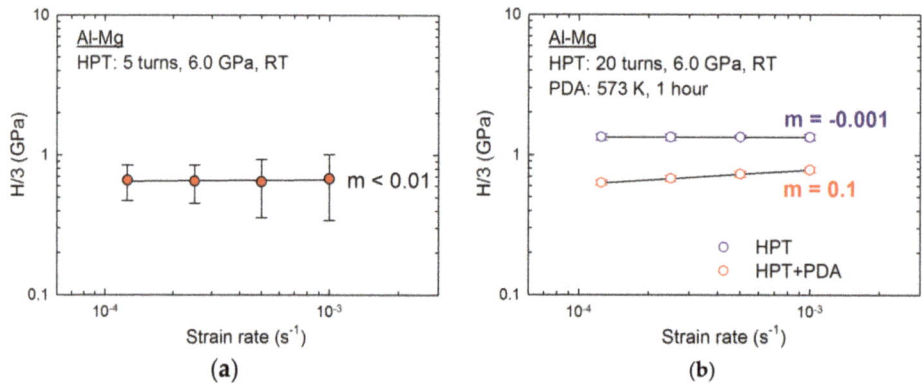

Figure 8. Variations of the strain rate sensitivity with increasing strain rate for the disk edges of the Al-Mg system after HPT for (**a**) five turns [42] and (**b**) 20 turns and after HPT and PDA [37].

There is a limited report of a negative strain rate sensitivity for an MMNC of a powder consolidated aluminum 6092/B_4C when tested at strain rates of <1.0 s^{-1} [45]. The report suggested the occurrence of dynamic strain ageing (DSA) due to the presence of the fast diffusion of solute atoms interacting with mobile dislocations. Thus, at the present MMNC in the Al Mg system immediately after HPT, decreasing plasticity demonstrated by the reduced strain rate sensitivity is reasonable because of the formation of dislocation junctions at the solute clusters, which is attributed to interaction of a significant number of dislocations created during HPT with the very rapid Mg solutes within Al matrix.

The improved *m* value of 0.1 for the PDA-treated MMNC in the Al-Mg system processed by HPT is even higher than the reported *m* values through nanoindentation testing of ~0.07 for a commercial purity Al after ECAP for 6–12 passes at RT [14,15,17,18,20] and after accumulative roll bonding (ARB) for eight cycles at RT [15] and ~0.035–0.050 for a ZK60 alloy after HPT for two turns at RT [21]. Thus, the MMNC after HPT followed by PDA demonstrates a significantly higher value of *m* compared with the initial materials when they are processed separately.

The present study demonstrated that a PDA treatment is feasible for enhancing the plasticity of the MMNCs in the Al-Mg system after HPT while maintaining reasonably high hardness as shown in Figure 4. In practice, this approach produces an ordering of the defect structures within the grain boundaries leading to an equilibrium state without any significant grain growth [46]. In addition, short-term annealing reduces the dislocation density in the grain interior of the UFG material after SPD so that the dislocation storage capability may increase and thus the strain hardening capability is enhanced. Thus, this leads to the possibility of high ductility in the SPD-processed material and there are several recent reports demonstrating the significance of PDA on mechanical properties at RT of HPT-processed materials [47–50].

4.2. Future Potential of a Nanoindentation Technique for UFG Metals

Recent developments in characterization techniques result in a better understanding of the enhancement in the mechanical properties of UFG materials processed by SPD. In particular, the novel technique of nanoindentation has become a common tool for the simultaneous measurement of a number of mechanical properties on the material surfaces at the submicron scale. [13]. Accordingly, there have been several studies to date demonstrating the use of the nanoindentation technique for examining the mechanical properties and parameters of SPD-processed metals [15], while numerous studies demonstrated the conventional procedures of mechanical testing in tension and compression using bulk UFG metals.

The attraction of a nanoindentation technique arises from the requirement of limited volumes, which is favorable for the SPD-processed materials where, with the current laboratory-scale studies on the SPD techniques, the tensile specimens machined from SPD-processed samples have often failed to meet the geometry defined by the American Society for Testing and Materials (ASTM) standard [51]. Moreover, several recent studies demonstrated a new scale of engineering materials processed by SPD techniques where the microstructure includes a gradient-type nanostructure according to grain size and composition [34,52]. This type of structural organization is defined as heterogeneous architecture materials [53] and is also demonstrated as a new type of structure in engineering materials leading to a significant potential for exhibiting excellent mechanical properties [54–57]. Thus, for understanding the local mechanical properties with microstructural gradations in length scales, the nanoindentation technique has been indispensable for measuring the specific mechanical response using a limited volume of material at arbitrarily selected local points.

Finally, it is reasonable to note that a recent review of nanoindentation describes the role, significance and feasibility of the nanoindentation technique [15]. Moreover, very recent reports describe current advances and capabilities of the novel nanoindentation techniques for measuring thermally-activated deformation mechanisms ranging from single crystalline to nanocrystalline metals through strain rate jump testing and long term creep testing [58] and for operating at high temperature in situ in a scanning electron microscope (SEM) [59].

5. Materials and Methods

A commercial purity Al-1050 alloy and a ZK60 magnesium alloy were used for the experiments. The extruded bars of the alloys having a diameter of 10 mm were cut into billets with lengths of ~65 mm and a set of disks was sliced from the billets and polished to achieve uniform thicknesses of ~0.83 mm. The direct bonding of the Al and Mg disks was performed through the application of the conventional HPT technique at RT following the general processing procedure described earlier [60] under a hydraulic pressure of 6.0 GPa for 1 to 20 turns at a rotational speed of 1 rpm. In practice, separate disks of the Al and Mg alloys were placed in the depression on the lower anvil on the order of Al/Mg/Al, where the Mg disk was positioned between the two Al disks but without using any glue or metal brushing treatment. Figure 9 shows (a) a picture of the HPT machine and (b) the stacked disks of Al and Mg between the conventional set-up of the HPT anvils [34,35]. In order to evaluate the effect of PDA, some of the HPT-processed disks after 20 turns were annealed at 573 K for 1 h.

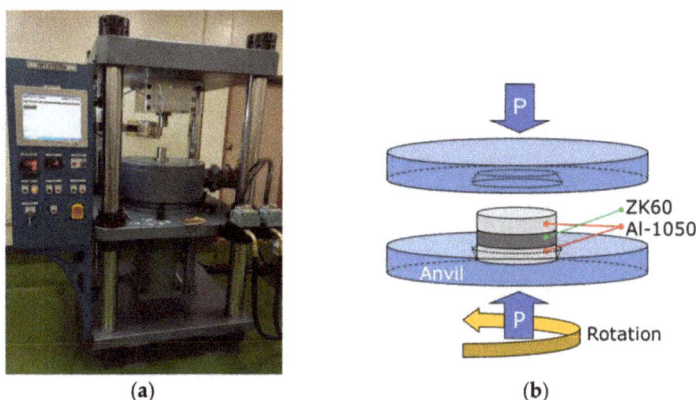

(a)   (b)

Figure 9. (a) Photograph of the HPT facility and (b) schematic illustration of the sample set-up for HPT processing [34,35].

Following processing, vertical cross-section from each processed disk was polished, chemically etched using Keller's etchant and examined by optical microscopy (OM). Subsequently, the values of Vickers microhardness, Hv, were recorded over the vertical cross-sections of the disks after HPT and HPT followed by PDA using a Shimazu HMV-2 facility (Kyoto, Japan) with a load of 50 gf. These individual microhardness values were recorded following a rectilinear grid pattern with an incremental spacing of 0.2 mm.

An elemental analysis was conducted using energy-dispersive X-ray spectroscopy (EDS) in a field emission scanning electron microscope (FE-SEM), FEI Nova NanoSEM 450 (Hillsboro, OR). The detailed microstructure was investigated by FE-SEM, JEOL JSM-6700F (Tokyo, Japan), in the peripheral region near the outer edge of the disks on the vertical cross-section after preparing using a broad ion beam cross-section polisher, JEOL IB-09020CP, with 6 kV Ar ion beam and 30° swing angle of specimen stage to minimize beam striations on the strain-free polished surface. An X-ray diffraction (XRD) analysis was performed using a Rigaku UltimaIV XRD (Tokyo, Japan) on the slightly polished surface of each disk. The examination used a CuKα radiation with a scanning speed of 3° min^{-1} and a step interval of 0.01°. Microstructural cell parameters and phase percentages were quantified by means of the XRD data analysis software, Materials Analysis Using Diffraction (MAUD), which is based on a full pattern fitting procedure (Rietveld method). Additional microstructural analysis was conducted by transmission electron microscopy (TEM) using a spherical aberration (Cs) corrected JEOL JEOM-2100 F with 200 kV accelerating voltage for a specimen prepared by an in situ lift-out technique using OmniProbe 200 (Oxfordshire, U.K.) and Omni gas injection system (GIS) in a focused ion beam (FIB), JEOL JIB-4300.

The micro-mechanical response was examined at the disk centers and the MMNC at the disk edges in the Al-Mg system after HPT for 5 and 20 turns and after PDA at RT using a nanoindentation facility, Nanoindenter-XP (formerly MTS; now Keysight, Santa Rosa, CA, USA), with a three-sided pyramidal Berkovich indenter having a centerline-to-face angle of 65.3°. More than 15 indentations were conducted at each specific phase at the measured locations to provide statistically valid data. All measurements were conducted under a predetermined peak applied load of P_{max} = 50 mN at constant indentation strain rates of 0.0125, 0.025, 0.05 and 0.1 s^{-1}, which are equivalent to general strain rates of 1.25×10^{-4}, 2.5×10^{-4}, 5.0×10^{-4} and 1.0×10^{-3} s^{-1} calculated through an empirical relationship [61,62].

6. Conclusions

There is a considerable potential for making use of HPT for the introduction of new alloy systems, especially for fabricating a wide range of hybrid materials. The nanoindentation technique provides a wide range of information including mechanical properties and the local microstructure. In addition to conventional tensile testing, this technique is promising for UFG materials processed by HPT, where the materials may have smaller overall dimensions and include gradient-type microstructures. Further investigations are needed to fully develop this approach.

Acknowledgments: The authors would like to thank Terence G. Langdon for technical discussion. This work was supported by the National Research Foundation Korea funded by Ministry of Education under Grant No. NRF-2016R1A6A1A03013422 and by Ministry of Science, ICT and Future Planning under Grant No. NRF-2016K1A4A3914691 (M.K.); and the NRF Korea funded by MSIP under Grant No. NRF-2015R1A5A1037627 and NRF-2017R1A2B4012255 (J.I.J.).

Author Contributions: Jae-il Jang supervised a series of experiments and Megumi Kawasaki analyzed the data and wrote the paper.

Conflicts of Interest: The authors declare no conflict of interest.

References

1. Valiev, R.Z.; Estrin, Y.; Horita, Z.; Langdon, T.G.; Zehetbauer, M.J.; Zhu, Y.T. Producing bulk ultrafine-grained materials by severe plastic deformation. *JOM* **2006**, *58*, 33–39. [CrossRef]

2. Zhilyaev, A.P.; Langdon, T.G. Using high-pressure torsion for metal processing: Fundamentals and applications. *Prog. Mater. Sci.* **2008**, *53*, 893–979. [CrossRef]

3. Langdon, T.G. Twenty-five years of ultrafine-grained materials: Achieving exceptional properties through grain refinement. *Acta Mater.* **2013**, *61*, 7035–7059. [CrossRef]

4. Korznikov, A.V.; Safarov, I.M.; Laptionok, D.V.; Valiev, R.Z. Structure and properties of superfine-grained iron compacted out of ultradisperse powder. *Acta Metall. Mater.* **1991**, *39*, 3193–3197. [CrossRef]

5. Stolyarov, V.V.; Zhu, Y.T.; Lowe, T.C.; Islamgaliev, R.K.; Valiev, R.Z. Processing nanocrystalline Ti and its nanocomposites from micrometer-sized Ti powder using high pressure torsion. *Mater. Sci. Eng. A* **2000**, *282*, 78–85. [CrossRef]

6. Edalati, K.; Horita, Z.; Fujiwara, H.; Ameyama, K. Cold consolidation of ball-milled titanium powders using high-pressure torsion. *Metall. Mater. Trans. A* **2010**, *41*, 3308–3317. [CrossRef]

7. Cubero-Sesin, J.M.; Horita, Z. Powder consolidation of Al–10 wt % Fe alloy by high-pressure torsion. *Mater. Sci. Eng. A* **2012**, *558*, 462–471. [CrossRef]

8. Zhilyaev, A.P.; Ringot, G.; Huang, Y.; Cabrera, J.M.; Langdon, T.G. Mechanical behavior and microstructure properties of titanium powder consolidated by high-pressure torsion. *Mater. Sci. Eng. A* **2017**, *688*, 498–504. [CrossRef]

9. Zhilyaev, A.P.; Gimazov, A.A.; Raab, G.I.; Langdon, T.G. Using high-pressure torsion for the cold-consolidation of copper chips produced by machining. *Mater. Sci. Eng. A* **2008**, *486*, 123–126. [CrossRef]

10. Edalati, K.; Yokoyama, Y.; Horita, Z. High-pressure torsion of machining chips and bulk discs of amorphous Zr$_{50}$Cu$_{30}$Al$_{10}$Ni$_{10}$. *Mater. Trans.* **2010**, *51*, 23–26. [CrossRef]

11. Beygelzimer, Y.; Estrin, Y.; Kulagin, R. Synthesis of hybrid materials by severe plastic deformation: A new paradigm of SPD processing. *Adv. Eng. Mater.* **2015**, *17*, 1853–1861. [CrossRef]

12. Bachmaier, A.; Pippan, R. Generation of metallic nanocomposites by severe plastic deformation. *Int. Mater. Rev.* **2013**, *58*, 41–62. [CrossRef]

13. Schuh, C.A. Nanoindentation studies of materials. *Mater. Today* **2006**, *9*, 32–40. [CrossRef]

14. Mueller, J.; Durst, K.; Amberger, D.; Göken, M. Local Investigations of the Mechanical Properties of Ultrafine Grained Metals by Nanoindentations. *Mater. Sci. Forum* **2006**, *503–504*, 31–36. [CrossRef]

15. Böhner, A.; Maier, V.; Durst, K.; Höppel, H.W.; Göken, M. Macro- and nanomechanical properties and strain rate sensitivity of accumulative roll bonded and equal channel angular pressed ultrafine-grained materials. *Adv. Eng. Mater.* **2011**, *13*, 251–255. [CrossRef]

16. Maier, V.; Durst, K.; Mueller, J.; Backes, B.; Höppel, H.W.; Göken, M. Nanoindentation strain-rate jump tests for determining the local strain-rate sensitivity in nanocrystalline Ni and ultrafine-grained Al. *J. Mater. Res.* **2011**, *26*, 1421–1430. [CrossRef]

17. Wheeler, J.M.; Maier, V.; Durst, K.; Göken, M.; Michler, J. Activation parameters for deformation of ultrafine-grained aluminium as determined by indentation strain rate jumps at elevated temperature. *Mater. Sci. Eng. A* **2013**, *585*, 108–113. [CrossRef]

18. Maier, V.; Merle, B.; Göken, M.; Durst, K. An improved long-term nanoindentation creep testing approach for studying the local deformation processes in nanocrystalline metals at room and elevated temperatures. *J. Mater. Res.* **2013**, *28*, 1177–1188. [CrossRef]

19. Chen, J.; Lu, L.; Lu, K. Hardness and strain rate sensitivity of nanocrystalline Cu. *Scr. Mater.* **2006**, *54*, 1913–1918. [CrossRef]

20. Maier, V.; Hohenwarter, A.; Pippan, R.; Kiener, D. Thermally activated deformation processes in body-centered cubic Cr–How microstructure influences strain-rate sensitivity. *Scr. Mater.* **2015**, *106*, 42–45. [CrossRef]

21. Choi, I.-C.; Lee, D.-H.; Ahn, B.; Durst, K.; Kawasaki, M.; Langdon, T.G.; Jang, J.-I. Enhancement of strain-rate sensitivity and shear yield strength of a magnesium alloy processed by high-pressure torsion. *Scr. Mater.* **2015**, *94*, 44–47. [CrossRef]

22. Alkorta, J.; Martínez-Esnaola, J.M.; Sevillano, J.G. Critical examination of strain-rate sensitivity measurement by nanoindentation methods: Application to severely deformed niobium. *Acta Mater.* **2008**, *56*, 884–893. [CrossRef]

23. Zhang, N.X.; Chinh, N.Q.; Kawasaki, M.; Huang, Y.; Langdon, T.G. Self-annealing in a two-phase Pb-Sn alloy after processing by high-pressure torsion. *Mater. Sci. Eng. A* **2016**, *666*, 350–359. [CrossRef]

24. Choi, I.-C.; Kim, Y.-J.; Ahn, B.; Kawasaki, M.; Langdon, T.G.; Jang, J.-I. Evolution of plasticity, strain-rate sensitivity and the underlying deformation mechanism in Zn–22% Al during high-pressure torsion. *Scr. Mater.* **2014**, *75*, 102–105. [CrossRef]

25. Lee, D.-H.; Choi, I.-C.; Seok, M.-Y.; He, J.; Lu, Z.; Suh, J.-Y; Kawasaki, M.; Langdon, T.G.; Jang, J.-I. Nanomechanical behavior and structural stability of a nanocrystalline CoCrFeNiMn high-entropy alloy processed by high-pressure torsion. *J. Mater. Res.* **2015**, *30*, 2804–2815. [CrossRef]

26. Lee, D.-H.; Seok, M.-Y.; Zhao, Y.; Choi, I.-C.; He, J.; Lu, Z.; Suh, J.-Y.; Ramamurty, U.; Kawasaki, M.; Langdon, T.G.; et al. Spherical nanoindentation creep behavior of nanocrystalline and coarse-grained CoCrFeMnNi high-entropy alloys. *Acta Mater.* **2016**, *109*, 314–322. [CrossRef]

27. Maier-Kiener, V.; Schuh, B.; Clemens, H.; Hohenwarter, A. Nanoindentation testing as a powerful screening tool for assessing phase stability of nanocrystalline high-entropy alloys. *Mater. Des.* **2017**, *115*, 479–485. [CrossRef]

28. Kawasaki, M.; Ahn, B.; Kumar, P.; Jang, J.-I.; Langdon, T.G. Nano- and Micro- mechanical properties of ultrafine-grained materials processed by severe plastic deformation techniques. *Adv. Eng. Mater.* **2017**, *19*, 1600578. [CrossRef]

29. Figueiredo, R.B.; Cetlin, P.R.; Langdon, T.G. Using finite element modeling to examine the flow processes in quasi-constrained high-pressure torsion. *Mater. Sci. Eng. A* **2011**, *528*, 8198–8204. [CrossRef]

30. Figueiredo, R.B.; Pereira, P.H.R.; Aguilar, M.T.P.; Cetlin, P.R.; Langdon, T.G. Using finite element modeling to examine the temperature distribution in quasi-constrained high-pressure torsion. *Acta Mater.* **2012**, *60*, 3190–3198. [CrossRef]

31. Oh-ishi, K.; Edalati, K.; Kim, H.S.; Hono, K.; Horita, Z. Diffusion and promoting solid-state reactions in aluminum-copper system. *Acta Mater.* **2013**, *61*, 3482–3489. [CrossRef]

32. Edalati, K.; Toh, S.; Iwaoka, H.; Watanabe, M.; Horita, Z.; Kashioka, D.; Kishida, K.; Inui, H.

33. Alfreider, M.; Jeong, J.; Esterl, R.; Oh, S.H.; Kiener, D. Synthesis and Mechanical characterisation of an ultra-fine grained Ti-Mg composite. *Materials* **2016**, *9*, 688. [CrossRef]

34. Ahn, B.; Zhilyaev, A.P.; Lee, H.-J.; Kawasaki, M.; Langdon, T.G. Rapid synthesis of an extra hard metal matrix nanocomposite at ambient temperature. *Mater. Sci. Eng. A* **2015**, *635*, 109–117. [CrossRef]

35. Kawasaki, M.; Ahn, B.; Lee, H.-J.; Zhilyaev, A.P.; Langdon, T.G. Using high-pressure torsion to process an aluminum-magnesium nanocomposite through diffusion bonding. *J. Mater. Res.* **2016**, *31*, 88–99. [CrossRef]

36. Lutterotti, L. Total pattern fitting for the combined size-strain-stress-texture determination in thin film diffraction. *Nucl. Instrum. Methods Phys. Res. Sect. B* **2010**, *268*, 334–340. [CrossRef]

37. Han, J.-K.; Lee, H.-J.; Jang, J.-I.; Kawasaki, M.; Langdon, T.G. Micro-mechanical and tribological properties of aluminum-magnesium nanocomposites processed by high-pressure torsion. *Mater. Sci. Eng. A* **2017**, *684*, 318–327. [CrossRef]

38. Kawasaki, M.; Alhajeri, S.N.; Xu, C.; Langdon, T.G. The development of hardness homogeneity in pure aluminum and aluminum alloy disks processed by high-pressure torsion. *Mater. Sci. Eng. A* **2011**, *529*, 345–351. [CrossRef]

39. Lee, H.-J.; Lee, S.K.; Jung, K.H.; Lee, G.A.; Ahn, B.; Kawasaki, M.; Langdon, T.G. Evolution in hardness and texture of a ZK60A magnesium alloy processed by high-pressure torsion. *Mater. Sci. Eng. A* **2015**, *630*, 90–98. [CrossRef]

40. Divinski, S.V.; Reglitz, G.; Rösner, H.; Estrin, Y.; Wilde, G. Ultra-fast diffusion channels in pure Ni severely deformed by equal-channel angular pressing. *Acta Mater.* **2011**, *59*, 1974–1985. [CrossRef]

41. Sauvage, X.; Wilde, G.; Divinski, S.V.; Horita, Z.; Valiev, R.Z. Review: Grain boundaries in ultrafine grained materials processed by severe plastic deformation and related phenomena. *Mater. Sci. Eng. A* **2012**, *540*, 1–12. [CrossRef]

42. Ahn, B.; Lee, H.-L.; Choi, I.-C.; Kawasaki, M.; Jang, J.-I.; Langdon, T.G. Micro-mechanical behavior of an exceptionally strong metal matrix nanocomposite processed by high-pressure torsion. *Adv. Eng. Mater.* **2016**, *18*, 1001–1008. [CrossRef]

43. Shim, S.; Jang, J.-I.; Pharr, G.M. Extraction of flow properties of single-crystal silicon carbide by nanoindentation and finite-element simulation. *Acta Mater.* **2008**, *56*, 3824–3832. [CrossRef]

44. Oliver, W.C.; Pharr, G.M. An improved technique for determining hardness and elastic modulus using load and displacement sensing indentation experiments. *J. Mater. Res.* **1992**, *7*, 1564–1583. [CrossRef]

45. Zhang, H.; Ramesh, K.T.; Chin, E.S.C. High strain rate response of aluminum 6092/B$_4$C composites. *Mater. Sci. Eng. A* **2004**, *384*, 26–34. [CrossRef]
46. Kumar, P.; Kawasaki, M.; Langdon, T.G. Review: Overcoming the paradox of strength and ductility in ultrafine-grained materials at low temperatures. *J. Mater. Sci.* **2016**, *51*, 7–18. [CrossRef]
47. Andreau, O.; Gubicza, J.; Zhang, N.X.; Huang, Y.; Jenei, P.; Langdon, T.G. Effect of short-term annealing on the microstructures and flow properties of an Al–1% Mg alloy processed by high-pressure torsion. *Mater. Sci. Eng. A* **2014**, *615*, 231–239. [CrossRef]
48. Maury, N.; Zhang, N.X.; Huang, Y.; Zhilyaev, A.P.; Langdon, T.G. A critical examination of pure tantalum processed by high-pressure torsion. *Mater. Sci. Eng. A* **2015**, *638*, 174–182. [CrossRef]
49. Huang, Y.; Lemang, M.; Zhang, N.X.; Pereira, P.H.R.; Langdon, T.G. Achieving superior grain refinement and mechanical properties in vanadium through high-pressure torsion and subsequent short-term annealing. *Mater. Sci. Eng. A* **2016**, *655*, 60–69. [CrossRef]
50. Shahmir, H.; He, J.; Lu, Z.; Kawasaki, M.; Langdon, T.G. Effect of annealing on mechanical properties of a nanocrystalline CoCrFeNiMn high-entropy alloy processed by high-pressure torsion. *Mater. Sci. Eng. A* **2016**, *676*, 294–303. [CrossRef]
51. ASTM. Available online: http://www.astm.org (accessed on 29 May 2017).
52. Kang, J.Y.; Kim, J.G.; Park, H.W.; Kim, H.S. Multiscale architectured materials with composition and grain size gradients manufactured using high-pressure torsion. *Sci. Rep.* **2016**, *6*, 26590. [CrossRef] [PubMed]
53. Bouaziz, O.; Bréchet, Y.; Embury, J.D. Heterogeneous and architectured materials: A possible strategy for design of structural materials. *Adv. Eng. Mater.* **2008**, *10*, 24–36. [CrossRef]
54. Fang, T.H.; Li, W.L.; Tao, N.R.; Lu, K. Revealing extraordinary intrinsic tensile plasticity in gradient nano-grained copper. *Science* **2011**, *331*, 1587–1590. [CrossRef] [PubMed]
55. Wu, X.L.; Jiang, P.; Chen, L.; Yuan, F.P.; Zhu, Y.T. Extraordinary strain hardening by gradient structure. *Proc. Natl. Acad. Sci. USA* **2011**, *111*, 7197–7201. [CrossRef] [PubMed]
56. Lu, K. Making strong nanomaterials ductile with gradients. *Science* **2014**, *345*, 1455–1456. [CrossRef] [PubMed]
57. Wu, X.L.; Jiang, P.; Chen, L.; Zhang, J.F.; Yuan, F.P.; Zhu, Y.T. Synergetic strengthening by gradient structure. *Mater. Res. Lett.* **2015**, *2*, 185–191. [CrossRef]
58. Durst, K.; Maier, V. Dynamic nanoindentation testing for studying thermally activated processes from single to nanocrystalline metals. *Curr. Opin. Solid State Mater. Sci.* **2015**, *19*, 340–353. [CrossRef]
59. Wheeler, J.M.; Armstrong, D.E.J.; Heinz, W.; Schwaiger, R. High temperature nanoindentation: The state of the art and future challenges. *Curr. Opin. Solid State Mater. Sci.* **2015**, *19*, 354–366. [CrossRef]
60. Kawasaki, M.; Langdon, T.G. The significance of strain reversals during processing by high-pressure torsion. *Mater. Sci. Eng. A* **2008**, *498*, 341–348. [CrossRef]
61. Wang, C.L.; Lai, Y.H.; Huang, J.C.; Nieh, T.G. Creep of nanocrystalline nickel: A direct comparison between uniaxial and nanoindentation creep. *Scr. Mater.* **2010**, *62*, 175–178. [CrossRef]
62. Choi, I.-C.; Yoo, B.-G.; Kim, Y.-J.; Seok, M.-Y.; Wang, Y.M.; Jang, J.-I. Estimating the stress exponent of nanocrystalline nickel: Sharp vs. spherical indentation. *Scr. Mater.* **2011**, *65*, 300–303. [CrossRef]

materials

MDPI

Article

Ultra High Strain Rate Nanoindentation Testing

Pardhasaradhi Sudharshan Phani [1,*] and Warren Carl Oliver [2]

[1] International Advanced Research Centre for Powder Metallurgy and New Materials (ARCI), Balapur PO, Hyderabad, Telangana 500005, India

[2] Nanomechanics Inc., 105 Meco Ln, Oak Ridge, TN 37830, USA; warren.oliver@nanomechanicsinc.com

* Correspondence: spphani@yahoo.com; Tel.: +91-40-2445-2418; Fax: +91-40-2444-3168

Received: 20 May 2017; Accepted: 14 June 2017; Published: 17 June 2017

Abstract: Strain rate dependence of indentation hardness has been widely used to study time-dependent plasticity. However, the currently available techniques limit the range of strain rates that can be achieved during indentation testing. Recent advances in electronics have enabled nanomechanical measurements with very low noise levels (sub nanometer) at fast time constants (20 µs) and high data acquisition rates (100 KHz). These capabilities open the doors for a wide range of ultra-fast nanomechanical testing, for instance, indentation testing at very high strain rates. With an accurate dynamic model and an instrument with fast time constants, step load tests can be performed which enable access to indentation strain rates approaching ballistic levels (i.e., 4000 1/s). A novel indentation based testing technique involving a combination of step load and constant load and hold tests that enables measurement of strain rate dependence of hardness spanning over seven orders of magnitude in strain rate is presented. A simple analysis is used to calculate the equivalent uniaxial response from indentation data and compared to the conventional uniaxial data for commercial purity aluminum. Excellent agreement is found between the indentation and uniaxial data over several orders of magnitude of strain rate.

Keywords: high strain rate; nanoindentation; aluminum alloy; dynamics

---

1. Introduction

Measuring the strain rate dependence of flow stress is of great interest to the materials community and has been a widely-studied research area [1]. The strain rate dependence of flow stress of bulk materials can be routinely measured over a wide range using many conventional techniques like uniaxial compression/tension for lower strain rates and Split-Hopkinson pressure bar for high strain rates [2]. However, these techniques are not readily applicable for small-scale structures or small volumes of materials, which has been a recent area of focus for the materials community. Several groups have used micro/nano impact testing [3–5] or dynamic indentation to understand the high strain rate behavior without necessarily using the depth sensing capability and they mostly fall under the microindentation regime [6–11]. Techniques based on nanoindentation such as constant strain rate test, strain rate jump test, constant rate of loading test, or a constant load and hold test have been widely used to measure the rate dependence of hardness of small volumes of materials over a range of strain rates [12–16]. These techniques are typically limited to the lower strain rate regimes (<1 1/s). In order to access high strain rates during indentation, a step load test can be performed, wherein, the load is ramped within a few micro seconds [14]. This results in sweeping across a wide range of strain rates especially in the high strain rate regime (>100 1/s) in a single test. This could be a powerful technique to measure the high strain rate response at small scales in a simple, quick, and cost-effective way.

While the step load test theoretically offers a great opportunity to probe the material response at high strain rates, there are several experimental challenges, such as the dynamic contribution of the instrument and the time constants of the measurement signals, which need to be carefully considered in order to make valid measurements. Recent advances in electronics have enabled nanomechanical measurements with very low noise levels (sub nanometer) at fast time constants (20 μs) and high data acquisition rates (100 KHz). These capabilities open the doors for a wide range of ultra-fast nanomechanical testing. In addition to having a measuring instrument with fast response, a comprehensive model for the dynamics of the instrument and measurement electronics is required to make accurate high strain rate measurements. In this work, we present a step load-based indentation high strain rate measurement technique that relies on fast response instrumentation and a comprehensive model for the instrument's dynamics and electronics. High strain rate tests are performed on annealed commercial purity aluminum alloy (1100 Al) to demonstrate the technique. A simple analysis is used to calculate the uniaxial equivalent response from the indentation results and is compared to the conventional high strain rate tests to assess the accuracy of this technique. The relative contribution of the instrument's dynamic response and the time constants of the measurement signals to the overall measurement are also discussed to demonstrate the importance of accurate instrument characterization for high strain rate indentation testing.

2. Experimental Procedure and Calculations

2.1. Measuring Strain Rate Dependence of Hardness

As mentioned earlier, indentation based techniques have been widely used to measure the strain rate dependence of hardness in the lower strain rate regime (<1 1/s). The strain rate during an indentation test is often defined as the ratio of the indenter velocity to the depth of indentation. Accessing higher strain rates requires higher indentation velocities at a given depth, or a lower depth for a given velocity or a combination of both. In order to minimize the contributions from indentation size effect (ISE) [17], it is preferable to access higher strain rates by achieving higher velocities at large depths. This can be accomplished by performing a step load test wherein the force is ramped as fast as the actuator used can physically accomplish the change. This results in sweeping a wide range of strain rates in the high strain rate regime in a single indentation test. The major requirements to perform these tests is a testing system with fast response actuators and sensors and a model for instrument's dynamics and electronics to accurately factor out the instrument's contribution from the measured response. These will be described in greater detail in the subsequent subsections.

Step load tests and the conventional constant load and hold (CLH) indentation tests were performed to a static load of 16 mN to cover a wide range of indentation strain rates. In both the cases, the tip is brought in contact with the sample at a slow approach rate of 200 nm/s. The CLH tests were performed by ramping the load to 16 mN at a loading rate of 5 N/s after contact and subsequently maintaining a constant force for 30 s. In the case of step load tests, a step force of 16 mN was input to the force actuator after contact. Note that the step load tests are not impact tests as the tip approaches the sample slowly before contact and the fast loading is only after contact. Unlike the CLH test, due to the fast loading in a step load test, there are significant inertial effects which result in actual load on sample being much higher than the applied step force of 16 mN for a short span. This is immediately followed by a decrease in the load on the sample due to the exhaustion of the dynamic forces, resulting in unloading of the contact even while the actuator applies the 16 mN force. This will be discussed in greater detail in Section 3.1. All the tests were performed on a 10 mm diameter commercial purity aluminum (1100 aluminum) sample which was polished and subsequently annealed at 350 °C for 4 h before testing. Ten repetitive step load and CLH tests were performed to ensure repeatability in the data. A diamond Berkovich tip (Micro Star Technologies, Huntsville, TX, USA) was used for all the tests. The load frame stiffness determination and tip area calibration was done using the results of constant strain rate tests (0.2 1/s) on fused silica. For these tests, the contact stiffness was continuously measured as a function of depth using a phase lock amplifier (Nanomechanics Inc., Oak Ridge, TN, USA) oscillating at 100 Hz frequency and a 2 nm displacement amplitude.

2.2. Measuring Instrumentation

In order to perform a step load test, the testing system requires a force actuator that can apply the desired force in a short time interval, which is typically less than a millisecond, and a displacement sensor that can accurately capture the rapid change in the displacement during that time. In addition, a high data acquisition rate is required. In a typical commercially-available nanoindentation system with an electromagnetic actuator, force is controlled by the current to the coil, which is a command signal, and the displacement is measured by a capacitance gage, which is a measured signal. To perform a step load test, a step function in current is sent as a command input to the actuator. Due to the finite time constant of the force signal, the actual force delivered by the actuator is not an instantaneous step function, but an exponential function with a finite rise time. The time constant of a signal is a parameter that characterizes the response of a signal to a unit step input. For first order linear-time invariant systems, it is the time required to reach 63% of its step input value or one third the time required to reach 95% of its step input. Hence, if the time scale of the test is comparable to the time constants of the measurement signals, corrections to the signals are required. In order to minimize these corrections, it is desirable to have measurement time constants much shorter than the time scale of measurement.

The step load tests in the current work were performed using a commercially available nanoindenter, iNano® from Nanomechanics Inc., Oak Ridge, TN, USA. It uses an electromagnetically-actuated InForce50 actuator with a force time constant of 290 μs, a displacement time constant of 20 μs, and data acquisition rate of 100 kHz. Force is the command signal and the displacement is the measured signal. The displacement sensor has sub nanometer noise levels even at a short time constant of 20 μs. This is critical for high strain rate testing as the velocity and acceleration are calculated by taking the first and second derivative of the displacement signal and any noise is amplified, especially by the second derivative.

2.3. Model for the Instrument's Dynamics and Electronics

At high strain rates, the instrument's dynamic contribution can dominate the measured response and can lead to inaccuracies in the measurement. In order to account for the instrument's dynamic contribution, a simple one degree of freedom (one DOF) damped harmonic oscillator model is proposed to model the electromagnetically-actuated indentation system. A schematic of the model is shown in Figure 1. The actuator is modeled as a single mass, spring, and dashpot system where the mass, m, is the moving mass of the coil and the indenter shaft, damping coefficient, b, is the damping generated due to resistance to the motion of the air in the capacitance gage and the eddy current damping in an electromagnetic actuator and the spring constant, k, is the spring constant of the leaf springs that support the indenter shaft. The sample or the contact is modeled as a spring and dashpot, which represents the contact stiffness and damping, respectively. As the moving mass of the sample is very small compared to the mass of the system, it is neglected. The load frame which holds the indentation system is modeled as a spring for simplicity.

The one DOF model shown in Figure 1 is one of the simplest possible dynamic models for an indentation system and given the complexity of most indentation systems, demonstrating that the testing system can be accurately described by this model is critical for high strain rate testing where the instruments contribution can dominate the measured response. Once the simple model for the actuator is validated its dynamic contribution can be simply factored out to accurately determine the response of the sample.

In order to validate the one DOF model for the actuator, a frequency sweep experiment is performed, wherein the actuator is excited dynamically at a fixed sinusoidal force oscillation amplitude over a wide range of frequency in free air (i.e., without a sample), The resultant dynamic displacement amplitude and the phase lag between the force and displacement signals are measured using a phase lock amplifier (PLA). Figure 2a shows the results of a typical frequency sweep experiment for an InForce50 actuator, wherein the measured dynamic compliance of the instrument in free air, which is the ratio of the dynamic displacement amplitude (h_0) to the dynamic force amplitude (F_0), is plotted as a function of the excitation frequency (ω). This is commonly referred to as the transfer function of the instrument. The dynamic compliance (C) and phase (ϕ) for a one DOF oscillator can be theoretically calculated using the following equations.

$$C = h_0 / F_0 = \sqrt{((k - m\omega^2)^2 + (b\omega)^2)} \tag{1}$$

$$\phi = \tan^{-1}\left(\frac{b\omega}{k - m\omega^2}\right) \tag{2}$$

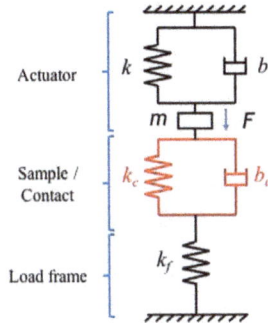

Figure 1. Schematic of the one DOF model for the indentation system showing the various dynamic elements used for the actuator, sample/contact, and the load frame.

The experimental data shown in the plot can be fit to a functional form given in Equation (1) to assess the suitability of using the one DOF model for the instrument. The solid red line in the plot shows the one DOF model fit to the experimental data. This plot clearly demonstrates that the actuator can be accurately modeled as a simple one DOF oscillator. The mass (m), damping coefficient (b), and spring constant (k) of the actuator used for the current work are 180.45 mg, 0.106 Ns/m, and 243 N/m, respectively. Figure 2b shows the experimental data and the model prediction for the phase angle between the displacement and force signals, reinforcing the excellent agreement in the results observed from the transfer function plot and also demonstrating the accuracy of the phase angle measurements. While the one DOF model presented here is simple, it can even be used under non-ambient conditions (vacuum or elevated temperature) for electromagnetic actuators as they do not require a medium for damping. This enables high strain rate experiments under vacuum or elevated temperatures.

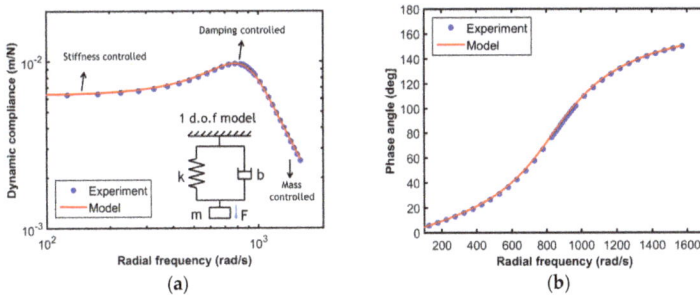

Figure 2. Comparison of experimental data and one DOF model data for (**a**) dynamic compliance, and (**b**) phase angle of the actuator, as a function of radial frequency for a typical InForce50 actuator.

In addition to the dynamic effects of the instrument, understanding and accounting for the time constants of the measurement signals is important for making accurate high strain rate measurements. As discussed in Section 2.2, time constant correction to the signals is required if the time scale of testing is comparable to the measurement time constant. A simple first order correction can be performed to the signal to account for the finite time constant, by the following equation:

$$y_{corr} = y_{msd} + \left(\dot{y}_{msd}\right)/2\tau \tag{3}$$

where y_{corr} is the corrected signal, y_{msd} is the measured signal, $\dot{y}_{msd}$ is the rate of change of the measured signal, and τ is the time constant.

While the transfer function plot demonstrates that the instrument can be modeled as a one DOF oscillator, it does not prove the accuracy of the model parameters m, b, and k when applied in conjunction with the time constants of the force and displacement signals which can have a significant effect during fast testing. In order to verify the accuracy of the time constant correction and dynamic model, a step load test is performed in free air wherein the force command signal is instantaneously stepped to 1 mN and the resultant displacement response is recorded at 100 kHz. Figure 3 shows the comparison of the experimental data for a typical InForce50 actuator and model prediction for a one DOF model at different time constants. There is excellent agreement between the experimental data and model predictions for a 20 µs time constant displacement signal. This demonstrates the validity of the model for the instrument's dynamics, as well as the electronics and, hence, can be used to accurately factor out the contribution of the instrument to the total measurement which will be discussed in greater detail in Section 2.4. The plot also shows the predicted response for 1 ms and 200 ms time constant displacement signals which show significant deviation from the actual response. Note that the data acquisition rate for all of the curves is 100 kHz and the difference is only in the time constant. While having a high data acquisition rate is important, this plot clearly demonstrates the need for a short time constant to accurately capture dynamic events.

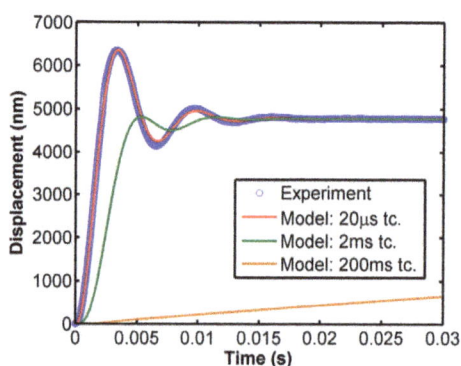

Figure 3. Comparison of experimental data and model predictions at different displacement time constants (tc.) for the step load response of the actuator in free air for a typical InForce50 actuator.

2.4. Calculation of the Indentation Strain Rate and Hardness

This section presents the procedure for calculating the indentation strain rate and hardness from the basic measurements, viz., force and displacement. As mentioned earlier, indentation strain rate is the ratio of indenter velocity ($\dot{h}$) to the depth of penetration (h). The depth of penetration is calculated by subtracting the displacement of the surface point from the measured displacement. The velocity ($\dot{h}$) and acceleration ($\ddot{h}$) are calculated by taking the analytical first and second derivative, respectively, of the spline fit to the displacement-time response. In order to calculate the load on the sample, the dynamic contribution of the instrument needs to be factored out. This can be done by using the simple one DOF model for the actuator which has been shown to accurately describe the system in Section 2.3. The equation to calculate the load on the sample (*P*) factoring out the dynamic contribution of the instrument, is as follows:

$$P = F - kh - b\dot{h} - m\ddot{h} \tag{4}$$

where *F* is the force output of the actuator and *h* is the depth of penetration into the surface. Note that the load on the sample can be very different from the force output of the actuator depending on the dynamic contributions of the instrument. This difference is quite significant for the step load tests where the inertial term ($m\ddot{h}$) dominates. For the case of the constant load and hold tests the contribution from the damping ($b\dot{h}$) and inertial terms is negligible as expected for static indentation.

Once the load on sample is calculated from Equation (4), the hardness is calculated using the conventional formula of load over the contact area. The contact area is calculated from the tip area function assuming a constant ratio of contact depth to total depth which, for this case, is found to be 0.99 based on the unloading data of the CLH tests.

2.5. Estimating Equivalent Uniaxial Response

The experimental procedure and calculations described in the earlier sections enable the determination of hardness as a function of the indentation strain rate. However, in order to compare the data to the conventional measurements, which are based on uniaxial testing, uniaxial equivalent parameters have to be estimated from the indentation data, which is challenging given the complexity of the stress fields during indentation. Recently, Su et al. [18] proposed a simple experimental technique based on the theoretical analysis of Bower et al. [19] to determine uniaxial creep parameters from indentation. The equations used to calculate the uniaxial equivalent strain rate and stress are presented here while the details can be found in Su et al. [18]. The equivalent uniaxial strain rate ($\dot{\varepsilon}$) and stress (σ) for a power-law creeping solid can be calculated from the basic indentation measurements (*h* and *P*) using the following equations:

$$\dot{\varepsilon} = \left(\frac{1}{c \tan \theta} \right) \left(\frac{1}{h} \right) \left(\frac{dh}{dt} \right) \tag{5}$$

$$\sigma = \left(\frac{1}{Fc^2} \right) \left[\frac{P}{\pi (h \tan \theta)^2} \right] \tag{6}$$

In the above equations, *θ* is the equivalent half cone angle, which is 70.3° for a Berkovich indenter and *F* and *c* are akin to constraint factor and pile-up/sink-in parameter, respectively. *F* and *c* are a function of the stress exponent and cone angle, and their functional dependence can be obtained from the recent work of Su et al. [18].

3. Results and Discussion

3.1. Step Load and Constant Load and Hold Tests: Basic Measurements

The depth of penetration into the sample as a function of the time on the sample in response to a 16 mN step force input is shown in Figure 4. Data from 10 tests recorded at 100 kHz is shown in the plot. Excellent repeatability can be observed from the plot. The depth increases to a maximum value and subsequently oscillates, wherein the contact is being unloaded and reloaded without causing further penetration. Hence, the depth up to the point of first maximum represents the elastoplastic regime which is of interest in the current analysis. The maximum depth is reached in less than 500 μs, after which the contact oscillates and hence having fast response and low noise displacement sensors is very important to perform these tests. The load on the sample during this time can be calculated by factoring out the dynamic contribution of the instrument from the force generated by the actuator using Equation (4).

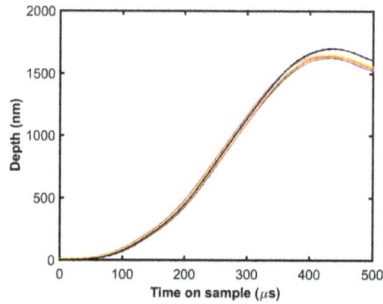

Figure 4. Indentation depth as a function of the time on the sample during the step force input of 16 mN for 10 different tests shown in different colors.

Figure 5 shows the load on sample calculated using Equation (4) for the step load tests. Good repeatability can be observed even in this case, in spite of having to use the first and second derivatives of the displacement signal for load calculations. Similar to the depth-time response shown in Figure 4, the load on sample reaches a maximum value which is at the peak depth. The maximum load on sample (~25 mN) is much higher than the force output of the actuator (16 mN) which is due to dynamic overload from deceleration of the tip. Due to the dynamic nature of this test and the experimental challenges described earlier, analyzing the results of these tests is quite challenging. There are contributions from ISE, the inertia of the instrument, the time constant of the command and measurement signals, and the data acquisition rate which need to be carefully considered. During the initial part of the step load (<100 µs), calculating the acceleration, which is the second derivative of displacement is difficult as there are only 10 data points (data points are 10 µs apart in time). Even though there is good repeatability in the results, data at such short time intervals may not be accurate. As we proceed further in time up to 250 µs, the strain rates are very high (>10^4 1/s) as shown in Figure 6, which is beyond the scope of the simple analysis presented in the current work, where a power law behavior is assumed for the strain rate dependence of the flow stress. There may be other physical phenomena that are operative at such high strain rates which are not modeled here. In addition, the ISE is found to exhaust after a depth of 1.2 µm, which corresponds to a time on sample of 340 µs. Furthermore, it is not ideal to use data less than the time constant of the load signal which is 290 µs. In view of the constraints discussed above, step load data beyond 340 µs, up to the point of reaching the maximum depth (440 µs) is useful for the current analysis. It may be noted that the contact unloads completely at ~500 µs and the load on the sample calculation based on Equation (4) is not valid beyond that point.

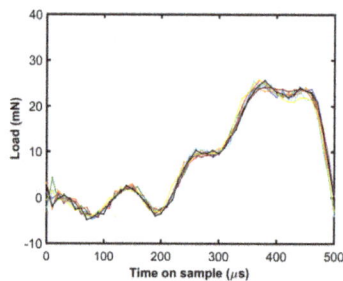

Figure 5. The load on the sample as a function of time on the sample calculated by accounting for the dynamic contribution of the instrument during the step load for 10 different tests shown in different colors.

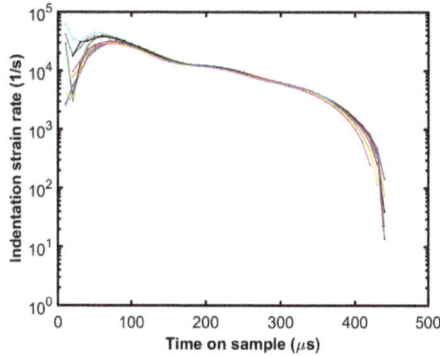

Figure 6. Indentation strain rate as a function of time during the step force input of 16 mN for 10 different tests shown in different colors.

Figure 7 shows the indentation depth as a function of time on sample for the CLH tests. The plot shows two distinct regions: the fast initial increase in depth corresponding to the fast load ramp (5 N/s) and the subsequent creep at a fixed force of 16 mN. The creep rate in the hold segment is almost insignificant as expected for aluminum at room temperature and the data may not correspond to the steady state creep as discussed in the previous work of the authors [20].

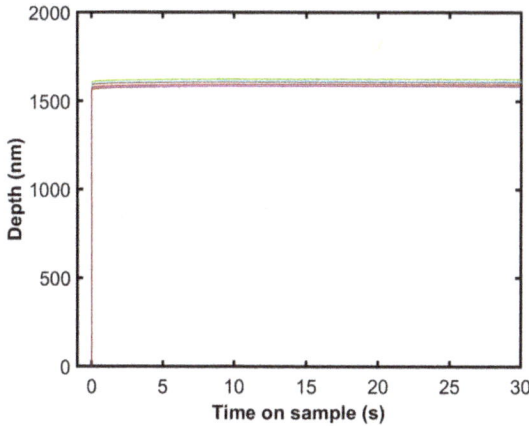

Figure 7. Indentation depth as a function of time during the CLH test for 10 different tests shown in different colors.

3.2. Strain Rate Effects

In this section, we present the strain rate dependence of hardness for the step load tests and the CLH tests. Figure 8 shows the hardness as a function of indentation strain rate for the step load tests. The hardness and the indentation strain rate are calculated using the procedure described in Section 2.4. The plot also shows the time on sample at the extremes of the data. As discussed in Section 3.1, data between 340 µs and 440 µs data is relevant for the current analysis. Within this window, the step load tests enable access to very high strain rates (4000 1/s). In addition, a range of strain rates at the higher end can be accessed in a single test which is simple and quick to perform compared to most conventional high strain rate tests.

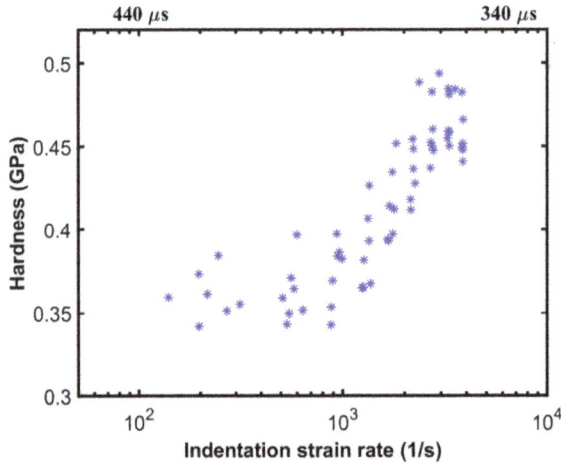

Figure 8. Indentation hardness as a function of indentation strain rate for step load tests.

Unlike the step load tests, the variation in hardness as a function of indentation strain rate for the CLH tests is calculated from the data in the hold segment. The calculation procedure is similar to the case of step load, but the load on the sample does not have any significant dynamic contribution from the instrument. Figure 9 shows the hardness as a function of indentation strain rate for the CLH tests. The data from the step load tests are also shown for comparison. The data from the CLH tests are above a depth of 1.2 μm where the ISE is not very significant. The plot shows the strength of the current testing methodology wherein a combination of step load tests and CLH tests is used to measure the strain rate dependence of hardness over seven orders of magnitude in the strain rate by a single instrument in a simple and quick manner. It is interesting to note that the strain rate sensitivity, which is the slope of the data, changes around a strain rate of 10^3 1/s. This will be discussed in greater detail in the next section.

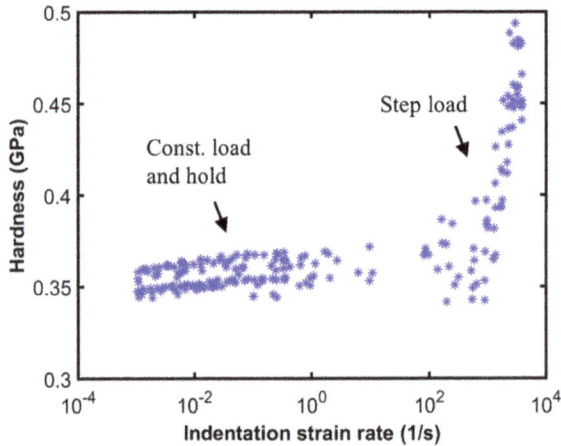

Figure 9. Indentation hardness as a function of strain rate during the constant load and hold (CLH) test and step load test.

3.3. Equivalent Uniaxial Response

In this section, we present the equivalent uniaxial strain rate and stress calculated from the indentation data, and the comparison with the uniaxial data from the literature. The procedure to calculate the equivalent uniaxial strain rate and stress is described in Section 2.5. Figure 10 shows the equivalent uniaxial stress as a function of equivalent uniaxial strain rate calculated from the indentation data and conventional uniaxial testing. Results from uniaxial testing for this alloy over the range of strain rates achieved in this work are obtained from Khan et al. [21]. The low strain rate data is from uniaxial compression testing, while the high strain rate data is from direct disc impact tests. In order to compare the results of this work where a Berkovich tip is used, which induces an equivalent strain of 8%, uniaxial results at two different strains (5% and 10%) are shown in the plot. The indentation data at 8% strain, lies within the region bounded by the uniaxial data at 5% and 10%, over a wide range of strain rates. Given the vast differences in test geometries and test protocols between the uniaxial and indentation tests, it is very interesting to note that the indentation data matches the uniaxial data over seven orders of magnitude in strain rate. The difference in the uniaxial data at two different strain levels indicates that steady state conditions may not have been achieved during indentation which induces 8% equivalent strain. This is similar to the observations of Luthy et al. [22] who performed torsion tests to very high strain (300–1000%) in order to achieve steady state creep in pure aluminum at room temperature. While the results of the current work may not represent steady state conditions, the experimental procedure and analysis presented here provide a generic framework to measure strain rate dependence of stress over a wide range of strain rates with a high degree of precision.

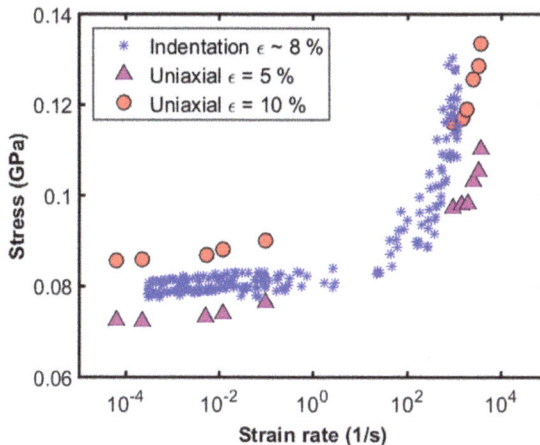

Figure 10. Comparison of equivalent uniaxial strain rate and stress calculated from the indentation data with the uniaxial results from conventional compression test and direct disc impact test (Khan et al. [21]).

4. Summary and Conclusions

Advances in instrumentation have enabled nanomechanical measurements with very low noise levels (sub nanometer) at fast time constants (20 µs) and high data acquisition rates (100 KHz). These capabilities in conjunction with a comprehensive model for instrument's dynamics and electronics is vital for accurate high strain rate measurements.

The indentation system used in this work is shown to be extremely well characterized by a simple one DOF harmonic oscillator model which can be readily extended to testing under non-ambient conditions (vacuum or high temperature).

A comprehensive model for the instrument's dynamics and electronics has been presented to accurately factor out the instrument's contribution to the measurement during a high strain rate test and validated with the step load experiments in free air. The importance of having fast measurement time constants for accurate high strain rate measurements has been demonstrated.

Step load tests have been performed on commercial purity aluminum (1100 Al), wherein the force is ramped to the desired level as fast as the instrument can physically accomplish the change, thereby enabling access to a range of high strain rates (>1000 1/s) in a single indentation test.

A simple procedure has been presented to determine the hardness and indentation strain rate during a step load test. The strain rate dependence of hardness has been measured over seven orders of magnitude by a combination of step load and CLH tests.

The uniaxial equivalent strain rate and stress calculated from the indentation data closely matches the uniaxial results over seven orders of magnitude. Given the vast differences in test geometries and test protocols between the uniaxial and indentation tests, it is very encouraging that the indentation data matches the uniaxial data over several orders of magnitude in strain rate.

The experimental procedure and analysis presented in this work provides a generic framework to measure strain rate dependence of stress over a wide range of strain rates with a high degree of precision in a simple, quick, and cost-effective way compared to the conventional methods. The simplicity of this experimental technique also enables it to be extended to high temperatures, thereby facilitating the measurement of a high strain rate response at elevated temperatures, which is a largely unexplored challenging area of research/experimentation.

Author Contributions: P.S.P. and W.C.O. conceived and designed the experiments; P.S.P. performed the experiments; P.S.P. and W.C.O. analyzed the data; P.S.P. wrote the paper.

Conflicts of Interest: The authors declare no conflict of interest.

References

1. Gray, G.T., III. High-Strain-Rate Deformation: Mechanical Behavior and Deformation Substructures Induced. *Ann. Rev. Mater. Res.* **2012**, *42*, 285–303. [CrossRef]
2. Ramesh, K.T. High Strain Rate and impact experiments. In *Springer Handbook of Experimental Solid Mechanics*; Sharpe, W., Ed.; Springer: New York, NY, USA, 2008; pp. 874–902. ISBN 978-0-387-26883-5.
3. Somekawa, H.; Schuh, C.A. High-strain-rate nanoindentation behavior of fine-grained magnesium alloys. *J. Mater. Res.* **2012**, *27*, 1295–1302. [CrossRef]
4. Kermouche, G.; Grange, F.; Langlade, C. Local identification of the stress-strain curves of metals at a high strain rate using repeated micro-impact testing. *Mater. Sci. Eng. A* **2013**, *569*, 71–77. [CrossRef]
5. Chen, J.; Shi, X.; Beake, B.D.; Guo, X.; Wang, Z.; Zhang, Y.; Zhang, X.; Goodes, S.R. An investigation into the dynamic indentation response of metallic materials. *J. Mater. Sci.* **2016**, *51*, 8310–8322. [CrossRef]
6. Varghese, J.; Radig, G.; Herkommer, D.; Dasgupta, A. Hybrid experimental and computational approach for rate dependent mechanical properties using indentation techniques. In Proceedings of the 6th International Conference on Thermal, Mechanial and Multi-Physics Simulation and Experiments in Micro-Electronics and Micro-Systems, Berlin, Germany, 18–20 April 2005; pp. 510–514.
7. Tirupataiah, Y.; Sundararajan, G. A dynamic indentation technique for the characterization of the high strain rate plastic flow behaviour of ductile metals and alloys. *J. Mech. Phys. Solids* **1991**, *39*. [CrossRef]
8. Subhash, G.; Koeppel, B.J.; Chandra, A. Dynamic indentation hardness and rate sensitivity in metals. *J. Eng. Mater. Technol. Trans. ASME* **1999**, *121*, 257–263. [CrossRef]
9. Shin, H.-S.; Ko, D.-K.; Oh, S.-Y. Investigation of deformation behavior in bulk amorphous metal under high strain rates using an indentation method. *J. Metastable Nanocryst. Mater.* **2003**, *15–16*, 167–172. [CrossRef]
10. Yamada, H.; Ogasawara, N.; Shimizu, Y.; Horikawa, K.; Kobayashi, H.; Chen, X. Effect of high strain rate on indentation in pure aluminum. *J. Eng. Mater. Technol. Trans. ASME* **2013**, *135*. [CrossRef]
11. Kumaraswamy, A.; Vasudeva Rao, V. High strain-rate plastic flow behavior of Ti-6Al-4V from dynamic indentation experiments. *Mater. Sci. Eng. A* **2011**, *528*. [CrossRef]

12. Raman, V.; Berriche, R. An investigation of the creep processes in tin and aluminum using a depth-sensing indentation technique. *J. Mater. Res.* **1992**, *7*, 627–638. [CrossRef]

13. Mayo, M.J.; Nix, W.D. A micro-indentation study of superplasticity in Pb, Sn, and Sn-38 wt% Pb. *Acta Metall.* **1988**, *36*, 2183–2192. [CrossRef]

14. Lucas, B. An Experimental Investigation of Creep and Viscoelastic Properties Using Depth-Sensing Indentation Techniques. Ph.D. Thesis, University of Tennessee, Knoxville, TN, USA, 1997.

15. Maier, V.; Durst, K.; Mueller, J.; Backes, B.; Höppel, H.W.; Göken, M. Nanoindentation strain-rate jump tests for determining the local strain-rate sensitivity in nanocrystalline Ni and ultrafine-grained Al. *J. Mater. Res.* **2011**, *26*, 1421–1430. [CrossRef]

16. Feldner, P.; Merle, B.; Göken, M. Determination of the strain-rate sensitivity of ultrafine-grained materials by spherical nanoindentation. *J. Mater. Res.* **2017**, *13*, 1466–1473. [CrossRef]

17. Pharr, G.M.; Herbert, E.G.; Gao, Y. The Indentation Size Effect: A Critical Examination of Experimental Observations and Mechanistic Interpretations. *Ann. Rev. Mater. Res.* **2010**, *40*, 271–292. [CrossRef]

18. Su, C.; Herbert, E.G.; Sohn, S.; LaManna, J.A.; Oliver, W.C.; Pharr, G.M. Measurement of power-law creep parameters by instrumented indentation methods. *J. Mech. Phys. Solids* **2013**, *61*, 517–536. [CrossRef]

19. Bower, A.F.; Fleck, N.A.; Needleman, A.; Ogbonna, N. Indentation of a Power Law Creeping Solid. *Proc. R. Soc. London. Ser. A Math. Phys. Sci.* **1993**, *441*, 97–124. [CrossRef]

20. Phani, P.S.; Oliver, W.C. A direct comparison of high temperature nanoindentation creep and uniaxial creep measurements for commercial purity aluminum. *Acta Mater.* **2016**, *111*, 31–38. [CrossRef]

21. Khan, A.S.; Huang, S. Experimental and theoretical study of mechanical behavior of 1100 aluminum in the strain rate range 1e-5–1e4 s-1. *Int. J. Plast.* **1992**, *8*, 397–424. [CrossRef]

22. Luthy, H.; Miller, A.K.; Sherby, O.D. The stress and temperature dependence of steady-state flow at intermediate temperatures for pure polycrystalline aluminum. *Acta Metall.* **1980**, *28*, 169–178. [CrossRef]

materials

MDPI

Article

Using Biotechnology to Solve Engineering Problems: Non-Destructive Testing of Microfabrication Components

Carla C. C. R. de Carvalho [1,*], Patrick L. Inácio [2], Rosa M. Miranda [2] and Telmo G. Santos [2]

[1] iBB-Institute for Bioengineering and Biosciences, Department of Bioengineering,
Instituto Superior Técnico, Universidade de Lisboa, Av. Rovisco Pais, 1049-001 Lisbon, Portugal

[2] UNIDEMI, Department of Mechanical and Industrial Engineering,
NOVA School of Science and Technology, NOVA University Lisbon, 2829-516 Caparica, Portugal;
p.inacio@campus.fct.unl.pt (P.L.I.); rmmdm@fct.unl.pt (R.M.M.); telmo.santos@fct.unl.pt (T.G.S.)

* Correspondence: ccarvalho@tecnico.ulisboa.pt; Tel.: +351-218419594

Received: 31 May 2017; Accepted: 8 July 2017; Published: 12 July 2017

Abstract: In an increasingly miniaturised technological world, non-destructive testing (NDT) methodologies able to detect defects at the micro scale are necessary to prevent failures. Although several existing methods allow the detection of defects at that scale, their application may be hindered by the small size of the samples to examine. In this study, the application of bacterial cells to help the detection of fissures, cracks, and voids on the surface of metals is proposed. The application of magnetic and electric fields after deposition of the cells ensured the distribution of the cells over the entire surfaces and helped the penetration of the cells inside the defects. The use of fluorophores to stain the cells allowed their visualisation and the identification of the defects. Furthermore, the size and zeta potential of the cells and their production of siderophores and biosurfactants could be influenced to detect smaller defects. Micro and nano surface defects made in aluminium, steel, and copper alloys could be readily identified by two *Staphylococcus* strains and *Rhodococcus erythropolis* cells.

Keywords: micro defects; NDT; indentation; microfabrication; *Staphylococcus*; *Rhodococcus*

---

1. Introduction

Current developments in microfabrication are demanding for reliable, economical, and ecologic non-destructive testing (NDT) techniques to detect unprecedented micro defects that conventional NDT cannot perceive. Usual defects in microfabrication are roughness, surface micro cracks, or voids created by micro size particle detachments, such as those observed e.g., in powder micro-injection [1]. New engineering materials such as nanostructured materials, functional surfaces and thermal barrier coatings, microelectronic and optical components, biomedical and orthodontic devices, and solar cells, may have their efficiency and reliability dependent on the existence of microcracks. These micro size surface defects are particularly critical, as it is there that the mechanical stresses are more intense, fatigue starts, and external damage (corrosion, wear, and thermal effects) exist.

Conventional NDT techniques include: (i) the eddy current technique, in which spatial resolution and size of the samples that may be analysed depend on the dimensions of the coil used [2,3]; (ii) optical transmission, by which the cracks that may be passed through by a broad-spectrum light or laser diode are detected by a CCD camera whilst those with odd shapes through which light cannot be transmitted will not be visible [4]; (iii) radiography and tomography, high-cost NDT methods, have difficulties in detecting defects smaller than 2% the thickness of the material, those perpendicular to the radiation beam, and defects in parts of complex three-dimensional geometry due to difficulties in image interpretation [5,6]; (iv) dye penetrant testing, in which the volume of the defect is more important than its area, because the actual crack depth is strongly related to the probability of detection [7]; and, magnetic particle inspection which may be used for the detection of defects on ferromagnetic materials [8]. Although most of these techniques are useful for the detection of defects during traditional manufacturing, they are no longer appropriate for microfabrication, as they were developed to answer other requirements and target scenarios, involving different materials, defect sizes, and morphologies. In micro-manufacturing, component size is small, preventing an effective coupling of ultrasound or eddy current probes. Dye penetrant testing allows the identification of some defects with ≈0.9 μm, but only when the defect depth is large as compared to its superficial open area, since its physical principle is exclusively based on the high capillarity and low viscosity of the penetrant. Magnetic particle inspection cannot be applied to most engineering materials, including polymers, ceramics, composites, and metals such as aluminium, copper, titanium, magnesium, and stainless steels alloys. Besides, parts with complex geometries may present areas with little or no magnetic flux [8].

The limit of detectability and sensitivity of conventional NDT techniques have improved during the last decades, but defects on the micrometre scale are still difficult to identify over background noise [9]. Several sophisticated characterisation techniques for microfabrication are available, such as Scanning Acoustic Microscopy (SAM), Near-field Scanning Optic Microscopy (MSOM), Scanning Electron Microscope (SEM) or Atomic Force Microscopy (AFM). However, they are limited to very small planar areas, are expensive, and require considerable time for analysis and interpretation of images [4,10], targeting a different scenario compared to our envisaged applications.

We have recently proposed a novel approach for the detection of micro defects and material characterisation based on the use of bacterial cells stained with fluorescent dyes [11,12]. The initial validation tests showed that *Rhodococcus erythropolis* cells allow the identification defects with a depth of 6.8, 4.3, and 2.9 μm in copper, aluminium, and steel [11]. The application of electric and magnetic fields to the samples when using *R. erythropolis* and *Staphylococcus* sp. cells, respectively, improved the movement of the cells over the surfaces which resulted in the observation of defects made by nanoindentation, and during micro-powder injection moulding and micro-laser welding [12]. However, the cells in the previous studies presented their intrinsic properties and no further exploitation of their ability to produce specialized compounds or to adapt to certain conditions was made.

In the present study, the properties of bacterial cells were influenced to improve the interactions with surfaces and the detectability threshold of small defects. By influencing the net surface charge of the cells and the natural production of biosurfactants, it was possible to help their penetration into micro-scale defects. The micro-indentations for the tests were made with side-lengths below the detection limit of eddy current techniques (ca. 50 μm [13]), with a very low ratio of depth/open area (unfavourable for dye penetrant testing), and on samples thicker than those that could be examined by radiography, tomography, and optical transmission techniques. Our study is an example of the advantages of multidisciplinary approaches to solve real problems created by new manufacturing techniques.

2. Results

2.1. Identifying Suitable Bacteria for NDT

To better observe micro-defects, the stained bacterial cells, which behave like solid particles in liquid suspensions, should cover the whole surface to be analysed and be able to enter and remain in the depressions. The application of cellular properties such as net surface charge, production of biosurfactants that decrease the surface tension of the liquid medium, and the production of siderophores to capture iron ions to increase the interaction between cell and magnetic and electric fields could represent a significant benefit when compared to the application of inert particles in NDT. Bacterial cells have evolved to form biofilms on the surface of most of the surfaces on Earth and bacterium–surface interactions involve electrostatic attraction or repulsion, Lifshitz-van der Waals and Brownian motion forces and Lewis acid–base interactions as described by the extended Derjaguin, Landau, Verwey, and Overbeek (xDLVO) theory [14,15].

Initially, we found that *Rhodococcus erythropolis* DCL14 moved in response to the direction of the applied electric field but no bacterium from our collection responded to magnetic fields. The first stage of this study was thus the screening and isolation of bacterial strains able to move as response to magnetic fields and the definition of the culture conditions leading to an increased electric/magnetic character of the cells. To favour the detection of defects in metallic surfaces, bacterial cells that could be influenced by magnetic or electric fields were sought in the space between the battery and the electronic board of cell phones. Since radio and microwave frequency electromagnetic fields (EMF) are emitted by mobile phones, bacterial cells living attached to these devices were expected to be influenced by EMF. Most of the aerobic bacterial species isolated using cultivation-based techniques from samples collected in the mobile phones were identified by the Sherlock® Microbial ID System from MIDI, Inc., (Newark, DE, USA) as belonging to the genus *Bacillus* and *Staphylococcus* (each represented ca. 40.0% of the total population; data not shown). These cells have been found in mobile phones by other research groups [16,17].

All isolated strains from the mobile phones were tested for their ability to respond to magnetic fields. Of the six strains that followed the movement of magnets under the microscope, two *Staphylococcus* strains were, surprisingly, the most responsive and were chosen for the following studies. Under the conditions used for growth, these *Staphylococcus* cells were spherical, their diameter ranging between 0.52 and 0.72 µm. One of the *Staphylococcus* isolates, designated "SA", shared a very high homology with several strains of *S. aureus*. The 16S rRNA-encoding gene sequence was aligned and found to be 100% identical to the 16S rRNA-encoding genes of *S. aureus* strains deposited in the GenBank database (data not shown). Strain "SA" was also identified as *S. aureus* by the Sherlock® Microbial ID System, and will therefore be designated as *S. aureus* SA in the following sections of this manuscript. The other *Staphylococcus* strain that responded to magnetic fields, which was designated as "SH", exhibited a sequence similar to that of the 16S rRNA-encoding gene of several other strains of the *Staphylococcus* genus. A homology of 99% was found between strain "SH" and *Staphylococcus* sp. HB2 (data not shown), which was isolated in China, where the mobile phone from which the SH strain was isolated was manufactured. Strain SH presented the highest similarity index with *S. hominis* in the Sherlock® Microbial ID System, but similarities of 98% were obtained with 16S rRNA-encoding gene sequences of different *Staphylococcus* species and thus this strain isolated from the mobile phone will be referred to as *Staphylococcus* sp. SH from now on.

2.2. Influencing Bacterial Properties to Help Cell Adhesion

Both *S. aureus* SA and *Staphylococcus* sp. SH isolated from the cell phones produced siderophores when placed in iron depleted media (Figure 1a). Siderophores are low molecular weight compounds used by bacterial cells to sequester Fe^{3+} under conditions of iron depletion [18,19]. *Staphylococcus* strains have also several iron-uptake ABC transporters comprising membrane-anchored lipoproteins on the external surface of the membrane, transmembrane permease proteins, and ATPases that provide energy for the process [20,21].

To test if the magnetic and electrostatic interactions between bacterial cells and metallic surfaces could be further enhanced, 0.005 g of iron per litre were added to the growth media of the bacteria. The cells grown in media supplemented with iron were compared to those grown on the same media without iron supplementation. Iron accumulation inside the cells was expected to stimulate their response to applied magnetic fields. In fact, non-crystalline magnetic inclusions have been reported in *Staphylococcus* species [22]. In the present study, the zeta potential of the cells, which may act as a measure of the surface charge of the cells, was influenced by both the carbon source used and concentration of iron(II) sulfate (Figure 2). When 0.005 $g \cdot L^{-1}$ of iron was supplemented to the growth media, the zeta potential of *S. aureus* SA decreased by 9.2% whilst that of *Staphylococcus* sp. SH decreased by 27.1%. A previous study showed a strong correlation between the chain length of n-alkanes used as carbon sources and the zeta potential of *R. erythropolis* DCL14 cells, with cells even becoming positive when grown on C14–C16 alkanes [23]. Changes in cell hydrophobicity and net surface charge that occurs during adaptation to the growth media affects the amount of cells that adhere to metallic and non-metallic surfaces [23–25].

In the current study, *R. erythropolis* DCL14 cells grown on ethanol exhibited a zeta potential 10% less negative when iron was added to the medium. However, these cells exhibited a zeta potential that was 8.5% more positive when grown with n-hexadecane as carbon source. Cells of strain DCL14 produced also glycolipids (Figure 1b), which acted as biosurfactants, thus decreasing the surface tension of the medium to 58.3 and 22.0 mN/m when the cells grew on ethanol and n-hexadecane, respectively. A decreased surface tension of the medium should aid the penetration of the cells in the defects.

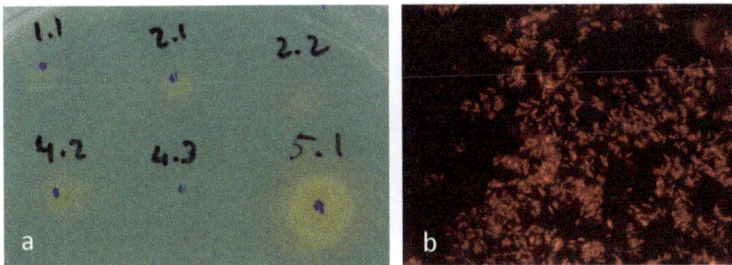

Figure 1. Production of siderophores and specialised lipids by bacteria. (**a**) CAS agar diffusion assay showing siderophore production (yellow halo) in the isolated *S. aureus* SA (marked as 1.1) and *Staphylococcus* sp. SH (indicated as 5.1); (**b**) *R. erythropolis* cells stained with Nile Red show storage lipids inside the cells and glycolipids excreted by the cells.

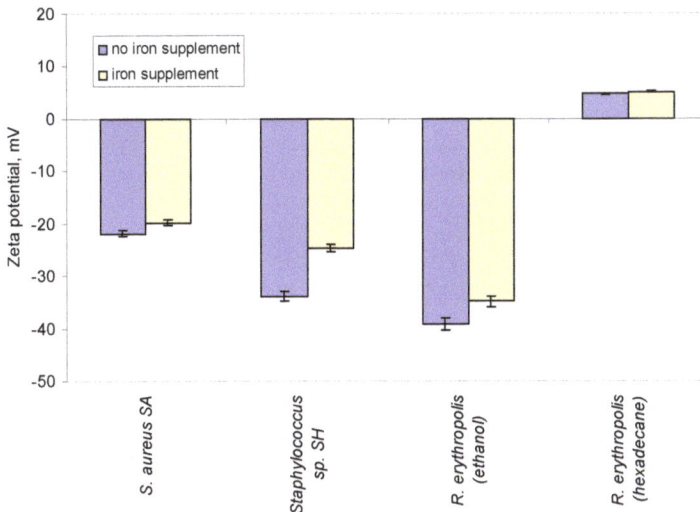

Figure 2. Influence of an iron supplement of 0.005 g/L iron(II) sulfate, in the growth media, on the zeta potential of the cells.

2.3. Application of Magnetic and Electric Fields to Help Bacterial Distribution

Bacterial cells in suspension behave like solid particles. To achieve a homogeneous distribution of the cells on the surfaces of the samples to be analysed and better penetration into the defects, the movement of the cells in horizontal and vertical planes was enhanced by the application of magnetic and electric fields. Images taken with 10.4 s exposure enabled the determination of the distance travelled by the cells and the selection of those cells that travelled the longest distance.

When exposed to magnetic fields created by electric currents with modified sine waves with frequencies of 1.25, 5 and 15 Hz, *S. aureus* SA cells travelled on average 38, 66, and 70 μm whilst *Staphylococcus* sp. SH cells travelled 29, 42, and 54 μm (Figure 3a). No significant differences (between 1.5% and 6.7%) were observed in the distance travelled by cells of either strain when modified sine waves with frequencies above 5 Hz were applied to create a magnetic field.

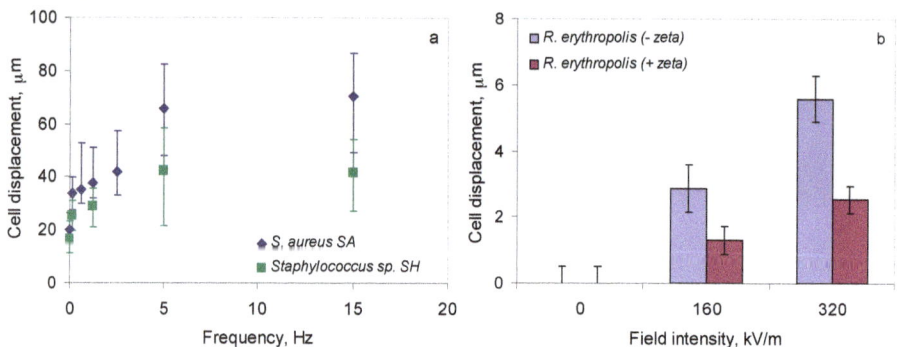

Figure 3. Cell displacement observed with an exposure of 10.4 s when *Staphylococcus* cells were placed on a glass surface subjected to magnetic fields (**a**) and when *R. erythropolis* cells presenting negative or positive zeta potential were placed under an electric field (**b**).

As previously mentioned, the zeta potential of *R. erythropolis* DCL14 cells can be influenced by the carbon source used for cell growth (Figure 2; [23]). When subjected to an electric field, cells grown on ethanol and presenting a negative zeta potential moved towards the anode, whilst cells grown on *n*-hexadecane and exhibiting a positive zeta potential moved towards the cathode. To favour the highest potential differences between cells and electrodes and the cell displacement, *R. erythropolis* cells were grown on ethanol without iron supplementation to obtain the highest negative zeta potential, whilst the medium used for cell growth using *n*-hexadecane as carbon source was supplemented with iron to increase the positive net surface charge of the cells (cf. Figure 2). The distance travelled by the cells was dependent on both their zeta potential and on the electric field applied: cells with negative net surface potentials were displaced 2.9 and 5.6 μm, whilst cells with positive potential were displaced 1.3 to 2.5 μm when electric field intensities of 160 and 320 kV/m were applied, respectively (Figure 3b). No displacement of the cells was observed in the absence of an electric field (Figure 3b).

2.4. Application of Bacterial Cells for the Identification of Standard Defects

A methodology for observing bacterial cells inside defects produced on the surface of metallic and polymeric materials was recently validated by our team [11,12]. It comprises the following steps:

i.   selection of the area to inspect;
ii.   surface cleaning;
iii. deposition of bacterial suspension;
iv.   penetration by and adherence of bacteria to defects;
v.   removal of the excess of bacterial suspension;

vi. inspection and evaluation of the surface using a microscope with fluorescent light;

vii. cleaning and sterilization of the material.

For validation purposes, after cleaning the surface and prior to the deposition of the bacterial suspension, the surface should be inspected under the microscope to ascertain that the surface is cell-free before the assays. Similarly, after cleaning and sterilization of the material at the end of the tests, the samples should be inspected under the microscope to assess if they are cell-free. Control samples should also be inspected by SEM.

To compare the ability of the bacterial cells to reveal each defect on the different metallic materials tested, 4 × 3 matrices containing surface indentations 300 μm apart were arranged in such a way that near a small defect there was always a larger one (Figure 4). The indentations were produced using a Vickers hardness testing machine from Mitutoyo with a square base pyramid with an angle between opposite faces of 136° (Figure 4a). Consequently, the depth of each pyramid is about 1/5 of its side length. This defect morphology presents a very low ratio of depth/open area, constituting one of the most adverse conditions in conventional dye penetrant NDT inspection, due to the possibility of removing the penetrant from the defect during the removal stage. The fluorescent dyes used allowed the easy observation of individual cells as they appeared quite bright over a dark field (Figure 4b,c).

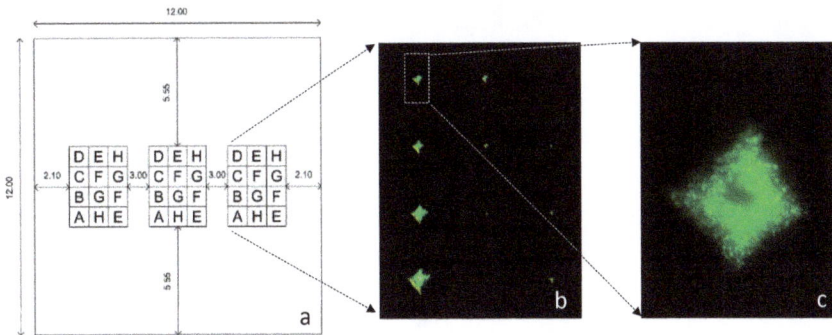

Figure 4. (**a**) Matrices of pyramidal defects created on the tested surfaces (units in mm). The letters represent the force applied to create the indentations: A—100 N, B—50 N, C—30 N, D—20 N; E—10 N; F—5 N; G—2.5 N and H—1 N. Units shown in mm; (**b**) Matrix of defects shown by the presence of green stained *R. erythropolis* cells; (**c**) Pyramidal defect filled with green stained cells.

The application of magnetic or electric fields in the horizontal direction may help the distribution of the cells across the surface to be analysed, whilst vertical fields may contribute to the entrance of the cells in the defects [12]. However, the fields may also influence the properties of the cells. It is known that cell–surface interactions involve e.g., hydrophobic and electrostatic interactions [26].

In the absence of an applied magnetic field, both *S. aureus* SA and *Staphylococcus* sp. SH helped the identification of the smaller defects in AISI 304L stainless steel (indentation pyramid with a depth of 3.7 μm) than in titanium or AA 1100 aluminium alloy (indentation pyramid with a depth of 4.6 μm; Figure 5). The application of a magnetic field in the 3 orthogonal spatial directions with 5 Hz allowed the entrance of both species in defects that were nearly half the size of those visible when no field was applied: in pyramidal defects with depths of 2.0 and 2.6 μm for steel and both titanium and aluminium, respectively. However, when the *Staphylococcus* cells were used to analyse defects made on copper samples, they only entered relatively large defects: with a side length of 30.3 μm and 6.2 μm depth when *S. aureus* SA was used, and a side length of 42.6 μm and 8.7 μm depth when cells of *Staphylococcus* sp. SH were applied (Figure 5). The application of a triaxial magnetic field with 5 Hz apparently further enhanced the known antimicrobial effect of copper [27,28]. With this metal, the *Staphylococcus* cells were only visible in the pyramids with a side length of 96.7 μm and depth of 19.7 μm.

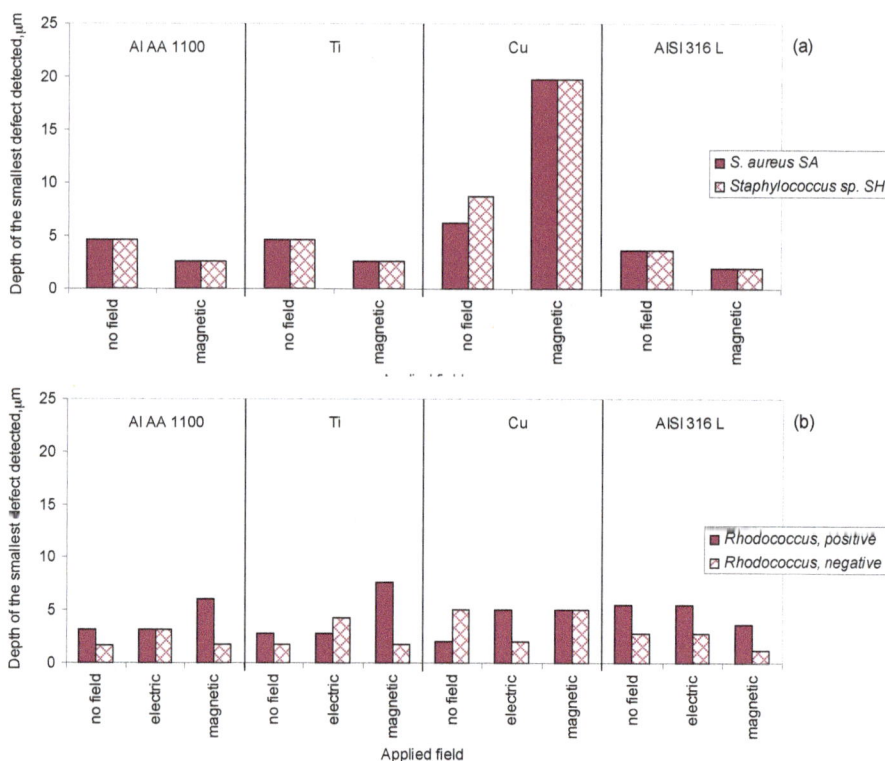

Figure 5. Influence of the applied field on the ability of the *Staphylococcus* (**a**) and *Rhodococcus* (**b**) cells to enter pyramidal indentations made as standard defects on metallic surfaces.

In the absence of an applied electric field, *R. erythropolis* cells grown on *n*-hexadecane as carbon source, and presenting positive zeta potential, allowed the identification of smaller defects in copper (side length of 9.7 μm and depth of 2 μm) than cells grown on ethanol and with negative net surface charge (side length of 24.6 μm and depth of 5 μm; Figure 5). In all other surfaces tested, when no field was applied, *R. erythropolis* cells presenting a negative zeta potential entered defects smaller than their positive counterparts. The application of a horizontal and vertical electric field with 8 kV during the assays with *R. erythropolis* cells with positive zeta potential allowed the identification of the same defects as when no field was applied, except in copper samples where only pyramidal defects with sides longer than 14.6 μm were shown (Figure 5). The reverse was observed with cells with negative net surface charge: smaller defects could be observed in copper samples but the detectability of the defects was worse in aluminium and titanium samples as compared to when no field was applied (Figure 5).

To assess if the cells presenting different zeta potentials could also be influenced by magnetic fields, *R. erythropolis* cells were tested under a triaxial magnetic field with 5 Hz. This allowed the observation of the smallest defects made in AISI 304L stainless steel, with only 1.7 μm of depth, with cells of both positive and negative zeta potential (Figure 5). However, in the other metallic surfaces, no changes were observed with cells with negative net charge whilst positive cells entered only defects that were larger than those observed without applied fields. The results thus indicate that the properties of the cells could be influenced, affecting the interaction between cells and surface. The application of magnetic and electric fields assured that all indentation matrices made on the surfaces were covered by cell suspension, but the existence of cells inside the pyramidal indentations during the observation stage is dependent on (i) the entrance of cells into the defect and (ii) the permanence of the cells in the defect after the removal of the excess of the culture suspension. This depends on the strength of the forces between cells and surfaces.

2.5. Application of Bacterial Cells for the Identification of Real Defects

The methodology proposed was not only efficient for identifying pyramidal indentations but also indicated scratches and other defects inadvertently introduced on the metal surfaces under study (Figure 6a–c). The cells could clearly indicate both the voids that were created on AISI 316L stainless steel microscrews made for dental implants during fabrication by Micro Powder Injection Molding (Figure 6c) and the scratches that were inadvertently made during manipulation of the microscrews in the laboratory (Figure 6a,b). Besides, the cells could efficiently show cracks formed on a pulsed laser weld of titanium thin sheets when stained with a fluorescent dye and observed under an optical microscope with fluorescent light (Figure 6d).

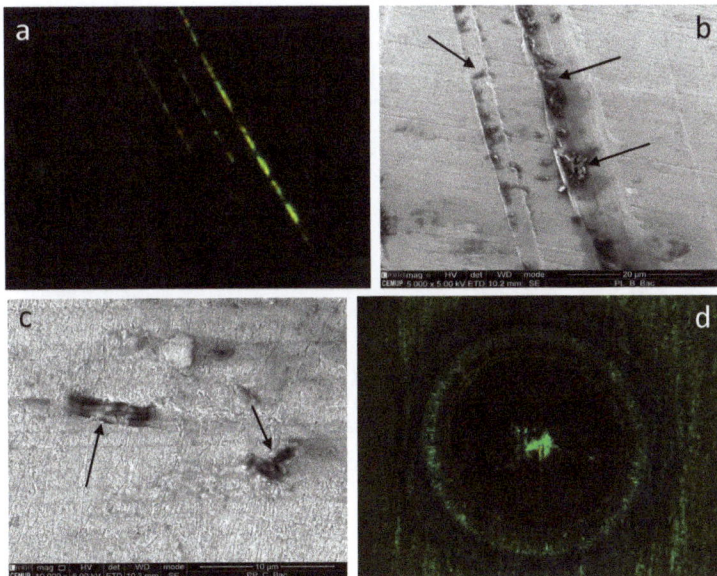

Figure 6. *Rhodococcus erythropolis* cells indicate the presence of scratches (**a,b**) created during the manipulation of microscrews for dental implants and voids (**c**) created during the manufacturing process by Micro Powder Injection Molding (μPIM). The cells were observed by fluorescence microscopy (**a,d**; magnification 30×) and SEM (**b,c**; magnification 5000× in **b** and 10,000× in **c**; some cells indicated by arrows). *R. erythropolis* stained with the green fluorescent dye SYTO®9 show radial cracks on a pulsed laser welding of titanium thin sheets (**d**).

When tested on engraved gold jewellery, the cells were able to accurately reveal both the topography and roughness of the sample, with the polished part of the surface remaining nearly cell free (Figure 7). Since the cells were deposited for less than 5 min on the surface of the metal samples to be examined, cell adhesion to the metal surfaces was reversible. Cleaning and disinfection of the samples could be achieved by any of the following methods: chemical sterilization, by application of 70% ethanol; irradiation with ultraviolet light (lamp Osram HNS 8 W G5, dominant wavelength

254 nm); and dry heat, or by placing the metallic samples in boiling water (data not shown). After the sterilization procedure, the samples were rinsed with distilled water, dried with tissue wipes, and observed under a microscope. No debris or whole cells were observed on the surfaces (data not shown).

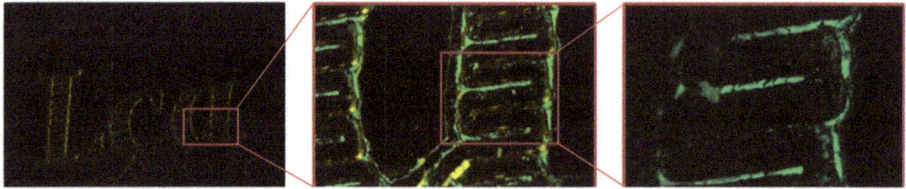

Figure 7. *Rhodococcus erythropolis* cells were able to highlight both the topography and roughness of a gold pendant (magnification from left to right: 30×; 150×; 300×).

To improve the stage of inspection and evaluation of the surface, and in particular to test the possibility of evaluating the surface under the naked eye, bacterial growth was promoted on the surfaces. It is known that bacterial cells may form biofilms on the surface of nearly all surfaces, including metals [25,29]. After the deposition stage, the excess liquid was removed and the sample was placed on sterile growth media overnight with temperature and agitation control. The cells inside the defects multiplied, allowing the observation of colonies under the naked eye which showed scratches made by sanding (Figure 8a) and matrices of defects made (Figure 8b).

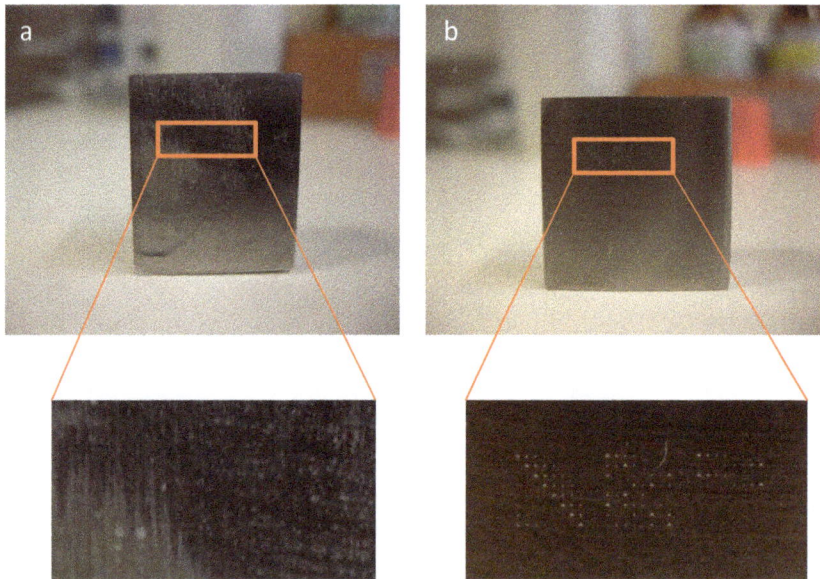

Figure 8. Colonies of *R. erythropolis* cells observed after letting the cells inside the defects grow overnight show scratches made by sanding (**a**) and matrices of indentations (**b**) under the naked eye.

3. Discussion

Non-destructive Testing (NDT) techniques for the identification of defects occurring during microfabrication of microstructures and microsystems are of paramount importance, especially following their application in e.g., biological, medical, and aeronautical systems. In the present study, we propose to push the frontier of NDT by introducing the use of biological phenomena in the scope of NDT techniques, showing the application of bacterial cells to help the visualisation of micro surface defects. In a previous validation study [11], the *R. erythropolis* cells were simply deposited on the surface to be analysed and the coverage of the whole surface was dependent on the concentration of cells in the suspension. Since most bacterial species that were good candidates for this new NDT methodology do not possess locomotion structures such as flagella or pili, we tested the hypothesis of promoting cell movement over the surfaces to be tested by the application of magnetic or electric fields. At the beginning of this study it was already known that the net surface charge of *R. erythropolis* DCL14 cells could be influenced by the carbon source used for growth [23], making this strain a good candidate to be tested in the assays involving electric fields. To find bacterial strains responsive to magnetic fields, we isolated bacteria from samples collected from mobile phones and found two *Staphylococcus* strains able to move over significant distances in a magnetic field when compared to their size.

The phenotypic characteristics of both *S. aureus* SA and *Staphylococcus* sp. SH, including the small size (around 600 nm) and the ability to move in magnetic fields aided the penetration of the cells into small defects. The unexpected response of these cells to the imposed magnetic fields could result from the capture of iron by siderophores and especially by lipoproteins positioned at the external surface of the cellular membrane. Besides, non-crystalline magnetic inclusions previously found in *Staphylococcus* sp. SH have been shown to enable movement of the cells, which could be brought about by magnetic fields [22].

Staphylococcal lipoproteins play a role in the acquisition of iron by serving as binding proteins for both heme and transferring iron, and as siderophore-binding components of ABC transporters [21]. Thus, when the Mueller–Hinton medium (iron concentration ca. $0.5–0.8$ mg·L^{-1} [30]) was supplemented with additional 0.005 g of $FeSO_4 \cdot 7H_2O$ per litre (iron added 1.0 mg·L^{-1}), the zeta potential of the cells decreased, suggesting that iron ions could be at the surface resulting in a less negative character of the cell surface. Furthermore, ferritin proteins have been identified in *Staphylococcus* species and shown to function as iron storage proteins [31]. When studying freshwater magnetotatic bacteria, Frankel and co-workers observed that besides magnetite, there was a substance that produced an extra quadrupole doublet which the authors hypothesised could result, among others, from ferritin, which at that time had not been recognized in prokaryotic cells [32]. In order to study iron-binding centres in bacterial ferritins, Pereira et al. showed that in the wild type of *Desulfovibrio vugaris* a diamagnetic diferrous species, a mixed valence $Fe^{2+}Fe^{3+}$ species and a mononuclear Fe^{2+} species were present, but a paramagnetic diferrous species was identified in a variant of the bacterium [33]. Although ferritin can only bind relatively small amounts of iron (a cell might bind ca. 7000 molecules of ferritin, containing a total of ca. 10^7 atoms of iron) and therefore the cells should be weakly paramagnetic, ferritin labelled particles (including polyacrylamide beads with size similar to whole cells) could be magnetically removed from a flowing suspension [34].

In the present study, the average distance travelled by *Staphylococcus* sp. SH and especially *S. aureus* SA cells when magnetic fields were applied was considerable when compared to their cell diameter (Figure 3a). *R. erythropolis* DCL14 cells, exhibiting both negative and positive zeta potentials, moved only 0.6–2.5 times the length of the cell when under an electric field (Figure 3b). Nevertheless, the horizontal and vibrational movement of the cells over the surface of the samples containing the indentations allows a homogeneous distribution of the cells over the area to be inspected. Other cell properties, including the ability of *R. erythropolis* to produce biosurfactants which lowers the surface tension of the medium, also helps the penetration of the cells by capillarity into the defects. Capillarity, together with wetting-contact angle, is of paramount importance in tests using liquid penetrants and is usually achieved by environmentally toxic chemical compounds [35]. Handling and disposal costs of synthetic and petroleum based oils used as carrier fluids for the fluorescent dye penetrants used in non-destructive inspections have been estimated at a few million dollars per year [36]. The proposed method using bacterial cells could thus constitute a cleaner and cheaper alternative to the existing liquid penetrants.

According to the extended Derjaguin–Landau–Verwey–Overbeek (DLVO) theory of colloid stability, microbial adhesion is governed by Lifshitz–Van der Waals forces, electrostatic attraction/repulsion, Lewis acid–base interactions and Brownian motion forces [15,26]. Since in the present study irreversible adhesion of cells to surfaces was prevented to maintain the non-destructive character of the procedure, the cells could only be observed inside the pyramidal defects if they could resist the hydraulic shear force caused by the cleaning of the surfaces prior to the observation under the microscope. The adhesion forces of bacteria to metals are influenced by electrostatic interactions and metal surface hydrophobicity, which in artificial sea water present the following zeta potential (in mV): Al—1754.0 ± 29.4; stainless steel—1364.4 ± 28.6; copper—449.4 ± 19.3 [37]. *Staphylococcus* sp. SH presented a more negative character (zeta potential of −33.9 mV) than *S. aureus* (zeta potential of −21.8 mV), but the same defects could be identified using both strains (Figure 5). The poor results observed with copper could result from lower cell adhesion forces due to its lower zeta potential which could lead to cells being wiped from the defects during the cleaning procedure. Although the exposure time during the assays was relatively short, the antimicrobial effect of metallic copper is also well known [27], and could have also influenced cell behaviour. The properties of the metals apparently had a higher influence on cell adhesion when magnetic and electric fields were applied to *R. erythropolis* cells and no general rule may be drawn (Figure 5). However, in general, smaller defects could be observed with *Rhodococcus* than with *Staphylococcus* cells.

The examples presented further demonstrate the feasibility of using bacterial cells on real samples made from different metals by different techniques as a valuable NDT to identify micro defects. From microscrews for dental implants to pulsed laser welding and engraved jewels, bacterial cells may be used to help the identification of defects produced during microfabrication. Furthermore, the cells also revealed very small scratches and holes that were inadvertently introduced on the metal surfaces of the samples during their manipulation. This suggests that further study could improve the detection limit of the technique in micro-manufactured and non-planar components (3D complex geometries) in different materials. The detection of micro and nano surface defects is a challenging area of NDT and is of paramount importance in current and arising applications such as transparent ceramics for optical applications and solar wafers where surface may also present an aesthetic, optical, or tribologic role.

One last comment goes to biosafety issues related to the procedure. *R. erythropolis* cells belong to biosafety level 1, and so standard microbiological practices are sufficient and the tests can be performed on an open laboratory bench. Although *Staphylococcus* species are part of the human flora (e.g., *S. aureus* is primarily found in the nose and on the skin of human beings), they belong to biosafety level 2, which includes microbes that are indigenous but associated with diseases of varying severity. If proper microbiological practices are followed, these microbes can be safely used in an open laboratory bench, as long as the potential for producing splashes or aerosols is low. Under such conditions, the use of laboratory coats and gloves as protective measures should be sufficient. At the beginning of this work, we tested the tolerance of both strains and found them tolerant to most commonly used antibiotics (data not shown). Currently, we are testing these bacterial strains in a lyophilised and dried state, and we are also testing commercially available level 1 bacteria known to respond to magnetic fields. It is envisaged that the cells may be provided in kits containing an appropriate medium, dye(s), and a disinfectant to be handled by technicians without microbiological training.

4. Materials and Methods

4.1. Strains and Bacterial Growth

Two *Staphylococcus* strains were isolated from samples collected from the space between the battery and the circuit board of mobile phones using sterile cotton swabs. The cells were screened for their ability to move in response to a magnetic field and were identified using the Sherlock® Microbial ID System (MIDI, Inc., Newark, DE, USA), as previously described [38,39]. In summary, tryptic soy agar plates containing each isolate were grown at 30 °C for 24 ± 1 h, after which the exponentially growing cells were harvested, and their fatty acids were extracted and methylated to fatty acid methyl esters (FAMEs) using the Instant FAME procedure according to the instructions provided by MIDI. FAMEs were analysed on a 6890N gas chromatograph from Agilent Technologies (Palo Alto, CA, USA).

The two isolates were also identified by 16S ribosomal RNA gene sequence analysis. The 16S rRNA-encoding genes of the two bacterial strains were determined according to previously described basic protocols [40]. Briefly, the genomic DNA was extracted from a 1 mL aliquot of an overnight grown culture using the commercial kit "High Pure PCR Template Preparation Kit" (Roche, Meylan, France), according to the manufacturer's instructions. The DNA concentration was estimated by assessing the absorbance at 260 nm, in a NanoDrop ND 1000 spectrophotometer (Thermo Fisher Scientific, Waltham, MA, USA). To amplify the 16S rRNA-encoding genes, a polymerase chain reaction (PCR) was performed with the Eubacteria universal primer pair E334F (5′CCAGACTCCTACGGGAGGCAGC3′) and E939R (5′ CTTGTGCGGGCCCCCGTCAATTC3′) [41]. Amplification mixtures were prepared in a total volume of 50 µL, containing 50 ng of total DNA as template, 200 µM of each deoxynucleotide, 0.5 µM of each primer, 1.5 mM $MgCl_2$, and 1U of Taq polymerase (Citomed). PCR conditions included an initial step of DNA denaturation at 95 °C for 5 min, followed by 30 cycles of 95 °C for 45 s, 58.2 °C for 60 s, and 72 °C for 45 s, and a final extension step at 72 °C for 7 min. Amplified fragments were visualised after electrophoresis in 0.8% (*w/v*) agarose gels and purified using the NZYGelpure kit (NZYTech, Lisbon, Portugal), according to the manufacturer's instructions. The DNA fragments were sequenced as a paid service by Eurofins MWG Operon (Ebersberg, Germany). The obtained sequences were aligned and compared with homologous sequences deposited in GenBank using the Basic Alignment Search Tool (BLAST, Bethesda MD, USA) [42].

For the different assays, the *Staphylococcus* cells were grown in Mueller–Hinton broth at 37 °C and 200 rpm in an Agitorb 200 (Aralab, Rio de Mouro, Portugal) incubator and collected by the end of the exponential phase.

Rhodococcus erythropolis DCL14 was isolated by the Division of Industrial Microbiology of the Wageningen University, in The Netherlands [43]. It is stored and maintained at iBB, IST, Portugal. The cells in the present study were grown in mineral medium [44] with 0.25% absolute ethanol or *n*-hexadecane as sole carbon sources at 28 °C and 200 rpm, and collected by the end of the exponential phase. When necessary (stated in the text), the mineral medium was supplemented with additional 0.005 $g \cdot L^{-1}$ of $FeSO_4 \cdot 7H_2O$, to reach a final concentration of 0.01 $g \cdot L^{-1}$.

4.2. Zeta Potential

The bacterial cells collected during the exponential phase were washed 3 times and suspended in 10 mM KNO_3. The zeta potential was calculated from the electrophoretic mobility (according to the method of Helmholtz-von Smoluchowski [45]) using a Doppler electrophoretic light scattering analyser (Zetasizer Nano ZS, Malvern Instruments Ltd., Worcestershire, UK). Calculations were made automatically using the Zetasizer Software 7.03, from Malvern Instruments Ltd.

4.3. Detection of Siderophores

Siderophore production was assessed in CAS agar plates, prepared according to Schwyn and Neilands [46], by inoculating the stock bacteria with the aid of a sterilised toothpick. The plates were incubated at 30 °C and monitored by image analysis every 24 h.

4.4. Surface Tension of Culture Supernatants

The surface tension of cell-free culture supernatants was measured by a plate device according to the Wilhelmy technique [47], using a K8 tensiometer from Krüss GmbH (Hamburg, Germany).

4.5. Application of Magnetic Fields under the Microscope

A customised functional prototype was developed to produce magnetic fields with peak intensities in horizontal and vertical directions, respectively. The magnetic fields were produced by a solenoid from a stator of a stepper motor when excited by an AC electrical current with modified sine wave at frequencies between 0.125 and 15 Hz. The stator was placed on the stage of an Olympus CX40 microscope (Olympus, Tokyo, Japan).

The cells were stained with a LIVE/DEAD® BacLight™ Bacterial Viability Kit from Molecular Probes (Life Technologies, Thermo Fisher Scientific, Waltham, MA, USA) containing a mixture of SYTO®9 green fluorescent nucleic acid stain (which stains all bacteria) and propidium iodide (which stains red only bacteria with damaged membranes). Cells were observed under fluorescent light provided by an Olympus U-RFL-T burner and an U-MWB mirror cube unit (excitation filter: BP450-480; barrier filter: BA515; Olympus, Tokyo, Japan) placed on the microscope.

To compare the movement of the different bacterial cells under the magnetic fields applied, photographs were captured during 10.4 s of exposure by an Evolution™ MP5.1 CCD colour camera using software Image-Pro Plus, both from Media Cybernetics, Inc. (Rockville, MD, USA). The distance traveled by each cell was calculated using Image-Pro Plus version 6.0. The images were calibrated using the TetraSpeck™ Fluorescent Microspheres Sampler Kit from Molecular Probes® (Life Technologies, Thermo Fisher Scientific, Waltham, MA, USA), which contains 0.1, 0.5, and 4.0 μm TetraSpeck™ beads.

4.6. Application of Electric Fields under the Microscope

A variable DC high voltage power supply was used to create electric fields between two parallel copper plates separated by 25 mm. Eight kV were applied between the plates, creating a constant electric field with an intensity of ca. 320 kV/m. The movement of the cells was followed as described for the magnetic fields.

Acknowledgments: C.C.C.R. de Carvalho acknowledges Fundação para a Ciência e a Tecnologia, I.P. (FCT), Portugal, for financial support under program "Investigador FCT 2013" (IF/01203/2013/CP1163/CT0002) and UID/BIO/04565/2013. TS and RM acknowledge FCT for PEst-OE/EME/UI0667/2014 and UID/EMS/00667/2013. The study was also partially supported by project PTDC/EME-TME/118678/2010. The authors would like to thank the R&D group of Teresa Vieira for the SEM images, and Jorge Leitão and Joana R. Feliciano for the 16S rRNA gene sequence analysis.

Author Contributions: C.C.C.R.C. and T.G.S. conceived and designed the experiments; C.C.C.R.C., P.L.I. and T.G.S. performed the experiments and analyzed the data; R.M.M. contributed with materials and scientific advising; C.C.C.R.C. wrote the paper.

Conflicts of Interest: The authors declare no conflict of interest.

References

1. Meng, J.; Loh, N.H.; Fu, G.; Tor, S.B.; Tay, B.Y. Replication and characterization of 316L stainless steel micro-mixer by micro powder injection molding. *J. Alloys Compd.* **2010**, *496*, 293–299. [CrossRef]
2. Ghoni, R.; Dollah, M.; Sulaiman, A.; Ibrahim, F.M. Defect characterization based on eddy current technique: Technical review. *Adv. Mech. Eng.* **2014**, *6*, 182496. [CrossRef]
3. Cherry, M.R.; Sathish, S.; Welter, J.; Reibel, R.; Blodgett, M.P. Development of high resolution eddy current imaging using an electro-mechanical sensor. *AIP Conf. Proc.* **2012**, *1430*, 324–331.
4. Abdelhamid, M.; Singh, R.; Omar, M. Review of microcrack detection techniques for silicon solar cells. *IEEE J. Photovolt.* **2014**, *4*, 514–524. [CrossRef]
5. Ibrahim, M.E. Nondestructive evaluation of thick-section composites and sandwich structures: A review. *Compos. Part A* **2014**, *64*, 36–48. [CrossRef]
6. Jolly, M.R.; Prabhakar, A.; Sturzu, B.; Hollstein, K.; Singh, R.; Thomas, S.; Foote, P.; Shaw, A. Review of non-destructive testing (NDT) techniques and their applicability to thick walled composites. *Procedia CIRP* **2015**, *38*, 129–136. [CrossRef]
7. McMaster, R.C. Liquid Penetrant Tests. In *Nondestructive Testing Handbook*, 3rd ed.; American Society for Nondestructive Testing: Columbus, OH, USA, 1982; Volume 2, p. 616.
8. Eisenmann, D.J.; Enyart, D.; Lo, C.; Brasche, L. Review of progress in magnetic particle inspection. *AIP Conf. Proc.* **2014**, *1581*, 1505–1510.
9. KTN. *A Landscape for the Future of NDT in the UK Economy*; Materials KTN: London, UK, 2014.

10. Hsu, J.W.P. Near-field scanning optical microscopy studies of electronic and photonic materials and devices. *Mater. Sci. Eng. R Rep.* **2001**, *33*, 1–50. [CrossRef]
11. Santos, T.G.; Miranda, R.M.; de Carvalho, C.C.C.R. A new NDT technique based on bacterial cells to detect micro surface defects. *NDT E Int.* **2014**, *63*, 43–49. [CrossRef]
12. Santos, T.G.; Miranda, R.M.; Vieira, M.T.; Farinha, A.R.; Ferreira, T.J.; Quintino, L.; Vilaça, P.; de Carvalho, C.C.C.R. Developments in micro- and nano-defects detection using bacterial cells. *NDT E Int.* **2016**, *78*, 20–28. [CrossRef]
13. García-Martín, J.; Gómez-Gil, J.; Vázquez-Sánchez, E. Non-destructive techniques based on eddy current testing. *Sensors* **2011**, *11*, 2525–2565. [CrossRef] [PubMed]
14. Van Oss, C.J. Hydrophobicity of biosurfaces—Origin, quantitative determination and interaction energies. *Colloids Surf. B Biointerfaces* **1995**, *5*, 91–110. [CrossRef]
15. Boks, N.P.; Norde, W.; van der Mei, H.C.; Busscher, H.J. Forces involved in bacterial adhesion to hydrophilic and hydrophobic surfaces. *Microbiology* **2008**, *154*, 3122–3133. [CrossRef] [PubMed]
16. Al-Abdalall, A.H.A. Isolation and identification of microbes associated with mobile phones in Dammam in eastern Saudi Arabia. *J. Fam. Community Med.* **2010**, *17*, 11–14. [CrossRef] [PubMed]
17. Gashaw, M.; Abtew, D.; Addis, Z. Prevalence and antimicrobial susceptibility pattern of bacteria isolated from mobile phones of health care professionals working in Gondar town health centers. *ISRN Public Health* **2014**, *2014*. [CrossRef]
18. Beasley, F.C.; Heinrichs, D.E. Siderophore-mediated iron acquisition in the staphylococci. *J. Inorg. Biochem.* **2010**, *104*, 282–288. [CrossRef] [PubMed]
19. De Carvalho, C.C.C.R.; Marques, M.P.C.; Fernandes, P. Recent achievements on siderophore production and application. *Recent Pat. Biotechnol.* **2011**, *5*, 183–198. [CrossRef] [PubMed]
20. Brown, J.S.; Holden, D.W. Iron acquisition by gram-positive bacterial pathogens. *Microb. Infect.* **2002**, *4*, 1149–1156. [CrossRef]
21. Schmaler, M.; Jann, N.J.; Ferracin, F.; Landolt, L.Z.; Biswas, L.; Götz, F.; Landmann, R. Lipoproteins in *Staphylococcus aureus* mediate inflammation by TLR2 and iron-dependent growth in vivo. *J. Immunol.* **2009**, *182*, 7110–7118. [CrossRef] [PubMed]
22. Vainshtein, M.; Suzina, N.; Kudryashova, E.; Ariskina, E. New magnet-sensitive structures in bacterial and archaeal cells. *Biol. Cell* **2002**, *94*, 29–35. [CrossRef]
23. De Carvalho, C.C.C.R.; Wick, L.; Heipieper, H. Cell wall adaptations of planktonic and biofilm *Rhodococcus erythropolis* cells to growth on C5 to C16 *n*-alkane hydrocarbons. *Appl. Microbiol. Biotechnol.* **2009**, *82*, 311–320. [CrossRef] [PubMed]
24. De Carvalho, C.C.C.R.; da Fonseca, M.M.R. Preventing biofilm formation: Promoting cell separation with terpenes. *FEMS Microbiol. Ecol.* **2007**, *61*, 406–413. [CrossRef] [PubMed]
25. Rodrigues, C.J.C.; de Carvalho, C.C.C.R. *Rhodococcus erythropolis* cells adapt their fatty acid composition during biofilm formation on metallic and non-metallic surfaces. *FEMS Microbiol. Ecol.* **2015**, *91*. [CrossRef] [PubMed]
26. Hermansson, M. The DLVO theory in microbial adhesion. *Colloids Surf. B Biointerfaces* **1999**, *14*, 105–119. [CrossRef]
27. De Carvalho, C.C.C.R.; Caramujo, M.J. Ancient procedures for the high-tech world: Health benefits and antimicrobial compounds from the Mediterranean empires. *Open Biotechnol. J.* **2008**, *2*, 235–246. [CrossRef]
28. Villapún, V.; Dover, L.; Cross, A.; González, S. Antibacterial metallic touch surfaces. *Materials* **2016**, *9*, 736. [CrossRef]
29. De Carvalho, C.C.C.R. Biofilms: New ideas for an old problem. *Recent Pat. Biotechnol.* **2012**, *6*, 13–22. [CrossRef] [PubMed]
30. Girardello, R.; Bispo, P.J.M.; Yamanaka, T.M.; Gales, A.C. Cation concentration variability of four distinct Mueller-Hinton agar brands influences polymyxin b susceptibility results. *J. Clin. Microbiol.* **2012**, *50*, 2414–2418. [CrossRef] [PubMed]
31. Morrissey, J.A.; Cockayne, A.; Brummell, K.; Williams, P. The staphylococcal ferritins are differentially regulated in response to iron and manganese and via PerR and Fur. *Infect. Immun.* **2004**, *72*, 972–979. [CrossRef] [PubMed]
32. Frankel, R.B.; Blakemore, R.P.; Wolfe, R.S. Magnetite in freshwater magnetotactic bacteria. *Science* **1979**, *203*, 1355–1356. [CrossRef] [PubMed]

33. Pereira, A.S.; Timóteo, C.G.; Guilherme, M.; Folgosa, F.; Naik, S.G.; Duarte, A.G.; Huynh, B.H.; Tavares, P. Spectroscopic evidence for and characterization of a trinuclear ferroxidase center in bacterial ferritin from *Desulfovibrio vulgaris* hildenborough. *J. Am. Chem. Soc.* **2012**, *134*, 10822–10832. [CrossRef] [PubMed]

34. Owen, C.S.; Lindsay, J.G. Ferritin as a label for high-gradient magnetic separation. *Biophys. J.* **1983**, *42*, 145–150. [CrossRef]

35. Tuğrul, A.B. Capillarity effect analysis for alternative liquid penetrant chemicals. *NDT E Int.* **1997**, *30*, 19–23. [CrossRef]

36. Sapienza, R.S.; Ricks, W.F.; Grunden, B.L.; Heater, K.J.; Badowski, D.E.; Sanders, J.W. *Environmentally Acceptable Alternatives for Non-Destructive Inspection with Fluorescent Penetrant Dyes*; METSS Corporation: Arlington, TX, USA, 2003.

37. Sheng, X.; Ting, Y.P.; Pehkonen, S.O. Force measurements of bacterial adhesion on metals using a cell probe atomic force microscope. *J. Colloid Interface Sci.* **2007**, *310*, 661–669. [CrossRef] [PubMed]

38. Kunitsky, C.; Osterhout, G.; Sasser, M. Identification of microorganisms using fatty acid methyl ester (fame) analysis and the midi sherlock microbial identification system. In *Encyclopedia of Rapid Microbiological Methods*; Miller, M.J., Ed.; Davis Healthcare International Publishing, LLC: Baltimore, MD, USA, 2005; Volume 3, pp. 1–17.

39. de Carvalho, C.C.C.R.; Caramujo, M.J. Bacterial diversity assessed by cultivation-based techniques shows predominance of *Staphylococccus* species on coins collected in Lisbon and Casablanca. *FEMS Microbiol. Ecol.* **2014**, *88*, 26–37. [CrossRef] [PubMed]

40. Nadais, H.; Barbosa, M.; Capela, I.; Arroja, L.; Ramos, C.G.; Grilo, A.; Sousa, S.A.; Leitão, J.H. Enhancing wastewater degradation and biogas production by intermittent operation of UASB reactors. *Energy* **2011**, *36*, 2164–2168. [CrossRef]

41. Rudi, K.; Skulberg, O.M.; Larsen, F.; Jakobsen, K.S. Strain characterization and classification of oxyphotobacteria in clone cultures on the basis of 16s rRNA sequences from the variable regions v6, v7, and v8. *Appl. Environ. Microbiol.* **1997**, *63*, 2593–2599. [PubMed]

42. Johnson, M.; Zaretskaya, I.; Raytselis, Y.; Merezhuk, Y.; McGinnis, S.; Madden, T.L. Ncbi blast: A better web interface. *Nucl. Acids Res.* **2008**, *36*, W5–W9. [CrossRef] [PubMed]

43. Van der Werf, M.J.; Swarts, H.J.; de Bont, J.A.M. *Rhodococcus erythropolis* DCL14 contains a novel degradation pathway for limonene. *Appl. Environ. Microbiol.* **1999**, *65*, 2092–2102. [PubMed]

44. Cortes, M.A.L.R.M.; de Carvalho, C.C.C.R. Effect of carbon sources on lipid accumulation in *Rhodococcus* cells. *Biochem. Eng. J.* **2015**, *94*, 100–105. [CrossRef]

45. Hiemenz, P.C.; Rajagopalan, R. *Principles of Colloid and Surface Chemistry*, 3rd ed.; Marcel Dekker Inc.: New York, NY, USA, 1997; p. 672.

46. Schwyn, B.; Neilands, J.B. Universal chemical assay for the detection and determination of siderophores. *Anal. Biochem.* **1987**, *160*, 47–56. [CrossRef]

47. Adamson, A.W. *Physical Chemistry of Surfaces*; John Wiley & Sons Inc.: New York, NY, USA, 1990; p. 777.

materials

MDPI

Article

A Novel Approach to Estimate the Plastic Anisotropy of Metallic Materials Using Cross-Sectional Indentation Applied to Extruded Magnesium Alloy AZ31B

Mingzhi Wang, Jianjun Wu *, Hongfei Wu, Zengkun Zhang and He Fan

School of Mechanical Engineering, Northwestern Polytechnical University, Xi'an 710072, China;
wangmz_nwpu@foxmail.com (M.W.); WuHongfei321@163.com (H.W.); 18292068849@163.com (Z.Z.);
henwpu@foxmail.com (H.F.)
* Correspondence: wujj@nwpu.edu.cn; Tel.: +86-29-8849-3101

Received: 21 July 2017; Accepted: 7 September 2017; Published: 11 September 2017

Abstract: In this paper, a methodology is presented for obtaining the plastic anisotropy of bulk metallic materials using cross-sectional indentation. This method relies on spherical indentation on the free edge of a specimen, and examining the out-of-plane residual deformation contour persisting on the cross-section after unloading. Results obtained from numerical simulation revealed that some important aspects of the out-of-plane residual deformation field are only sensitive to the extent of the material plastic anisotropy, and insensitive to strain hardening, yield strain, elastic anisotropy, and the selected displacement threshold value. An explicit equation is presented to correlate the plastic anisotropy with the characteristic parameter of the bottom shape of residual deformation contour, and it is used to uniquely determine the material plastic anisotropy in cross-sectional indentation. Effectiveness of the proposed method is verified by application on magnesium alloy AZ31B, and the plastic anisotropy parameter obtained from indentation and uniaxial tests show good agreement.

Keywords: indentation; plastic anisotropy; parameter identification; metallic materials

---

1. Introduction

Indentation testing has long been used as a simple and effective method to extract the mechanical properties of materials, e.g., hardness [1–3]. Recent advances in depth-sensing instrumented indentation equipment and optical profiling techniques have greatly stimulated the tremendous interests in characterization of various mechanical properties of materials, e.g., elastic modulus [4], stress strain curves [5,6], fracture toughness [7,8], wear properties [9–11], as well as the material anisotropy [12–17].

In the nature and synthetic material systems, the anisotropic materials are often observed and widely used in industrial products, such as rolled sheets, composites, extrusions, and so on [18]. The plastic anisotropy is a very important factor that should be considered in the material design and application stages [19–21]. On the one hand, the plastic anisotropy has non-negligible influence on the final shapes of metal products made by pressure forming, e.g., the in-plane plastic anisotropy is tightly related to the tendency of rolled sheets to form ears during drawing [18,22]. As one example, magnesium alloys have received significant attention in industrial applications because of the excellent physical properties, e.g., light weight/high strength [23,24]. However, it was reported that this material has relatively poor ductility and obvious plastic anisotropy at room temperature because of the pronounced basal-type texture, e.g., magnesium alloy AZ31 [24,25]. Therefore, a large number of prior works were dedicated to the improvement of the formability and performance of magnesium alloys, such as the usage of different rolling techniques [26–28], and the introduction of rare earth [29,30].

On the other hand, the anisotropy of static mechanical properties is tightly related to the micro-texture and micro-structure of materials [31–34]. For the lithium aluminum alloys extrusion, the deformation texture created during shaping usually causes high levels of anisotropy, such as the asymmetric flow and in-plane plastic anisotropy [32–34]. In aerospace applications, a lower strength in the off-axis direction will neglect the benefits of high strength in the working direction, e.g., the wing skins and fuselage stringers [32,33]. Therefore, it is especially important to pay attention to the mechanical characterization of material plastic anisotropy.

The traditional method for the measurement of material plastic anisotropy relies on the uniaxial experiments along different orthogonal directions, e.g., uniaxial tensile/compression tests. However, this testing method is not applicable for the specimens with finite volumes [16]. Therefore, researchers are seeking alternative ways to extract the material plastic anisotropy, e.g., using indentation, by examining the non-uniform surface displacement distribution around indenter [16,17,35]. Additionally, many researchers [13–15] resorted to the sophisticated numerical optimization algorithms to calibrate the prescribed anisotropic model, e.g., Hill's criterion [36], by minimizing a suitably-defined error norm between experiment and simulation. However, the inverse parameters identification processes in these methods usually involve complicated parameters regression and the extensive off-line numerical simulations [16]. It is noted that the stress field induced by indentation is essentially under multi-axial state. Therefore, the indentation actually reflects an "averaged" material response [12,37]. For the anisotropic materials, the indentation response is determined by all the constitutive parameters of the postulated constitutive law. Thus, it is challengeable to correlate solely the plastic anisotropy parameter with a specific material response amount, without taking account of the influence of the other constitutive parameters.

In this paper, we proposed a novel method to estimate the plastic anisotropy parameter of metal materials by examining the out-of-plane residual deformation field in the cross-sectional indentation. The advantage of this method is that the shape of the out-of-plane deformation contour induced by cross-sectional indentation is only sensitive to the material plastic anisotropy, and insensitive to strain hardening, yield strain, and elastic anisotropy. Therefore, the extent of material plastic anisotropy can be uniquely correlated with the indentation response, e.g., out-of-plane residual deformation field.

2. Experiment Investigation

2.1. Material

The material studied here is the magnesium alloy extrusion AZ31B. This material is widely used in industrial applications for its light weight/low density, and good mechanical properties [23,24]. The chemical composition of this alloy is listed in Table 1, and it refers to the mass percentage. The original shape of the material sample is rounded bar, and the extrusion direction is along the axis. Due to the extrusion process, this material exhibits strong plastic anisotropy. Uniaxial compression experiments were used to investigate the plastic anisotropy of this alloy so that the plastic anisotropy parameter estimated from indentation and uniaxial tests can be comparable.

Table 1. Chemical composition of magnesium alloy AZ31B.

Chemistry	Al	Zn	Mn	Si	Fe	Cu	Ni	Mg
wt %	2.5–3.5	0.6–1.4	0.2–1.0	≤0.80	≤0.003	≤0.01	≤0.001	Balance

The compression experiment was implemented on the universal material test machine (CIMACH, Changchun, China), respectively, along the longitudinal and transverse directions. Specimens used in the uniaxial tests were cylindrical, with diameters of 10 mm and heights of 12 mm. Strain during the compression process was measured with the assistance of strain gages and data acquisition devices. The maximum compression ratio was about 10%. Stress–strain curves in uniaxial compression processes were fitted using the Hollomon hardening law. The uniaxial experimental result is listed in Table 2. In this table, the symbols "1, T" and "3, T" represent the transverse direction, while "2, L" represents the longitudinal direction. Results show the yield stress of this material exhibits strong anisotropy along the longitudinal and transverse directions, and the yield stress ratio between longitudinal and transverse directions is about 1.51, while the elastic modulus and strain hardening exponent show very slight anisotropy.

Table 2. Uniaxial compression experimental data of magnesium alloy AZ31B.

AZ31B	E (GPa)	σ_y (MPa)	n
1, T	33.5	64.2	0.287
3, T	34.5	68.5	0.260
2, L	32.3	100.5	0.268

2.2. Cross-Sectional Indentation

Figure 1a shows the material coordinate and specimen used in cross-sectional indentation experiment. The specimen was cut along the diagonal direction (45° to the transverse direction) in the T-L plane. Then the two cross-sectional surfaces were carefully polished to a mirror finish and glued together. The rapid glue used in the indentation experiment was a soft type, and the glue thickness was less than 20 μm. The top and bottom surfaces of the specimen were flat, in order to avoid the tilt of the specimen in the indentation loading process. Spherical indentation was implemented on the hardness tester at room temperature, and the indenter was a tungsten ball with a diameter of 5 mm. The indenter was vertically pressed against the free edge along the bounded interface, so that the material under the indenter is able to flow freely along the out-of-plane (vertical to the cross-sectional surface) direction. Before indentation, the surface of the specimen was carefully polished to a mirror finish. Two different maximum indentation loads, Load-1: P_{max} = 400 N and Load-2: P_{max} = 500 N were, respectively, used in the indentation experiments. The holding time was 15 s. Figure 1b shows the residual imprint left on the surface of specimen (Load-1), the out-of-plane flowing direction (the red arrows), and the interface line.

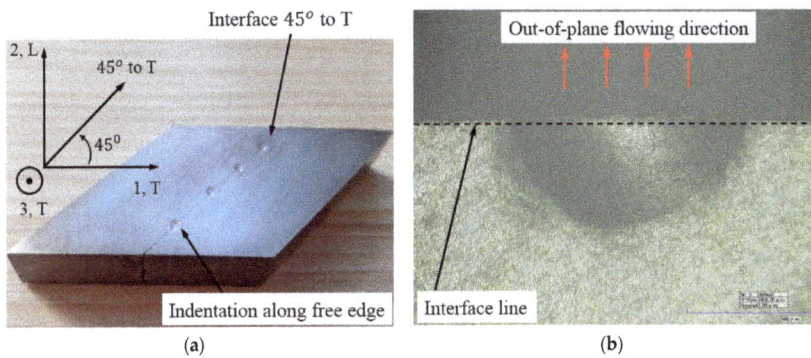

Figure 1. (a) Specimen used in cross-sectional indentation experiment; and (b) the residual imprint left on the surface of specimen.

The out-of-plane deformation field persisted on the interface surface after unloading is measured using the optical profiling systems (WYKO NT1100, Veeco, Champaign, IL, USA) and the results are, respectively, shown in Figure 2a for Load-1 and Figure 2b for Load-2. This measuring process is based on the white light interferometry theory, and high-resolution 3D surface measurements can be obtained. The orientation of the deformation plane in the experiment can be described by the material coordinate defined in Figure 1. The red dotted line in Figure 2 represents the indentation center axis.

It shows that the out-of-plane residual deformation contour deviates from the indentation center axis obviously. We recall that the geometry of the indenter-specimen system is essentially symmetric along the indentation center axis. Therefore, it may be deduced that the asymmetric deformation field persisting on the interface surface is caused by the material plastic anisotropy. This problem will be further investigated in Sections 3 and 4.

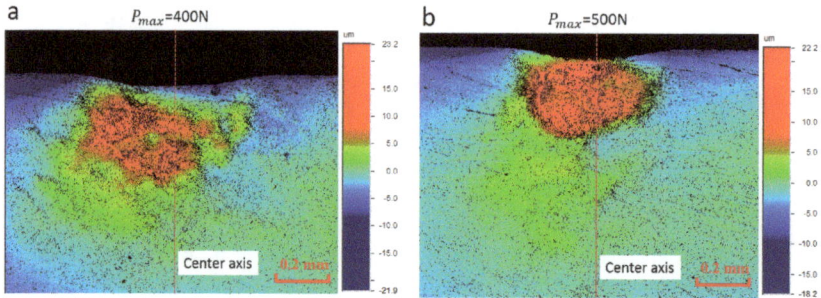

Figure 2. Out-of-plane residual deformation field obtained from indentation experiments using different prescribed loads: (**a**) the prescribed indentation load is 400 N; and (**b**) the prescribed indentation load is 500 N.

3. Methods

3.1. Simulation Setup

Figure 3 shows the schematic of the cross-sectional indentation model for the anisotropic materials. The anisotropic material considered in the present study exhibits different plastic properties, (e.g., yield stress) along the two orthogonal directions (e.g., longitudinal and transverse), as shown by the material coordinate defined in Figure 3. This material coordinate is used throughout the entire numerical study. The cross-section is designed along the diagonal direction (e.g., 45° to the transverse direction). The indenter is pressed against the free edge of the cross-section, and the loading direction is vertical to the T-L plane. In the indentation process, the cross-section is a free surface with no boundary constraints.

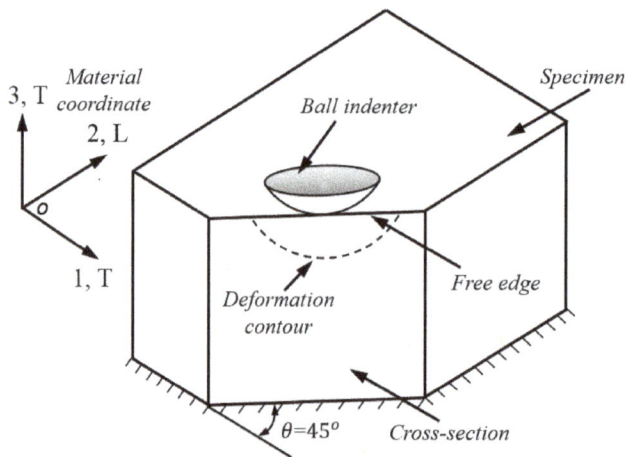

Figure 3. Schematic of the cross-sectional indentation model.

Hill's plasticity theory [36] is used to describe the deformation behaviors of anisotropic materials in cross-sectional indentation simulation. The general state of this yield function is expressed in Equation (1):

$$f(\sigma) = \sqrt{F(\sigma_{22} - \sigma_{33})^2 + G(\sigma_{33} - \sigma_{11})^2 + H(\sigma_{11} - \sigma_{22})^2 + 2L\tau_{23}^2 + 2M\tau_{31}^2 + 2N\tau_{12}^2} \tag{1}$$

where, F, G, H, L, M, and N are anisotropic parameters of the current state of anisotropy [16], and these parameters can be determined by using Equation (2) [38]. The normal and shear yield stress along three orthogonal axes (e.g., 1, 2, and 3 in the material coordinate) are defined as σ_{11}, σ_{22}, σ_{33} and τ_{12}, τ_{31}, τ_{23}, respectively:

$$F = \frac{1}{2}\left(\frac{1}{R_{22}^2} + \frac{1}{R_{33}^2} - \frac{1}{R_{11}^2}\right); G = \frac{1}{2}\left(\frac{1}{R_{33}^2} + \frac{1}{R_{11}^2} - \frac{1}{R_{22}^2}\right); H = \frac{1}{2}\left(\frac{1}{R_{11}^2} + \frac{1}{R_{22}^2} - \frac{1}{R_{33}^2}\right); L = \frac{3}{2R_{23}^2}; M = \frac{3}{2R_{13}^2}; N = \frac{3}{2R_{12}^2} \tag{2}$$

In Equation (2), the six anisotropic yield stress ratios, R_{11}, R_{22}, R_{33}, R_{12}, R_{13}, and R_{23} in, respectively, three normal (R_{11}, R_{22}, and R_{33}) and three shear (R_{12}, R_{13}, and R_{23}) directions are used to quantify the orthogonal anisotropic plasticity, as shown by the material coordinate defined in Figure 3. These six anisotropic constants are inputted by using the POTENTIAL sub-option in ABAQUS software (Dassault, Paris, France) [38]. The R-values are defined by the reference yield stress σ_Y, and the reference shear yield stress τ_Y is defined as $\tau_Y = \sigma_Y / \sqrt{3}$ according to von Mises criterion. It is noted that the R-value defined here is different from the strain ratio r, and the latter is usually called the Lankford index [22]. For the anisotropic materials studied here, the other five R-values are maintained at unity, and only R_{22} is varied to simulate the in-plane anisotropic plasticity along the 2, L direction. More details about the anisotropic material model studied here can be found in [16,18,38]. The transverse direction is along 1, 3, T axis, and the longitudinal direction is along the 2, L axis. The stress-strain curve along transverse direction is defined as the reference input amount, and the yield stress along the longitudinal axis is determined by using the relation $\sigma_{YL} = R_{22}\sigma_{YT}$. The strain hardening behavior of material is described using the Hollomon hardening law, with a single strain hardening exponent "*n*" for both the longitudinal and transverse directions [18].

3.2. Finite Element Modelling

The ABAQUS/standard commercial codes [38] were used in cross-sectional indentation simulation. The finite element (FE) model, meshes, and boundary conditions are shown in Figure 4. The spherical indenter was assumed as discrete rigid using the R3D4 element type. Its diameter was 5 mm. The specimen was modeled using the C3D8R element type, and refined meshes were created around the local contact regions between indenter and specimen. Minimum element size in this refined region was 50 μm. Poisson's ratio of specimen was fixed at 0.3, because Poisson's ratio is a minor factor in indentation studies [39,40]. The height and radius of specimen was 9.6 mm, and this value is large enough to avoid the influence of outer boundary effects. The total number of elements were 20,232 for the specimen and 4000 for the indenter. Contact friction on the surfaces between the indenter and specimen was defined at 0.1, because the contact friction between metals and diamond is around this value [41,42]. Displacement of the bottom nodes of the specimen was fixed. No boundary constraints were applied on the free surface, so that material during indentation is able to flow freely along the out-of-plane direction (vertical to the free surface, as shown in Figure 3). Vertical displacement of the indenter was controlled by the force increments, up to a prior defined maximum value, and then the withdrawal of the indenter was simulated in one step.

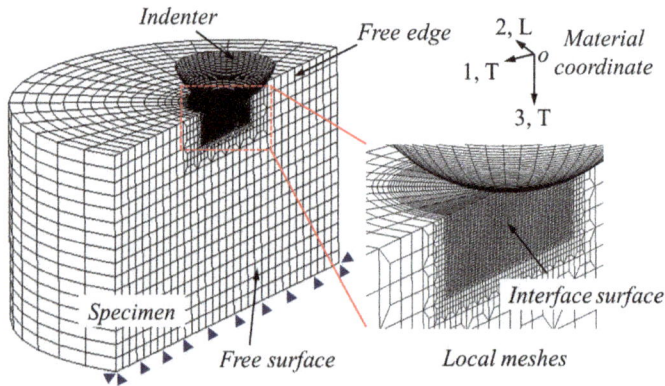

Figure 4. FE model, meshes, and boundary conditions used in the cross-sectional indentation simulation.

3.3. Characterization of the Out-of-Plane Deformation Contour

We first performed the FE simulation on a test case, where E = 400 GPa, σ_{YT} = 200 MPa, n = 0.3, and R_{22} = 1.5. The prescribed maximum indentation load is fixed at 800 N. The out-of-plane (vertical to the cross-section) deformation contour of this tested material is shown in Figure 5. The red dotted line is the indentation center axis. Here, a critically-selected displacement threshold value, U_{thr} = 2 μm is used. Figure 5 shows clearly that the bottom shape of the out-of-plane deformation contour is not symmetrical with respect to the indentation center axis, and it deviates toward a specific direction (T-side).

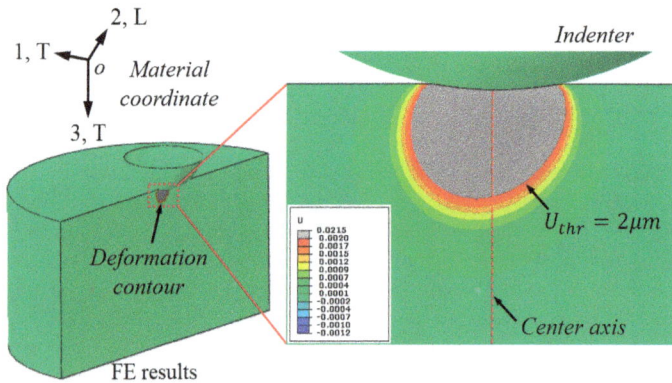

Figure 5. Out-of-plane deformation contour persisting on the cross-section after unloading.

Figure 6 shows the bottom shape of the out-of-plane deformation contour can be well approximated by using two circles, with diameters d_T and d_L, respectively. This approximation process is very simple. On each side of the deformation contour, select three points on the displacement threshold line. Then, these three points can be used to uniquely identify one approximation circle, as shown in Figure 6. Given the asymmetric shape of the out-of-plane deformation contour persisted on the cross-section after unloading, we hypothesized that the deviation of the bottom shape of the deformation field is caused by the difference of yield stress σ_{YT} and σ_{YL} in indentation simulation. Additionally, the deformation contour deviates toward an "easier" side. In the study, the relative position of the two approximation circles is described by the ratio of their corresponding diameters

R_d, as $R_d = d_L/d_T$. Therefore, R_d can be used to reflect the deviation extent of the bottom shape of the out-of-plane residual deformation contour in cross-sectional indentation.

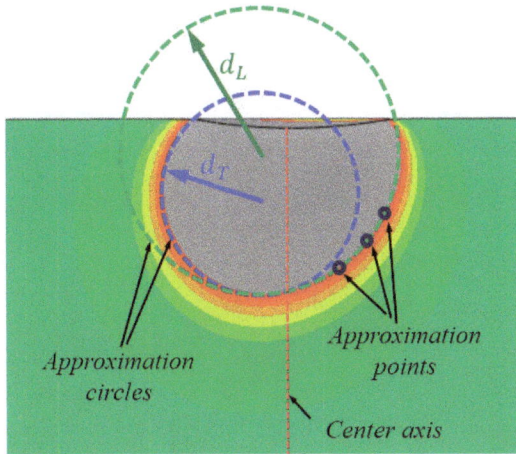

Figure 6. Description of the bottom shape of the out-of-plane deformation contour using two circles.

4. Fundamental Relationship between R_{22} and R_d

4.1. Results Obtained from Numerical Simulation

In this section, the possible factor that influences the out-of-plane deformation contour is fully investigated. Figure 7 shows the FE simulation results of a test case, where $E = 400$ GPa, $\sigma_{YT} = 200$ MPa, $n = 0.3$, and the R_{22} value is varied from 0.6 to 2.0. The prescribed indentation load is 800 N, and the selected displacement threshold value is fixed at $U_{thr} = 2$ μm. It is clearly shown in Figure 7 that the deviation direction (see the red arrows) of the out-of-plane deformation contour changes monotonously with R_{22} increases. When $R_{22} < 1$, the contour line deviates toward the L side, while its deviation direction changes gradually to the T side when $R_{22} > 1$. Additionally, when $R_{22} = 1$, the contour line is symmetric. Therefore, the deformation contour always deviates toward an "easier" side, where the yield stress is lower.

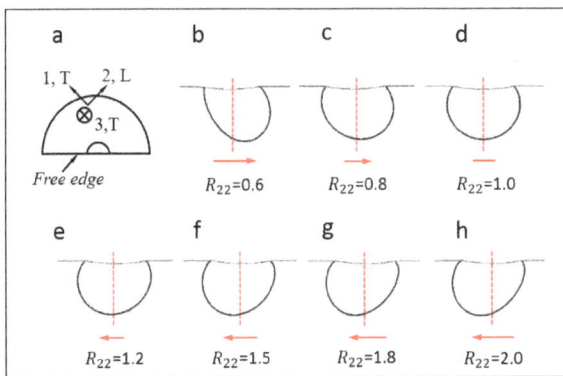

Figure 7. Influence of the R_{22} value on the shape of the out-of-plane deformation contours.

Figure 8 shows the evolution of the relative position of two approximation circles with R_{22} increases. It shows clearly that when $R_{22} < 1$, d_T is larger than d_L ($R_d < 1$) while d_T becomes smaller than d_L ($R_d > 1$) when $R_{22} > 1$. Additionally, when the material is isotropic ($R_{22} = 1$), the two approximation circles are nearly coincident ($d_T = d_L$), and $R_d = 1$. Therefore, there exists a strong relevance between the bottom shape of the out-of-plane deformation contour (R_d value) and the plastic anisotropy parameter R_{22}.

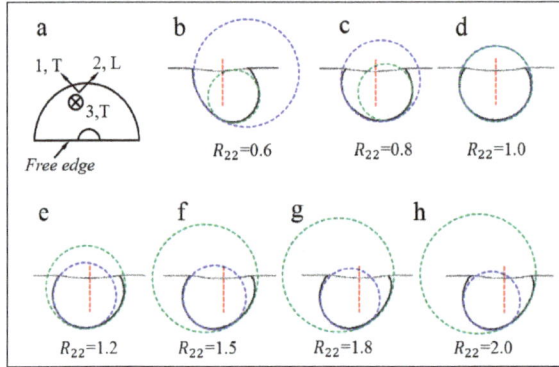

Figure 8. Evolution of the relative position between two approximation circles as R_{22} increases.

Figure 9 shows the FE simulation results of another test case, where the influence of the strain hardening exponent, yield strain, elastic anisotropy, and the selected displacement threshold value, on the shape of out-of-plane deformation contour is systematically investigated. In all the simulations, the material is assumed as isotropic ($R_{22} = 1.0$). In Figure 9, it shows all the deformation contours are symmetric, as it was depicted in Figure 7d. Although, the magnitudes of these contour lines are different, the two approximation circles are nearly coincident.

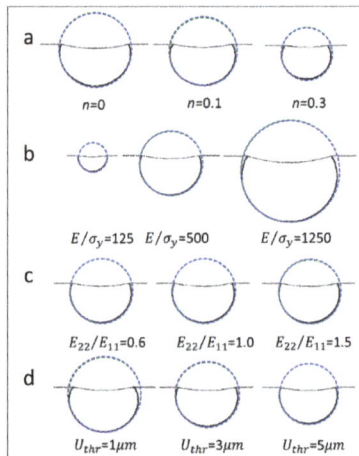

Figure 9. Influence of the strain hardening exponent, yield strain, elastic anisotropy, and the selected displacement threshold value on the shape of contour lines (for all the simulations in this figure, R_{22} is fixed at 1.0).

Similar result can be found in Figure 10, where the R_{22} value in all these simulations is fixed at 1.5. Figure 10 shows that the magnitudes of these contour lines are different, while the relative position of the two approximation circles are nearly the same, as it was depicted in Figure 8f. Therefore, the results indicate the strain hardening exponent, yield strain, elastic anisotropy, and the selected displacement threshold value hardly influence the bottom shape of the out-of-plane deformation contour. It is noted that the orientation of the deformation planes shown in Figures 7–10 are the same.

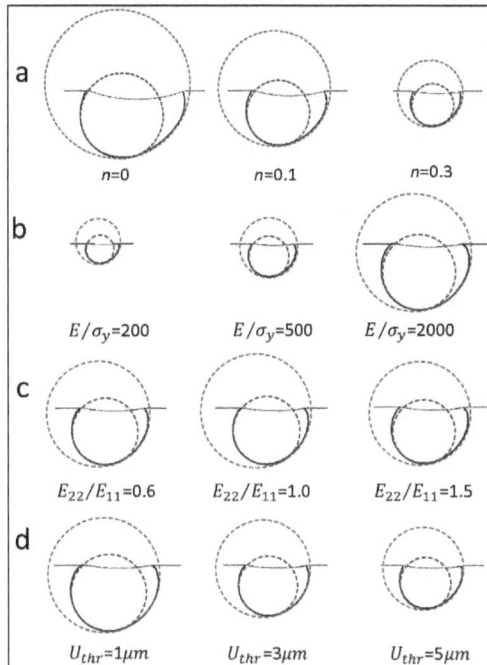

Figure 10. Influence of the strain hardening exponent, yield strain, elastic anisotropy, and the selected displacement threshold value on the relative position between two approximation circles (for all the simulations in this figure, R_{22} is fixed at 1.5).

Figure 11 shows the extracted R_d values using all the contour lines in Figure 10a–d, respectively, under four different situations. Figure 11 shows the R_d values are very close. The small difference of R_d values in Figure 11 is probably caused by some uncertain factors, e.g., numerical oscillations in the FE simulation and the approximation error. Therefore, the result indicates the bottom shape of the out-of-plane deformation contour is only dependent on the R_{22} value, and independent of the strain hardening exponent, yield strain, elastic anisotropy and the selected displacement threshold value.

Figure 12 shows the relationship between R_{22} and R_d. The result is extracted from Figure 8, and its relationship is approximated by using a second polynomial function, as expressed in Equation (3). Its shows the relationship between R_{22} and R_d is monotonic. Therefore, Equation (3) can be used to uniquely estimate the anisotropic parameter R_{22}, when the R_d value is obtained from the cross-sectional indentation.

$$R_{22} = 0.1972R_d^2 + 0.3639R_d + 0.4137 \tag{3}$$

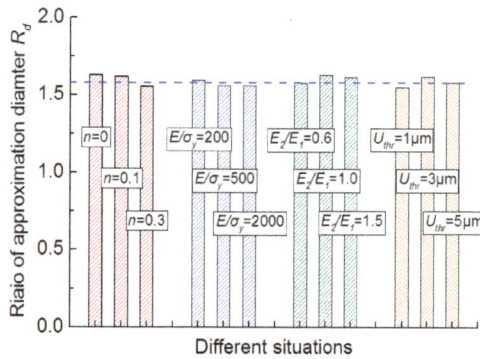

Figure 11. Influence of the strain hardening exponent, yield strain, elastic anisotropy, and the selected displacement threshold value on the diameter ratio, R_d of the two approximation circles.

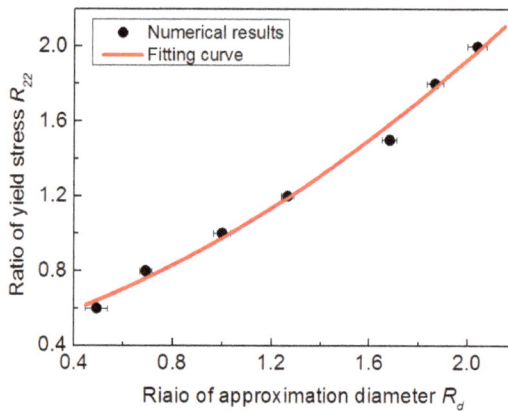

Figure 12. Relationship between R_{22} and R_d.

4.2. Comparison with the Experiment

Figure 13 shows the approximation results of the out-of-plane residual deformation field using two circles. Here, the selected U_{thr} value is about 10 µm. Results show the R_d values are about 1.56 for Load-1, and 1.65 for Load-2. Figure 14 shows the estimated R_{22} value using Equation (3), and its comparison with the uniaxial test data. A good agreement can be found, as shown in Figure 14. The error of the R_{22} value between indentation and uniaxial tests are about −3.31% under Load-1, and +2.65% under Load-2. Results indicate the proposed method in the present study is very effective, and the reproducibility is good. Additionally, an indentation load of 500 N provides a more accurate result. In the proposed methodology it is very important to obtain the effective experimental data, e.g., the residual deformation field since the fitting of the circle is dependent on the selected three approximation points. This approximation is reliable when the displacement contour is well-defined. In the numerical simulation, the displacement contours are obtained from the FE simulation and they are well-defined while, in the experiment, the fitting result may be influenced by the quality of the experimental data, because the displacement contour can be influenced by some uncertainty factors, such as the prescribed indentation load, surface evenness and roughness, the glue thickness, and so on. Thus, it is essential to improve the quality of the experimental data.

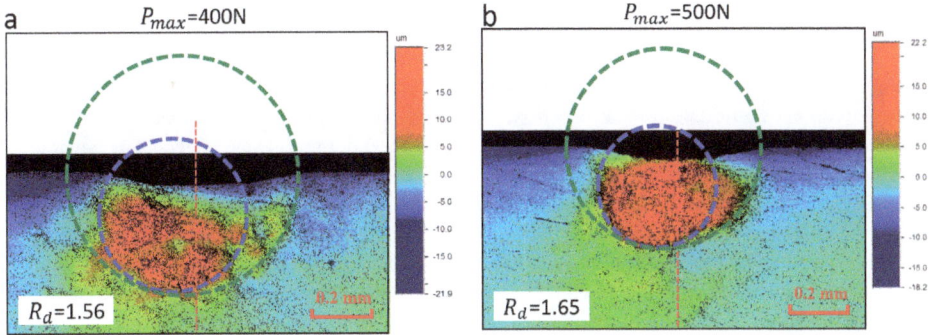

Figure 13. Out-of-plane deformation contour: (a) The prescribed indentation load is 400 N; (b) The prescribed indentation load is 500 N.

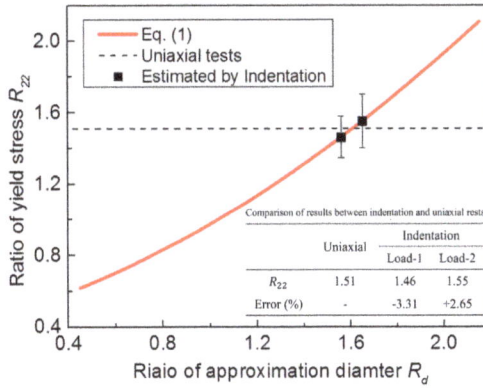

Figure 14. Comparison of the estimated R_{22} value between indentation and uniaxial tests.

In the experiment, the indentation load should be properly selected. On the one hand, elastic plastic transition [5] will occur with load increases. Thus, the load should be larger, so that the deformation is plasticity dominates, while too higher a load may result in the occurrence of some uncertain factors, such as the fracture along the edge. Therefore, it is suggested here that the load can be optimized in the real experiment using the trial and error method to obtain a better result, e.g., a well-defined displacement contour. On the other hand, it should be noted that the quality of the experimental data depends on the combined effects between the load and the other uncertain experiment factors, e.g., surface roughness and glue thickness. In the experiments, two indentation loads, 400 N and 500 N, were selected, because they gave good experimental data. When the experiment error (e.g., surface evenness and roughness) is fixed at a certain level, a higher indentation load may induce a larger out-of-plane displacement value, thus it can reduce the influence of these uncertain factors on the quality of the displacement contour. However, this effect may be simultaneously influenced by the glue thickness when the displacement value is too large. From this aspect, the current experiment condition has a lot of room for the improvement, e.g., higher experiment precision, or a more properly designed glue thickness value. These are open questions, and will be further studied in our future work.

5. Discussion

In the present study, a novel approach is established to estimate the plastic anisotropy of bulk metallic materials by examining the bottom shape of the out-of-plane residual deformation contour in cross-sectional indentation. It is noted that the indentation size in the present study is within the macro-scale (e.g., the indenter radius is 2.5 mm), so the indentation size-effect is not considered. Additionally, the top and bottom surfaces of the specimen should be flat in order to avoid the tilt of the specimen during the indentation loading process. The influence of tilt angle, indenter offset, glue thickness, and the other possible factors, e.g., size-effect, grain size, on the shape of the out-of-plane deformation field will be further investigated in our future work. In numerical computations, a relatively simple yield criterion, e.g., Hill's plasticity theory [36], with isotropic strain hardening behaviors was used. Although more complex yield criterion may bringing more accurate numerical results, e.g., Barlat 91 criterion [43], this problem is out of the current research scope. In addition, the influence of element size on the simulation output was systematically investigated. Results indicate the more refined meshes in the simulation model are able to give a more precise description of the residual deformation field, while this improvement is very limited, and the computation burden will increase accordingly. Additionally, the selected element size in the present study is sufficient to give a very precise description of the material deformation field.

The established methodology in the present study can be used to estimate the plastic anisotropy of specimens with finite volumes, where the uniaxial tests are not applicable. Although the cross-sectional indentation technique has been used to characterize the mechanical behaviors of a large number of materials, e.g., fracture toughness of thermal barrier coatings [8] and the shear band of metallic glasses [44], its application on the characterization of mechanical behaviors of in-plane anisotropic materials is rare. Perhaps, the experiment process of this method may be as laborious as the conventional uniaxial tests. However, it does provide a novel method to uniquely correlate the plastic anisotropy parameter with the measurable material response, e.g., the out-of-plane deformation field in cross-sectional indentation. For the research purposes, it can be used to probe the relationship between anisotropy, microtexture, microstructure, and processing behaviors of materials [17,31–33], particularly in the situations where the information of the entire indentation loading history (e.g., the load-displacement curve data) is not available.

6. Conclusions

We presented in this paper a methodology for obtaining the plastic anisotropy of metal materials using cross-sectional indentation. The study focuses on the bulk metallic materials which possess different yield stress along the longitudinal and transverse directions. A computational model was built to simulate the spherical indentation on the free edge of the cross-section. Results obtained from numerical simulation revealed that the bottom shape of the out-of-plane deformation field persisted on the cross-section after unloading is only sensitive to the extent of material plastic anisotropy, and insensitive to strain hardening exponent, yield strain, elastic anisotropy and the selected displacement threshold value. Based on the parametric FE study, an explicit equation is established to determine the material plastic anisotropy using the characteristic parameter of the out-of-plane residual deformation contour in the cross-sectional indentation. The effectiveness of this method is verified by application on magnesium alloy AZ31B, and the anisotropic parameter estimated from indentation and uniaxial tests show good agreement.

Acknowledgments: This project is supported by National Natural Science Foundation of China (Grant No. 51675431).

Author Contributions: Mingzhi Wang and Jianjun Wu conceived and designed the experiments; Mingzhi Wang performed the experiments; Hongfei Wu and Zengkun Zhang analyzed the data; He Fan contributed analysis tools; and Mingzhi Wang wrote the paper.

Conflicts of Interest: The authors declare no conflict of interest.

References

1. Tabor, D. *The Hardness of Metals*; Oxford University Press: London, UK, 1951.
2. Wang, W.; Wu, J.; Hui, Y.; Kun, Z.; Zhan, X.; Guo, R. Identification of elastic-plastic properties of metal materials by using the residual imprint of spherical indentation. *Mater. Sci. Eng. A* **2016**, *679*, 143–154. [CrossRef]
3. Zambaldi, C.; Raabe, D. Plastic anisotropy of γ-TiAl revealed by axisymmetric indentation. *Acta Mater.* **2010**, *58*, 3516–3530. [CrossRef]
4. Oliver, W.C.; Pharr, G.M. An improved technique for determining hardness and elastic-modulus using load and displacement sensing indentation experiments. *J. Mater. Res.* **1992**, *7*, 1564–1583. [CrossRef]
5. Patel, D.K.; Kalidindi, S.R. Correlation of spherical nanoindentation stress-strain curves to simple compression stress-strain curves for elastic-plastic isotropic materials. *Acta Mater.* **2016**, *112*, 295–302. [CrossRef]
6. Moussa, C.; Hernot, X.; Bartier, O.; Delattre, G.; Collin, J.M.; Mauvoisin, G. Mechanical characterization of carbonitrided steel with spherical indentation using the average representative strain. *Mater. Des.* **2016**, *89*, 1191–1198. [CrossRef]
7. Roy, T.K. Assessing hardness and fracture toughness in sintered zinc oxide ceramics through indentation technique. *Mater. Sci. Eng. A* **2015**, *640*, 267–274. [CrossRef]
8. Wang, X.; Wang, C.J.; Atkinson, A. Interface fracture toughness in thermal barrier coatings by cross-sectional indentation. *Acta Mater.* **2010**, *60*, 6152–6163. [CrossRef]
9. Attar, H.; Ehtemam-Haghighi, S.; Kent, D.; Okulov, I.V.; Wendrock, H.; Bnisch, M.; Volegov, A.S.; Calin, M.; Eckert, J.; Dargusch, M.S. Nanoindentation and wear properties of Ti and Ti-TiB composite materials produced by selective laser melting. *Mater. Sci. Eng. A* **2017**, *688*, 20–26. [CrossRef]
10. Ehtemam-Haghighi, S.; Cao, G.; Zhang, L.C. Nanoindentation study of mechanical properties of Ti based alloys with Fe and Ta additions. *J. Alloys Comp.* **2017**, *692*, 892–897. [CrossRef]
11. Ehtemam-Haghighi, S.; Prashanth, K.G.; Attar, H.; Chaubey, A.K.; Cao, G.H.; Zhang, L.C. Evaluation of mechanical and wear properties of Ti-xNb-7Fe alloys designed for biomedical applications. *Mater. Des.* **2016**, *111*, 592–599. [CrossRef]
12. Vlassak, J.J.; Nix, W.D. Measuring the elastic properties of anisotropic materials by means of indentation experiments. *J. Mech. Phys. Solids* **1994**, *42*, 1223–1245. [CrossRef]
13. Bocciarelli, M.; Bolzon, G.; Maier, G. Parameter identification in anisotropic elastoplasticity by indentation and imprint mapping. *Mech. Mater.* **2005**, *37*, 855–868. [CrossRef]
14. Nakamura, T.; Gu, Y. Identification of elastic-plastic anisotropic parameters using instrumented indentation and inverse analysis. *Mech. Mater.* **2007**, *39*, 340–356. [CrossRef]
15. Bolzon, G.; Talassi, M. An effective inverse analysis tool for parameter identification of anisotropic materials. *Int. J. Mech. Sci.* **2013**, *77*, 130–144. [CrossRef]
16. Yonezu, A.; Yoneda, K.; Hirakata, H.; Sakihara, M.; Minoshima, K. A simple method to evaluate anisotropic plastic properties based on dimensionless function of single spherical indentation-Application to SiC whisker-reinforced aluminum alloy. *Mater. Sci. Eng. A* **2010**, *527*, 7646–7657. [CrossRef]
17. Kalkhoran, S.M.; Choi, W.B.; Gouldstone, A. Estimation of plastic anisotropy in Ni–5% Al coatings via spherical indentation. *Acta Mater.* **2012**, *60*, 803–810. [CrossRef]
18. Wang, M.; Wu, J.; Zhan, X.; Guo, R.; Hui, Y.; Fan, H. On the determination of the anisotropic plasticity of metal materials by using instrumented indentation. *Mater. Des.* **2016**, *111*, 98–107. [CrossRef]
19. Garmestani, H.; Kalidindi, S.R.; Williams, L.; Bacaltchuk, G.M.; Fountain, C.; Lee, E.W.; Es-Said, O.S. Modeling the evolution of anisotropy in Al-Li alloys: Application to Al-Li 2090-T8E41. *Int. J. Plast.* **2002**, *18*, 1373–1393. [CrossRef]
20. Yoshida, F.; Hamasaki, H.; Uemori, T. A user-friendly 3D yield function to describe anisotropy of steel sheets. *Int. J. Plast.* **2013**, *45*, 119–139. [CrossRef]
21. Zhang, H.; Diehl, M.; Roters, F.; Raabe, D. A virtual laboratory using high resolution crystal plasticity simulations to determine the initial yield surface for sheet metal forming operations. *Int. J. Plast.* **2015**, *80*, 111–138. [CrossRef]
22. Lankford, W.T.; Snyder, S.C.; Bauscher, J.A. New criteria for predicting the press performance of deep drawing sheets. *Trans. Am. Soc. Met.* **1950**, *42*, 1197–1231.

23. Wang, Y.N.; Huang, J.C. The role of twinning and untwining in yielding behavior in hot-extruded Mg-Al-Zn alloy. *Acta Mater.* **2007**, *55*, 897–905. [CrossRef]

24. Yan, H.; Chen, R.S.; Han, E.H. Room-temperature ductility and anisotropy of two rolled Mg–Zn–Gd alloys. *Mater. Sci. Eng. A* **2010**, *527*, 3317–3322. [CrossRef]

25. Yi, S.; Bohlen, J.; Heinemann, F.; Letzig, D. Mechanical anisotropy and deep drawing behavior of AZ31 and ZE10 magnesium alloy sheets. *Acta Mater.* **2010**, *58*, 592–605. [CrossRef]

26. Tang, W.; Huang, S.; Li, D.; Peng, Y. Mechanical anisotropy and deep drawing behaviors of AZ31 magnesium alloy sheets produced by unidirectional and cross rolling. *J. Mater. Process. Technol.* **2015**, *215*, 320–326. [CrossRef]

27. Kim, W.J.; Yoo, S.J.; Chen, Z.H.; Jeong, H.T. Grain size and texture control of Mg–3Al–1Zn alloy sheet using a combination of equal-channel angular rolling and high-speed-ratio differential speed-rolling processes. *Scr. Mater.* **2009**, *60*, 897–900. [CrossRef]

28. Li, X.; Al-Samman, T.; Gottstein, G. Mechanical properties and anisotropy of ME20 magnesium sheet produced by unidirectional and cross rolling. *Mater. Des.* **2011**, *32*, 4385–4393. [CrossRef]

29. Stanford, N.; Barnett, M.R. The origin of "rare earth" texture development in extruded Mg-based alloys and its effect on tensile ductility. *Mater. Sci. Eng. A* **2008**, *496*, 399–408. [CrossRef]

30. Stanford, N.; Atwell, D.; Beer, A.; Davies, C.; Barnett, M.R. Effect of microalloying with rare-earth elements on the texture of extruded magnesium-based alloys. *Scr. Mater.* **2008**, *59*, 772–775. [CrossRef]

31. Crooks, R.; Wang, Z.; Levit, V.I.; Shenoy, R.N. Microtexture, microstructure and plastic anisotropy of AA2195. *Mater. Sci. Eng. A* **1998**, *257*, 145–152. [CrossRef]

32. Alexander, B.-B.; Carl, B.; Franck, A.T.G.; Daniel, L. Modelling of anisotropy for Al–Li 2099 T83 extrusions and effect of precipitate density. *Mater. Sci. Eng. A* **2016**, *673*, 581–586. [CrossRef]

33. Alexander, B.-B.; Carl, B.; Franck, A.T.G.; Daniel, L.; Julien, B.; Mathieu, B. Characterization of Al-Li 2099 extrusions and the influence of fiber texture on the anisotropy of static mechanical properties. *Mater. Sci. Eng. A* **2014**, *597*, 62–69.

34. Rioja, R.J. Fabrication method to manufacture isotropic Al-Li alloys and products for space and aerospace applications. *Mater. Sci. Eng. A* **1998**, *257*, 100–107. [CrossRef]

35. Yonezu, A.; Kuwahara, Y.; Yoneda, K.; Hirakata, H.; Minoshima, K. Estimation of the anisotropic plastic property using single spherical indentation-An FEM study. *Comp. Mater. Sci.* **2009**, *47*, 611–619. [CrossRef]

36. Hill, R. A theory of the yielding and plastic flow of anisotropic metals. *Proc. R. Soc. A* **1948**, *193*, 281–297. [CrossRef]

37. Donohue, B.R.; Ambrus, A.; Kalidindi, S.R. Critical evaluation of the indentation data analyses methods for the extraction of isotropic uniaxial mechanical properties using finite element models. *Acta Mater.* **2012**, *60*, 3943–3952. [CrossRef]

38. *ABAQUS*; Analysis User's Manual Version 6.9; Software for Finite Element Analysis and Computer-Aided Engineering; ABAQUS Inc.: Providence, RI, USA, 2009.

39. Zhao, M.; Ogasawara, N.; Chiba, N.; Chen, X. A new approach to measure the elastic-plastic properties of bulk materials using spherical indentation. *Acta Mater.* **2006**, *54*, 23–32. [CrossRef]

40. Hernot, X.; Moussa, C.; Bartier, O. Study of the concept of representative strain and constraint factor introduced by Vickers indentation. *Mech. Mater.* **2014**, *68*, 1–14. [CrossRef]

41. Bowden, F.P.; Tabor, D. *The Friction and Lubrication of Solids*; Oxford University Press: Clarendon, VT, USA, 2001.

42. Bucaille, J.L.; Stauss, S.; Felder, E.; Michler, J. Determination of plastic properties of metals by instrumented indentation using different sharp indenters. *Acta Mater.* **2003**, *51*, 1663–1678. [CrossRef]

43. Barlat, F.; Lege, D.J.; Brem, J.C. A six-component yield function for anisotropic materials. *Int. J. Plast.* **1991**, *7*, 693–712. [CrossRef]

44. Zhang, H.; Jing, X.; Subhash, G.; Kecskes, L.J.; Dowding, R.J. Investigation of shear band evolution in amorphous alloys beneath a Vickers indentation. *Acta Mater.* **2005**, *53*, 3849–3859. [CrossRef]

materials

MDPI

Article

Mechanical Contact Characteristics of PC3 Human Prostate Cancer Cells on Complex-Shaped Silicon Micropillars

Brandon B. Seo [1], Zeinab Jahed [2], Jennifer A. Coggan [3], Yeung Yeung Chau [4], Jacob L. Rogowski [1], Frank X. Gu [1], Weijia Wen [4], Mohammad R. K. Mofrad [2] and Ting Yiu Tsui [1,*]

[1] Department of Chemical Engineering, University of Waterloo, 200 University Avenue West, Waterloo, ON N2L 3G1, Canada; bsbseo@uwaterloo.ca (B.B.S.); jrogowsk@uwaterloo.ca (J.L.R.); frank.gu@uwaterloo.ca (F.X.G.)
[2] Departments of Bioengineering and Mechanical Engineering, University of California Berkeley, 208A Stanley Hall, Berkeley, CA 94720, USA; zjahed@berkeley.edu (Z.J.); mofrad@berkeley.edu (M.R.K.M.)
[3] Department of Chemistry, University of Waterloo, 200 University Avenue West, Waterloo, ON N2L 3G1, Canada; jcoggan@uwaterloo.ca
[4] Department of Physics, The Hong Kong University of Science and Technology, Clear Water Bay, Kowloon, Hong Kong, China; cyyab@connect.ust.hk (Y.Y.C.); phwen@ust.hk (W.W.)
* Correspondence: tttsui@uwaterloo.ca; Tel.: +1-519-888-4567 (ext. 38404)

Received: 16 May 2017; Accepted: 26 July 2017; Published: 2 August 2017

Abstract: In this study we investigated the contact characteristics of human prostate cancer cells (PC3) on silicon micropillar arrays with complex shapes by using high-resolution confocal fluorescence microscopy techniques. These arrays consist of micropillars that are of various cross-sectional geometries which produce different deformation profiles in adherent cells. Fluorescence micrographs reveal that some DAPI (4′,6-diamidino-2-phenylindole)-stained nuclei from cells attached to the pillars develop nanometer scale slits and contain low concentrations of DNA. The lengths of these slits, and their frequency of occurrence, were characterized for various cross-sectional geometries. These DNA-depleted features are only observed in locations below the pillar's top surfaces. Results produced in this study indicate that surface topography can induce unique nanometer scale features in the PC3 cell.

Keywords: nanoindentation; contact mechanics; deformation; silicon; pillars; PC3 cells

1. Introduction

The recent development of vertically-aligned micro- and nano-pillar arrays for a range of biological applications, such as circulating tumor cell (CTC) capturing devices [1–11], biosensors [12–16], and neural microprobe implants [17–21], have helped to stimulate active research initiatives in understanding how cells interact with these micro- and nanometer-scale pillars. In addition to overall cell responses to these small features, several recent works were focused on the effects of nuclear deformation induced by such pillars. Pan et al. [22] cultured neonatal Sprague Dawley (SD) rat bone marrow stromal cells on poly (lactide-co-glycolide) (PLGA) square-shaped micropillars and observed pillar-induced shape changes of the nuclei. They reported that the nuclei were severely deformed and would conform to the contours of the PLGA surface features. Their results indicate that gravitational force is not the primary driving factor that causes the cells to deform. Instead, Pan et al. suggested that the deformation is due to cytoskeletal stresses. Furthermore, they revealed that cells with severely deformed nuclei continue to survive and maintain the ability to differentiate. Hanson et al. [4] evaluated the nuclear deformation of various cell types that showed different nuclear stiffness characteristics on nanopillars. They showed that both the cell and nuclear membranes conformed to the shape of the patterned structures. These cells survived even with severe pillar-induced nuclear deformations and continued to proliferate and migrate on the patterned surface. However, as expected, the amount of nuclear deformation depends on the stiffness of the nucleus, with more compliant organelles exhibiting greater strain. Davidson et al. [23] studied nuclear deformation behaviors of two cancerous osteosarcoma cell lines (SaOs-2 and MG-63) surrounding poly-L-lactic acid (PLLA) micropillars, and compared them with healthy human osteoprogenitor cells. Their results showed that the nuclei of cancerous osteosarcoma cells deformed more than those of healthy cells. These deformations were already visible only six hours after cell deposition. In addition, the severely deformed cancer cells continued to undergo mitosis and cell division. Their results suggested that the primary driving force behind nuclear deformation is the pressure generated by the actin cap above the nucleus which pushes the nucleus downwards onto the nanopillars.

While these works produced valuable information on how cell and nuclear shape changed with various patterned surfaces, there have been few insights into how pillar cross-sectional geometries affect the nuclear deformation characteristics and their internal structure. Experiments conducted in previous works were focused on solid core structures with high orders of symmetry, such as rounded rectangular [23,24] or cylindrical pillars [25]. As a result, they lack sharp points or edge features that can produce high mechanical stresses, and do not represent certain complex shapes that are often seen in platforms designed for biological applications [4,9,14,16]. Furthermore, the pillar material commonly used in previous nucleus deformation studies [23,24] was PLLA which has an elastic modulus significantly lower than silicon (~4.1 GPa [26] vs. ~190 GPa [27]). This polymeric material is also weak and soft with a tensile strength ~163 times lower than silicon (~43 MPa [26] vs. ~7000 MPa [27]). Therefore, PLLA pillars with a given geometry may not induce as significant a mechanical stress on the adhered cells as their silicon counterpart due to localized Hertizan deformation or plastic yielding [28]. A good analogy is a stiff and strong diamond spherical tip producing larger mechanical contact stress when pressed on an elastic half-space in comparison to the contact stress of a PLLA tip with the same geometry under the same loading force. This difference exists because the compliant PLLA tip deforms readily at the contact point, thereby increasing the contact area and reducing the localized stress. In addition, the pressure applied by the cells may alter the polymeric pillar's geometries by bending [25,29], a possibility which brings uncertainty when interpreting cell deformation results. Jahed et al. recently demonstrated that a nanometer-scale *Staphylococcus aureus* bacterial cell network can produce sufficient forces to mechanically deform polydimethylsiloxane (PDMS) micropillars [29], and even high-strength nanocrystalline nickel nanopillars [30].

Determining the exact stress profiles within cells surrounding the pillars is difficult due to complex topographic features of the pillars, uncertainties in the magnitude/direction of external forces applied by the cytoskeletal elements on the adherent cells, the mechanical properties of live cells, and changes in cell contour as they spread around pillars. However, it may still be possible to appreciate the effect of pillar geometries on contact pressure by using simplified contact mechanic models as examples. Consider two Hertzian contact models for spherical and cylindrical tips on an elastic half-space [28] as schematically illustrated in Figure 1a,b, respectively. The maximum contact pressure (p_{max}) produced by the external force (F) are given by the following equations:

$$p_{max}(spherical\ tip) = \left(\frac{6FE^{*2}}{\pi^3 R_s{}^2}\right)^{1/3} \tag{1}$$

$$p_{max}(cylinder) = \left(\frac{FE^*}{\pi L R_c}\right)^{1/2} \tag{2}$$

where parameters R_s and R_c correspond to the radius of the spherical and cylindrical tips, respectively. The length of the cylinder is represented by parameter L. The effective modulus (E^*) is related to the elastic moduli (E) and the Poisson's ratio (v) of the spherical/cylindrical tip material (t) and the elastic half-space (h) by the following equation:

$$\frac{1}{E^*} = \frac{1 - v_t^2}{E_t} + \frac{1 - v_h^2}{E_h}$$

(3)

Equations (1) and (2) show that the tip radius maximum contact pressure relationship varies with $R^{-2/3}$ and $R^{-1/2}$, respectively. These models reveal that the maximum contact pressure (p_{max}) increases as the spherical and cylindrical tip radii reduce. Hence, pillars with sharp corners or edges may induce significant contact pressure on the adherent cells.

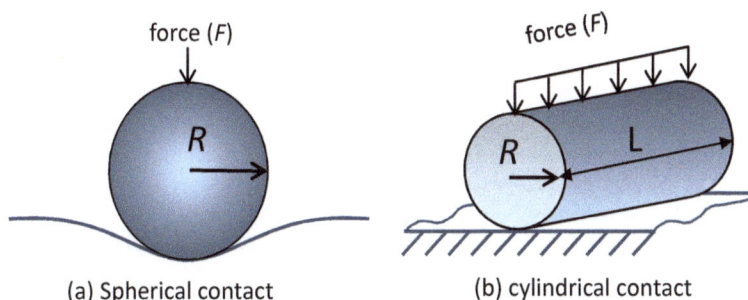

(a) Spherical contact   (b) cylindrical contact

Figure 1. Schematic drawings of (**a**) spherical and (**b**) cylindrical tip contact geometries.

The objective of this work is to gain an understanding of how prostate cancer (PC3) cell nuclei respond to sharp, mechanically stiff, and hard silicon surface structures with various cross-sectional profiles. Herein, complex C-shaped and hollow micropillars with different outer diameters were fabricated on hard silicon substrates. Each C-shaped pillar contains two sharp points and edges which may amplify the local mechanical stresses imposed on the cells by the cytoskeletal forces. This resembles the indentation contacts of sharp tips on viscoelastic materials. Silicon was chosen because of the fact that its elastic modulus, hardness, and yield strength are significantly higher than monolithic polymers, such as PLLA. This work is uniquely different from previous studies that used rectangular or cylindrical polymeric pillars: (1) silicon pillar cross-sectional geometries contain sharp and pointed features with tip radii in the nanometer-scale; (2) silicon pillars are ~45 times stiffer and their mechanical strength is ~160 times larger than polymeric pillars with the same geometry. This allows the mechanical stress to be concentrated on the cells without the uncertainty that the pillars will deform locally or bend when in contact with the cells. Prostate cancer cells were chosen because they are one of the most common cancers and leading causes of cancer death among men [31]. Each PC3 cell covers a large surface area in tens or hundreds of square microns when it is fully adherent. This allows a large number of silicon pillars to make contact with individual cells. DNA and cytoskeletal elements were subsequently stained and inspected with high-resolution confocal fluorescence microscopy. Results show that the DNA of PC3 cancer cells that have been incubated for at least 24 h would spread and cover the complex-shaped pillars. More importantly, DAPI (4′,6-diamidino-2-phenylindole)-stained nuclear micrographs reveal a new type of topography-induced feature at the nuclei locations. This feature appears as nanometer-scale slits emanating from the pillars, particularly near the sharp corners of the C-shaped pillars. To the best of our knowledge, this is the first time such small slits have been observed within human prostate cancer cells. Careful analysis showed that C-shaped pillars with smaller outer diameters resulted in a more frequent appearance of these slit features. A few slit structures were observed for hollow pillars that do not contain any sharp corners. Further inspection of the nuclei revealed that these slits were only observed at focal planes between the pillar top and the substrate surface. The lack of DAPI fluorescence signal from these slits suggests that they do not contain significant amounts of DNA.

2. Materials and Methods

2.1. Substrate Preparation

Patterned silicon pillar arrays were prepared using standard microfabrication ultraviolet (UV) lithography and silicon etching techniques performed at the Hong Kong University of Science and Technology. The thickness of the AZ®7908 photoresist (AZ Electronic Materials, Merck KGaA, Darmstadt, Germany) was ~0.9 µm. After a soft-bake on a hotplate at 90 °C for 60 s, patterns were transferred to the substrate by UV light in an ASML Stepper 5000 (PHT-S1) (Veldhoven, The Netherlands) at an intensity of 500 mW for 0.66 s. The wafers were developed in FHD-5 (Fujifilm Electronic Materials Co. Ltd., Tokyo, Japan) for 60 s, followed by a hard-bake process at 120 °C for 30 min. Pillar patterns were transferred onto the silicon substrate using a Bosch anisotropic etch process [32] to render pillars of approximately 2.2 µm in height. Three different pillar geometries and dimensions were successfully manufactured. These include C-shaped pillars with outer diameters of approximately 1.2 and 2.7 µm, as well as hollow-shaped columnar structures with outer and inner diameters of approximately 5.6 µm and 3 µm. Detailed physical dimensions and spacing of these pillars are summarized in Table 1. These silicon substrates were rinsed with ethanol (70%) followed by phosphate-buffered saline (PBS) (Bio-Rad Laboratories, Mississauga, ON, Canada) prior to cell deposition.

2.2. Cell Culture and Staining

The human prostate cancer cell line PC3 was obtained from the American Type Culture Collection (ATCC, Manassas, VA, USA). Cells were passaged every three to five days upon reaching 80% confluence using 0.25% trypsin/0.05% ethylenediaminetetraacetic acid (EDTA) solution (ATCC). They were cultured in F-12K medium (ATCC) and incubated at 37 °C with 5% CO_2 atmosphere in a Thermo Fisher Scientific (Rochester, NY, USA) incubator. Kaighn's modification of Ham's F-12 medium, containing 2 mM L-glutamine, 1500 mg/L sodium bicarbonate, 10% fetal bovine serum, and 1% penicillin-streptomycin (100 U/mL penicillin and 100 ug/mL streptomycin), was used.

Passage number eight was used for all experiments. After the targeted cell concentrations were achieved, the prostate cancer cells were deposited on the patterned silicon substrates in BioLite 35 mm tissue culture dishes. Three batches of cells with incubation periods of 30 min, 24 h, and 72 h were performed. The cell culture medium was subsequently removed and the silicon chips with adhered cells were washed twice with an equal volume of PBS (Thermo Scientific BupH phosphate-buffered saline, Waltham, MA, USA). Unless otherwise stated, the concentration used for these experiments was approximately 105 cells/mL. Cell concentrations were determined by using a hemocytometer (Hausser Scientific Co., Horsham, PA, USA) with cells stained with Trypan Blue (Lonza Walkersville Inc., BioWhittaker, Walkersville, MD, USA).

Cell fixing was conducted at room temperature by submerging chips in 4% paraformaldehyde solution for one hour. After washing the chips with PBS, they were kept submerged in PBS at 4 °C until staining. This is a well-established fixation process developed for the nuclear morphology studies in the literature [8,22,23,33,34]. Fixed PC3 cells were fluorescently stained for F-actin (phalloidin, CytoPainter F-actin Staining Kit, abcam®, Cambridge, MA, USA) and nuclei (4′,6-diamidino-2-phenylindole, DAPI, Life Technologies, Carlsbad, CA, USA). First, red fluorescent phalloidin conjugate was prepared according to the manufacturer's specified protocol. Cells were then rinsed twice with PBS prior to permeabilization with 0.1% Triton X-100 (Sigma-Aldrich Corporation, St. Louis, MO, USA) in PBS for five minutes at room temperature. After permeabilization, cells were rinsed twice with PBS and incubated in the dark with red fluorescent phalloidin conjugate at room temperature for 45 min. After incubation with phalloidin, the chips were rinsed two more times with PBS. DAPI counterstain was prepared according to the manufacturer's specified protocol. Following F-actin staining, PC3 cells were incubated in the dark with 300 nM DAPI in PBS for three minutes at room temperature. Chips were then rinsed twice in PBS and remained submerged in PBS until imaging. Optical inspections were conducted with a Leica SP5 laser scanning confocal microscope (Leica Microsystems GmbH, Wetzlar, Germany) at the University of Guelph. The specimens were submerged in room temperature PBS during the imaging process. This instrument can acquire multiple channels of fluorescent signals simultaneously.

3. Results and Discussion

3.1. Silicon Pillars

Representative scanning electron micrographs of silicon pillars with three different cross-sectional geometries are displayed in Figure 2a–c. They were fabricated on the same silicon wafers simultaneously and have an identical height of ~2.2 μm. C-shaped pillars with outer diameters of ~1.2 and ~2.7 μm are shown in Figure 2a,b. Each C-shaped pillar consists of two sharp corners and curved edges as labeled in Figure 2. Hollow-shaped columnar pillars were also examined and their micrographs are shown in Figure 2c. The pillars were positioned in an orthogonal orientation. Physical dimensions of these structures are listed in Table 1. Each of these pillar arrays covers a square area of ~3.5 mm × ~3.5 mm. The pillar's center-to-center distances of the three specimen groups—namely 1, 2, and 3—are approximately 6.2, 7.6 and 10.5 μm, respectively.

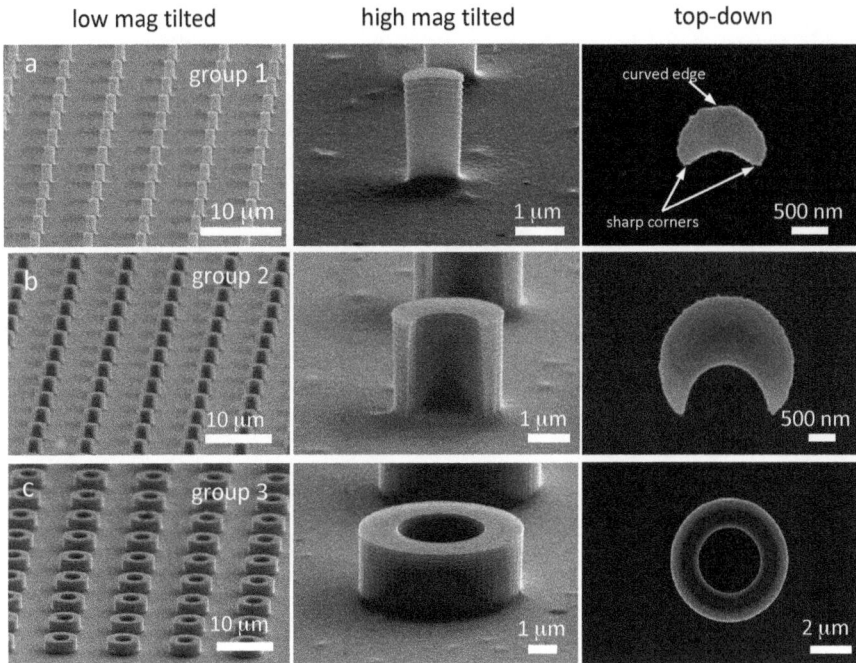

Figure 2. SEM micrographs of silicon pillars with complex cross-sectional geometries examined in this work: (**a**) group 1, small C-shaped pillars; (**b**) group 2, large C-shaped pillars; and (**c**) group 3, hollow pillars.

Table 1. Physical dimensions of micron and sub-micron scale silicon pillars. Statistical distributions of PC3 cells that show fine line features after incubation on silicon substrates with various pillar geometries for approximately 72 h. Data spreads correspond to one standard error.

Group	Pillar Shape	Outer Diameter (μm)	Wall Thickness (μm)	Center-to-Center Distance (μm)	Nuclei Inspected	Nuclei with Line Features	% Nuclei with Line Features
1	small C-shape	1.2	–	6.2	55	35	64 ± 6
2	large C-shape	2.7	–	7.6	38	17	45 ± 8
3	hollow	5.6	1.3	10.5	28	5	18 ± 7

More importantly, the gaps between adjacent pillars—regions that allow portions of cells to extend into during the spreading process—are identical for all pillars (~5 μm). Badique et al. [24] and Wang et al. [8] have suggested that pillar spacing is an important parameter for nuclear deformation. Maintaining identical gap spacing between the pillars will reduce uncertainties of flow dynamic variations among pillars with different cross-sectional geometries and allow for a direct comparison of results from these three pillar groups. Unlike cylindrical pillars with axisymmetric geometry, the C-shaped pillars fabricated in this work provide a unique surface topography that produces non-axisymmetric mechanical stress states on the attached cells. Measured from the top-down views, the tip radii at the corners of the small and large C-shaped pillars are approximately 80 nm and 136 nm, respectively. Since the mechanical stress concentration factor increases with reduced tip radius [28], the smaller C-shaped pillars with sharper corners and edges are expected to induce greater stress on the cells. In contrast, the 5.6 μm outer diameter hollow pillars are axisymmetric structures with low stress concentration and are provided as baselines for comparison with the C-shaped pillars.

The precise contact pressure profiles produced by these three pillar structures on the adherent cells are difficult to determine because they require detailed information on the pillar-cells' three-dimensional contact profile, the mechanical properties of the live cells, and the directions and magnitudes of applied forces—which are hard to define as they change during the dynamic cell-spreading process. However, by using Equations (1)–(3) we can describe the effects of tip radius on contact pressure under simple loading geometries, such as spherical and cylindrical contacts on an elastic half-space. By reducing the spherical tip radius from 5.6 μm to 136 nm and 80 nm, the maximum contact pressure increases ~10.9 and ~16.0 times, respectively. The maximum contact pressures for smaller cylindrical contacts are ~5.4 and ~7.4 times greater, respectively. In both contact cases, the maximum pressure increases with the reduced radius of the structure. As mentioned above, it is important to note that the actual stress profiles experienced by the adherent cells can be more complex than the two simple models evaluated here; however, they do serve as a demonstration of the pillar corner and edge radius effects.

3.2. Cells on Bare Silicon Substrates

A typical top-down high-resolution confocal micrograph of a PC3 cell incubated for approximately 30 min on a flat, smooth bare silicon substrate without pillar patterns is shown in Figure 3a. This micrograph shows that the cell is compact with near circular geometry and the nucleus is located at approximately the center of the cell. This is expected as this cell has only been deposited on the rigid surface for a short period of time. Additionally, cytoskeletal protein F-actin projection was observed only at the cell periphery where the cells contact the substrate. The average diameter of ten randomly-selected 30-min incubated cells was 16.6 ± 2.9 μm. Their sizes are statistically indistinguishable from results published in the literature of 15.1 ± 2.6 μm [35] and reported by Nexcelom Bioscience LLC of 18.08 ± 2.69 μm [36]. This demonstrates that the cell morphology was not grossly altered by the fixation processes. Representative confocal inspections were also conducted on PC3 cells adhered to flat, smooth, bare silicon substrates after approximately 72 h of incubation; these are shown in Figure 3b.

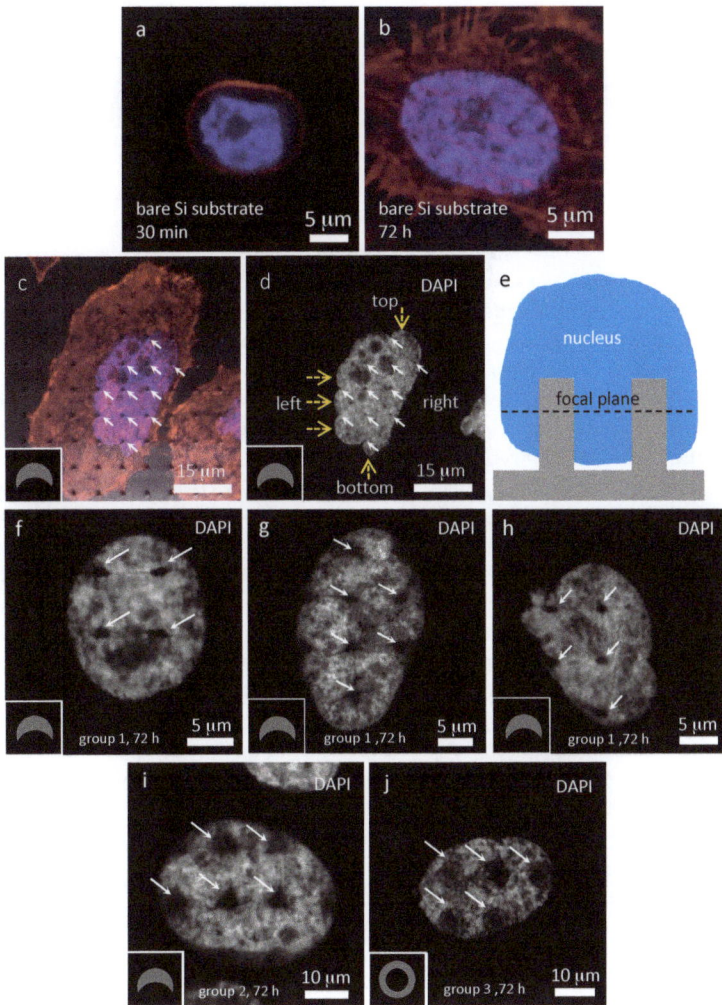

Figure 3. Confocal fluorescence micrographs of PC3 cells incubated for (**a**) 30 min and (**b**) 72 h on bare Si substrate; (**c**) Composite and (**d**) DAPI-only micrographs of PC3 cells incubated for 72 h on small C-shaped pillars (group 1). Pillar locations are indicated by solid arrows while bulged edges of the DAPI-stained materials are highlighted by dashed arrows in (**c**). A schematic drawing of the focal plane used in these micrographs is shown in (**e**). Representative micrographs of PC3 cells incubated for 72 h reveal fine line structures emanating from the small C-shaped pillars are displayed in (**f**–**h**). Lines in (**f**) are connected with the adjacent pillars while lines in (**g**) and (**h**) are not. Confocal micrographs of 72 h incubated PC3 cells on large C-shaped and hollow-shaped pillars are displayed in (**i**,**j**), respectively. Inset drawings indicate the pillar orientations.

This DAPI and fluorescent phalloidin-stained composite micrograph shows that the cell has spread and successfully attached to the flat, smooth, bare silicon substrate. This cell contains a large number of F-actin-rich filopodia and lamellipodia. Some areas within the nucleus, as shown in Figure 3a,b, were not stained as intensely by DAPI and appear dim or dark. These regions are randomly distributed with varying sizes and shapes, and most likely correspond to different nuclear components, such as nucleoli or euchromatin compartments.

3.3. Cells on Patterned Silicon Substrates

Representative confocal micrographs of PC3 cells incubated on small (group 1) and large C-shaped (group 2) pillars for 72 h are shown in Figure 3. An example of PC3 cells incubated on small C-shaped pillars (group 1) for 72 h is displayed in Figure 3c,d. These micrographs were collected at a focal plane below the top of the pillar, as schematically illustrated in Figure 3e. A composite micrograph of blue DAPI, red phalloidin, and gray-scale optical reflection images of the silicon-patterned structures is displayed in Figure 3c. The inset schematic drawing indicates the orientation of the small C-shaped pillars on the substrate. Pillar locations are highlighted with solid arrows unless otherwise stated. This micrograph shows many filopodia and lamellipodia are observed on this cell, with the majority of these being attached to the substrate. It is interesting to note that some appear preferentially bonded to the pillars. Such selective attachment behaviors of actin structures on small pillars have been previously reported by Albuschies and Vogel [37] with human dermal foreskin fibroblast cells. In addition, Jahed et al. [38] showed 3T3 Swiss Albino fibroblasts cells preferentially sensing and remaining attached to metallic nanopillars. Figure 3d shows the DAPI-stained areas of this cell in gray-scale. Detailed inspections of this image reveal the locations of C-shaped pillars as small solid dark spots (highlighted with white solid arrows). This micrograph also showed other features with a varying intensity of DAPI fluorescence which may represent the locations of different sub-nuclear organelles. At least eleven silicon pillars can be identified surrounded by the nucleus displayed in Figure 3d with six located at the center of the stained region, while the rest are positioned along the edges. Figure 3d shows this nucleus to have distinctly bulged edges (highlighted with yellow dashed arrows) at the top, bottom, and left portions of the nucleus. They are likely created by the flow of membrane-bound nuclear material in between the stiff silicon pillars during the cell spreading process. The nucleus and its membrane are expected to experience highly-localized mechanical stress at the pillar locations where the material flows are restricted and the nuclear membrane is severely deformed in order to conform to the shape of the silicon pillars.

Another important characteristic feature observed in some of the cells deposited on the small C-shaped pillars (group 1) are distinct fine dark line structures revealed in the DAPI-labeled micrographs collected at focal planes below the pillar top, as schematically illustrated in Figure 3e. Figure 3f shows examples of such fine dark lines in a DAPI-stained area radiating outward from the pillar locations and the inset drawing indicates the orientation of the small C-shaped pillars on the substrate. The locations of four pillars are highlighted with solid arrows. These dark lines, less than a micron wide, can be clearly seen emanating from all four pillars located near the center of the nucleus, as displayed in Figure 3f. They are uniquely different from the randomly-distributed and irregularly-shaped dim background features produced by other sub-nuclear organelles. The lack of DAPI fluorescence signal in these fine line structures indicates a low concentration of nuclear DNA in these features. Some nuclei, as shown in Figure 3g,h, contain line features that are not connected to adjacent pillars and have lengths in the micron scale. Additional micrographs are revealed in Supplementary Figure S1 to demonstrate the reproducibility of these dark line features. The DAPI-labeled micrographs shown in Figure 3f–h also reveal that the pillar's dark spots resembling filled semi-circles rather than their true C-shaped geometry. This indicates that the pillars are not in direct contact with the DNA. A possible reason is that the nuclear envelope and the plasma membrane remain intact at the cell-pillar interface and provide a gap between the DNA and the pillar. These pillars are unable to rupture the cell membranes and make direct contact with DNA materials. Instead, cells putatively engulf and wrap around the pillars. This is expected as several previous studies [39–42] have shown that micropillars are unlikely to penetrate the cell membranes due to their large dimensions and small height to diameter aspect ratios. The narrow dark spacing between the pillar and the DAPI stained DNA shown in Figure 3f,h may be filled with other cellular components, such as the plasma membrane, cytoplasmic components, the nuclear envelope, or other sub-nuclear organelles. Hence, the dark spots do not resemble pillar true geometries.

Random inspection of 55 cells deposited on small C-shaped pillars (group 1) revealed that 35 cells (64 ± 6%) exhibited these dark line structures where the data spread corresponds to one standard error. These results are plotted in Figure 4a. The length distributions of 132 dark lines surrounding small C-shaped silicon pillars were measured and displayed in Figure 4b. This cumulative probability plot shows the line lengths are in the range of ~1.3 and ~7.3 μm, with the average value of 3.8 ± 1.4 μm. The data spread represents one standard deviation. On average there are ~3.8 lines formed in each cell that contains dark lines. Among the 132 lines inspected, 104 of them (79% of the line population) extended from the sharp corners, while 28 lines (21%) emanated from the curved edges. This demonstrates that the sharp corners of the C-shaped pillars are the more likely locations to form dark line structures.

Careful inspection of PC3 cells incubated for approximately 72 h on large C-shaped pillars (group 2) revealed similar sub-micron-scale dark line structures, as shown in Figure 3i. The large C-shaped pillars surrounded by the nucleus are labeled with arrows and the separation distance between the pillars is identical to the small C-shaped pillars (~5 μm). This confocal micrograph shows that the dark spots where the pillars are located do not bear a resemblance to a true pillar C-shape but instead have filled-triangular profiles. Interestingly, several of the fine dark line features extend from the tips of these triangles—an indication that they may be structurally connected. Random inspections of 38 PC3 cells attached on large C-shaped pillars (group 2) showed that 17 of them (45 ± 8%) exhibit line features as shown in Figure 4a. These results show that fewer PC3 cell nuclei (45% vs. 64%) exhibit line structures when they are deposited on the large C-shaped pillars in comparison to the smaller counterparts.

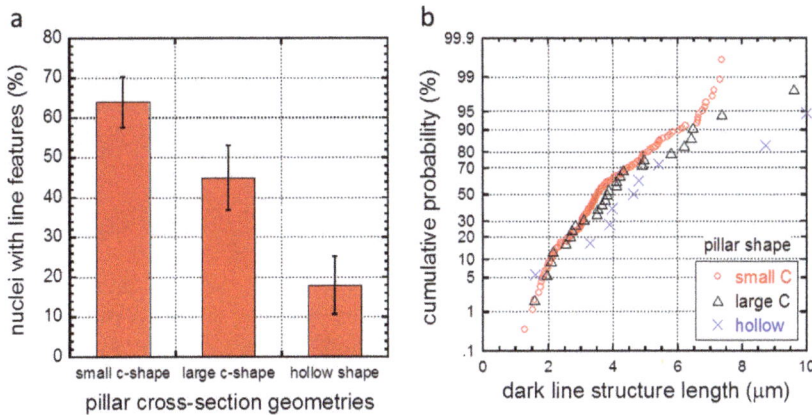

Figure 4. (a) Population of cell nuclei with slit features for three different pillar cross-section geometries. Slit length distributions on various shaped pillars are displayed in (b).

The length distributions of 27 dark line structures surrounding the large C-shaped pillars are plotted in Figure 4b. The average dark line length in these cells is 4.2 ± 1.9 μm where the data spread corresponds to one standard deviation. These results are statistically indistinguishable from the small C-shaped pillars results. The average number of dark lines observed in cells contacting large C-shaped pillars is ~1.6—two times fewer than those observed in small C-shaped pillars of 3.8. One primary reason is a reduced number of dark lines originated from the curved edges of the large C-shaped pillars. Inspection of cells on large C-shaped pillars show 24 out of 27 lines (89%) originated from the two sharp corners of pillars but only three lines (11%) were extended from the curved edges. The lack of dark lines emanating from large C-shaped pillar curved edges may suggest that their formation may be related to the dimension of the pillars. Large C-shaped pillar diameters are more than two times larger than the small C-shaped counterpart (1.2 vs. 2.7 µm).

Results shown in Figures 3 and 4 indicate that the dark line formation process depends on the pillar's cross-sectional geometry and may be influenced by the sharp corners and curved edges of the small C-shaped pillars. These sharp features are expected to magnify the mechanical forces produced by the cytoskeletal structures pushing downward on the nuclei [4,28]; hence, greater indentation forces are applied to the cells. To test this hypothesis, cells were deposited on hollow-shaped pillars fabricated on the same substrate as the C-shaped pillars. The separation distance between the hollow pillars is identical to the C-shaped structures of ~5 µm in order to reduce the uncertainties related to dynamic flow of material between the pillars. Since these hollow pillars have axisymmetric geometry with smooth, curved surfaces, the amount of induced stress is expected to be low when compared to C-shaped pillars with sharp corners.

A representative image of a PC3 cell that was incubated on hollow pillars for approximately 72 h is shown in Figure 3j and does not show distinct line structures. Random sampling of 28 cells indicated that only five nuclei (18 ± 7%) showed dark line structures, as shown in Figure 4a. These results signify that the important factor for the dark line formation is pillar's cross-sectional geometry. The line length distributions on these cells are shown in Figure 4b and show the dark lines observed on hollow-shaped pillars have average lengths of 5.1 ± 2.6 µm. While the average length is longer than those observed in C-shaped pillars, the differences are statistically insignificant due to the large data standard deviation. Furthermore, the average number of lines observed surrounding each cell is 1.8 which is approximately the same as those on the large C-shaped specimen. Experiments were repeated with cells incubated for approximately 24 h and showed similar dark line structures. Examples of 24 h cell incubation on small C-shaped pillars (group 1) and large C-shaped pillars (group 2) are shown in Supplementary Figure S2a,b respectively.

To understand the three-dimensional configurations of these line features, additional inspections of the PC3 cells were conducted at multiple focal planes. High-magnification DAPI-stained micrographs of the fine line structure from the large nucleus displayed in Figure 5a were collected and are shown in Figure 5b–e. The locations of these focal planes are schematically illustrated in the accompanied diagram. The z-section image sequence begins at a focal plane slightly above the top of small C-shaped pillars, and sequentially downward to the pillar base. White solid arrows in the micrographs indicate where the two pillars are located. No fine line structures or pillar dark spots are observed in the micrographs taken at focal planes above the pillar tops (Figure 5b). Faint impressions of two C-shaped pillars appear in Figure 5c with the focal plane located near the top of the pillars, but no dark line is observed in this micrograph. When the focal plane is located below the pillar top level, as the micrograph shown in Figure 5d, a faint fine dark line that connects the pillars is clearly visible. In addition, the two pillar dark spots displayed in Figure 5d show they do not resemble the true pillar C-shaped profile but instead a tear-drop shaped geometry where the tail portions of the feature narrows to form fine dark lines connecting adjacent pillars. This indicates that the pillar's dark spots and the fine dark lines are structurally connected.

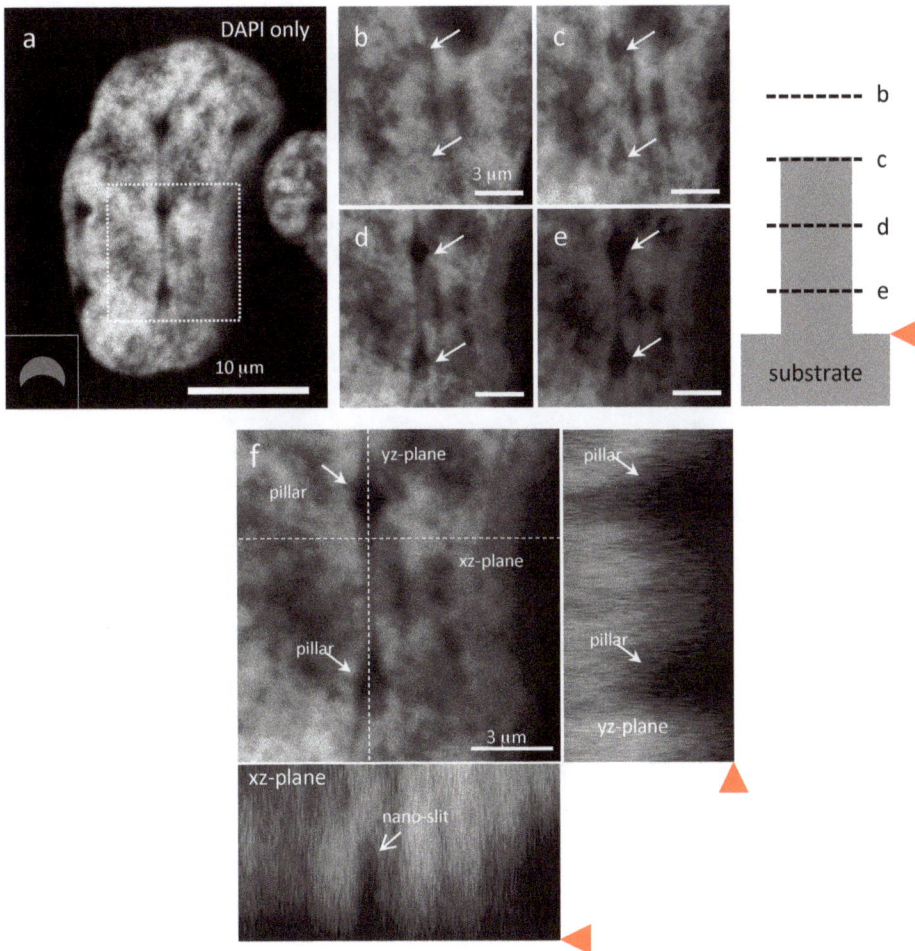

Figure 5. (**a**) DAPI only micrograph revealing fine line structures in cells incubated on small C-shaped pillars. High magnification images of a representative nucleus surrounding two small C-shaped pillars imaged in various focal planes are shown in (**b–e**). The corresponding focal plane locations are illustrated in the schematic drawing. Orthogonal views of this nucleus are shown in (**f**) revealing two pillars and a nano-slit. Red triangle pointers indicate substrate surface locations.

Furthermore, the tear-drop shaped and fine dark line structures are observed at other focal planes further down toward the substrate (see Figure 5e). This suggests that these observed features are cross-sectional views of a thin continuous vertical slit connecting adjacent pillars. It begins below the top of the pillars and extends downward. The geometric structure of this nanometer scale slit is further confirmed by the orthogonal views of this cell nucleus as shown in Figure 5f where the cross-section locations are labeled with dashed lines. A nano-slit is clearly visible on the xz-plane.

Schematic cross-sectional drawings of nuclear DNA elements surrounding two C-shaped pillars are shown in Figure 6. The drawings indicate that the DNA elements do not make direct contacts with the pillars. Instead, tear-drop shaped gaps are formed between the DNA and the pillars where they may be filled with the plasma membrane, nuclear envelope, or other sub-nuclear organelles that were not stained by DAPI. To the best of our knowledge, this is the first literature report of silicon micropillar-induced nano-slit structures within human prostate cancer cells. Confocal micrographs reveal that these continuous slits of materials are depleted of DNA.

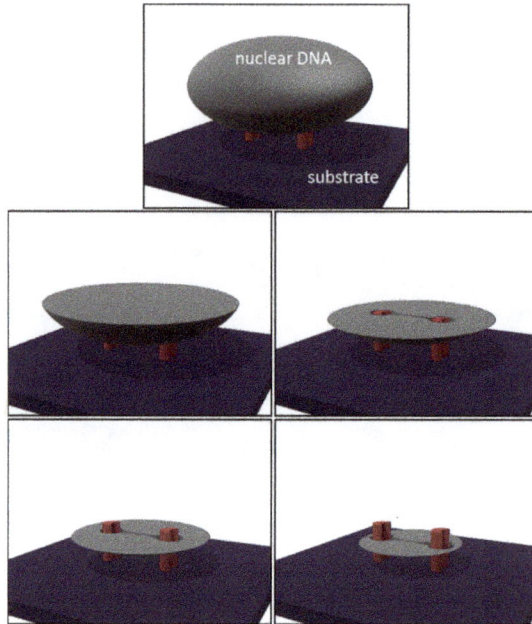

Figure 6. Schematic drawings of nuclear DNA elements (gray) surrounding two C-shaped pillars (red) at different cross-sectional planes above and below the pillars. Note the DNA materials are not in direct contact with the pillars. The tear drop-shaped openings in between the pillars indicate the slit location.

The formations of these nanometer-scale thick slits are not only observed in cells deposited on a uniform array of C-shaped pillars, but have also been confirmed in cells that are simultaneously contacting pillars with different cross-sectional geometries, as revealed in Figure 7a,b. These confocal fluorescence micrographs show the DAPI and composite images of a cell that is adhered concurrently to two different arrays of hollow silicon pillars (group 3) and small C-shaped (group 1) geometries. The majority of the cell nucleus is surrounding the small C-shaped pillars (highlighted with arrows) while the remaining part of the cell is extended to the hollow pillars. Even though the cell contacts drastically different surface topographies, the DAPI-strained DNA materials still show slits as those shown above. It is remarkable that slits can be observed in the two C-shaped pillars that are less than 5 μm away from the hollow pillar arrays (marked with dash arrows). However, it is unclear if the orientations and the lengths of these slits have been influenced by the nearby hollow pillar arrays. The evidence presented here indicates that the slit formation mechanism may be driven by a localized effect at the micron-scale that is determined by the pillar geometries.

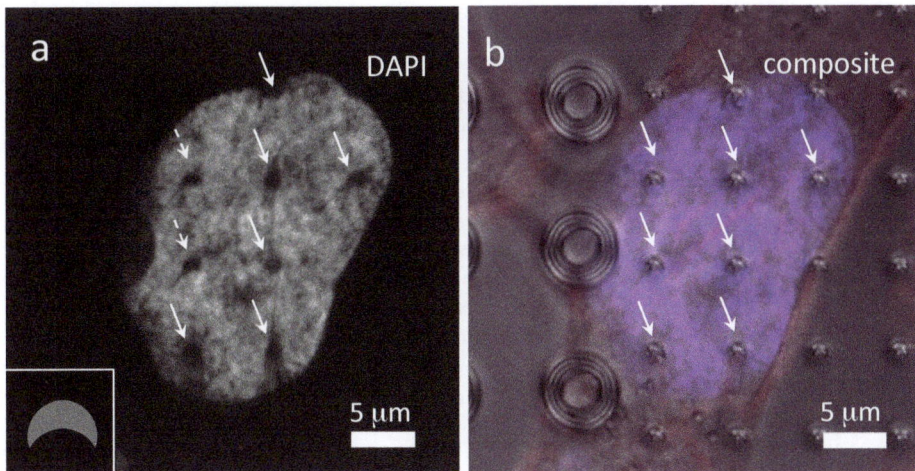

Figure 7. (a) DAPI only and (b) composite top-down micrographs of a cell adhered on small C-shaped and hollow pillars. The locations of the pillars are highlighted with arrows. Cells were incubated on the pattern substrate for 72 h. Inset drawing indicates the pillar orientation.

The confocal micrographs presented above show that sub-micron-scale stiff silicon pillar structures with C-shaped geometries produce nanometer-scale slit features in the PC3 cells. Results also revealed that not only can the nuclei be reshaped, but they may also be sensitive to the cross-sectional geometries of the surface topography. The exact origin and mechanisms of these slit formations have yet to be identified. One possible explanation may be related to the pillar/cell contact geometries. As cells spread on textured silicon surfaces their cytoskeletons push the nuclei and other sub-cellular organelles downward and press against the pillars [22,23]. This resembles nano-indentations of viscoelastic materials (cytoplasm and nuclei) contained within elastic membranes (plasma membranes and nuclear membranes) by a series of complex-shaped silicon cylinders. Hanson et al. showed that nuclear deformation on nanopillars is determined by nuclear stiffness as well as cytoskeletal forces, and that the geometry of nanopillar arrays highly influences nuclear deformation. As live cells press against the silicon pillars, the plasma membrane, cytoplasm, and nuclear envelope deform and occupy the space in between the pillars. For asymmetric hollow pillars, deformations near the contact rims were evenly distributed. However, membrane deformations on complex-shaped pillars, such as C-shaped structures, were more severe near the two sharp corners. This may cause the plasma membrane and nuclear envelope to fold inward and form creases near the corners. Since the membranes do not contain DNA, the creases formed resemble thin dark lines or slits, as illustrated in Figures 3 and 5.

This work demonstrates that PC3 cell mechanical contact responses to stiff silicon micro-pillars are more complex than previously understood and open a new direction of investigation in this field. Future research should include in situ live cell time-lapse microscopy imaging studies of these cancer cells with different pillar geometries. This focus could gain temporal information about how the slits are formed and help develop an understanding of whether the presence of the nano-slits will prevent or restrict the transport of the nuclear material between different parts of the nucleus. The effects of these structures on basic cell biological functions, such as proliferation, metabolism, mitosis, and cell division, should be investigated and compared between cancer and normal cells. During typical cell division processes, DNA molecules are replicated and chromatin molecules are being condensed into chromosomes (prophase). Eventually, the chromosomes or sister chromatids are being pulled by the microtubules to the opposite ends of the cells (metaphase and anaphase). However, it is unclear how the slits, which may act as physical barriers, affect the chromosome replication processes and microtubule motions. Even if cells were successfully divided, it is unclear if the presence of slits will induce chromosome replication errors. To address these questions, further experiments are needed to compare the DNA sequences among cells grown on the patterned surfaces with different passages. Other essential macro molecules, such as ribonucleic acids (RNAs) and peptide chains, should also be analyzed from different passages to understand if the cell functions are compromised.

Additional investigation is also required to determine how the response of PC3 cells to silicon micropillars compares with other cancer and healthy cell types. One type of the potentially healthy cells to be investigated are human fibroblast cells. Their average nuclear dimensions are similar to the PC3 cells, on the order of tens of microns, when they adhered and spread on the patterned silicon surfaces. This type of study will provide a comparison of surface topographic responses on normal and cancerous cells. Investigations of slit formations should also be conducted on other immortalized cancerous cell lines, such as HeLa cervical cancer cells, to determine if the slit formation is a mechanical contact phenomenon unique to cancerous cells or PC3 cells. Finally, other pillar cross-sectional geometries with higher stress concentration points should also be investigated to understand the effect of introducing a stress field to sub-nuclear organelles.

4. Conclusions

The response of human prostate cancer cell line PC3 nuclei on nanometer-scale textured silicon micropillars with complex cross-sectional geometries was studied using laser scanning confocal fluorescence microscopy. Results show the formation of nanometer scale slit structures in the DAPI-stained elements near some pillars where the number of slits formed is related to pillar size and cross-sectional morphology. These features develop most frequently in small C-shaped pillars, followed by large C-shaped pillars and hollow circular pillars. DAPI fluorescence micrographs reveal a low concentration of DNA in these slit structures, which are only observed in the portion of the nucleus located below the pillar top surfaces.

Supplementary Materials: The following are available online at www.mdpi.com/1996-1944/10/8/892/s1. Figure S1: (a) show a DAPI only micrograph with dark line structures radiating out at the corners of the small c-shaped pillars (group 1). Composite image of the same cell is shown in (b). This cell was incubated on patterned silicon substrate for 72 h. Inset indicates the pillar orientation, Figure S2: Representative DAPI only and composite confocal micrographs of dark line structures radiating out at the corners of (a) small C-shaped pillars (group 1) and (b) large C-shaped pillars (group 2) respectively. Cells were incubated on patterned substrate for 24 h. Insets indicate the orientation of the pillars.

Acknowledgments: The authors would like to acknowledge Abdulilah Alofi, Lilian Ho, Carlie Kong, Shweta Lad, and Junhua Gu for their support of this work. Ting Yiu Tsui thanks the Hong Kong University of Science and Technology for the support of his 2014 sabbatical. Ting Yiu Tsui would also like to thank Marc Aucoin and Professor Moira Glerum for useful discussions on the topics of cell biology. Ting Yiu Tsui thanks the Natural Sciences and Engineering Research Council of Canada Discovery Grants (grant no. RGPIN-355552).

Author Contributions: Brandon B. Seo and Ting Tsui designed the silicon pillar patterns, performed confocal experiments, data analysis, and prepared the manuscript. Zeinab Jahed and Mohammad R. K. Mofrad developed the idea to study nuclei deformation and prepared the manuscript. Jennifer A. Coggan prepared and cultured PC3 cells on patterned silicon substrates. Yeung Yeung Chau and Weijia Wen designed and fabricated the patterned silicon substrates. Jacob L. Rogowski and Frank X. Gu developed the cell fixation and staining processes on cells after the incubation processes, as well as manuscript preparations.

Conflicts of Interest: The authors declare no conflicts of interest.

References

1. Bettinger, C.J.; Langer, R.; Borenstein, J.T. Engineering Substrate Micro- and Nanotopography to Control Cell Function. *Angew. Chem. Int. Ed. Engl.* **2009**, *48*, 5406–5415. [CrossRef] [PubMed]
2. Bettinger, C.J.; Orrick, B.; Misra, A.; Langer, R.; Borenstein, J.T. Microfabrication of poly (glycerol-sebacate) for contact guidance applications. *Biomaterials* **2006**, *27*, 2558–2565. [CrossRef] [PubMed]
3. Bucaro, M.A.; Vasquez, Y.; Hatton, B.D.; Aizenberg, J. Fine-tuning the degree of stem cell polarization and alignment on ordered arrays of high-aspect-ratio nanopillars. *ACS Nano* **2012**, *6*, 6222–6230. [CrossRef] [PubMed]

4. Hanson, L.; Zhao, W.; Lou, H.-Y.; Lin, Z.C.; Lee, S.W.; Chowdary, P.; Cui, Y.; Cui, B. Vertical nanopillars for in situ probing of nuclear mechanics in adherent cells. *Nat. Nanotechnol.* **2015**, *10*, 554–562.
5. Hu, W.; Crouch, A.S.; Miller, D.; Aryal, M.; Luebke, K.J. Inhibited cell spreading on polystyrene nanopillars fabricated by nanoimprinting and in situ elongation. *Nanotechnology* **2010**, *21*, 385301. [CrossRef] [PubMed]
6. Kim, D.-H.; Provenzano, P.P.; Smith, C.L.; Levchenko, A. Matrix nanotopography as a regulator of cell function. *J. Cell Biol.* **2012**, *197*, 351–360. [CrossRef] [PubMed]
7. Liu, X.; Chen, L.; Liu, H.; Yang, G.; Zhang, P.; Han, D.; Wang, S.; Jiang, L. Bio-inspired soft polystyrene nanotube substrate for rapid and highly efficient breast cancer-cell capture. *NPG Asia Mater.* **2013**, *5*, e63. [CrossRef]
8. Wang, S.; Wan, Y.; Liu, Y. Effects of nanopillar array diameter and spacing on cancer cell capture and cell behaviors. *Nanoscale* **2014**, *6*, 12482–12489. [CrossRef] [PubMed]
9. Wang, L.; Asghar, W.; Demirci, U.; Wan, Y. Nanostructured substrates for isolation of circulating tumor cells. *Nano Today* **2013**, *8*, 347–387. [CrossRef] [PubMed]
10. Kim, W.; Ng, J.K.; Kunitake, M.E.; Conklin, B.R.; Yang, P. Interfacing silicon nanowires with mammalian cells. *J. Am. Chem. Soc.* **2007**, *129*, 7228–7229. [CrossRef] [PubMed]
11. Fujie, T.; Shi, X.; Ostrovidov, S.; Liang, X.; Nakajima, K.; Chen, Y.; Wu, H.; Khademhosseini, A. Biomaterials Spatial coordination of cell orientation directed by nanoribbon sheets. *Biomaterials* **2015**, *53*, 86–94. [CrossRef] [PubMed]
12. Lu, Y.; Peng, S.; Luo, D.; Lal, A. Low-concentration mechanical biosensor based on a photonic crystal nanowire array. *Nat. Commun.* **2011**, *2*, 576–578. [CrossRef] [PubMed]
13. Gupta, B.; Zhu, Y.; Guan, B.; Reece, P.J.; Gooding, J.J. Functionalised porous silicon as a biosensor: Emphasis on monitoring cells in vivo and in vitro. *Analyst* **2013**, *138*, 3593–3615. [CrossRef] [PubMed]
14. Xie, C.; Lin, Z.; Hanson, L.; Cui, Y.; Cui, B. Intracellular recording of action potentials by nanopillar electroporation. *Nat. Nanotechnol.* **2012**, *7*, 185–190. [CrossRef] [PubMed]
15. Xie, C.; Hanson, L.; Xie, W.; Lin, Z.; Cui, B.; Cui, Y. Noninvasive neuron pinning with nanopillar arrays. *Nano Lett.* **2010**, *10*, 4020–4024. [CrossRef] [PubMed]
16. Spira, M.E.; Hai, A. Multi-electrode array technologies for neuroscience and cardiology. *Nat. Nanotechnol.* **2013**, *8*, 83–94. [CrossRef] [PubMed]
17. Nguyen, J.K.; Park, D.J.; Skousen, J.L.; Jorfi, M.; Skousen, J.L.; Weder, C.; Potter, K.A.; Buck, A.C.; Self, W.K.; Mcconnell, G.C.; et al. Mechanically adaptive intracortical implants improve the proximity of neuronal cell bodies. *J. Neural Eng.* **2011**, *8*, 066011.
18. Potter, K.A.; Buck, A.C.; Self, W.K.; Mcconnell, G.C.; Rees, H.D.; Levey, A.I.; Nguyen, J.K.; Park, D.J.; Skousen, J.L.; Prasad, A.; et al. Comprehensive characterization and failure modes of tungsten microwire. *J. Neural Eng.* **2012**, *9*, 56015.
19. Harris, J.P.; Hess, A.E.; Rowan, S.J.; Weder, C.; Zorman, C.A.; Tyler, D.J.; Capadona, J.R. In vivo deployment of mechanically adaptive nanocomposites for intracortical microelectrodes. *J. Neural Eng.* **2011**, *8*, 046010. [CrossRef] [PubMed]
20. Okun, M.; Lak, A.; Carandini, M.; Harris, K.D. Long term recordings with immobile silicon probes in the mouse cortex. *PLoS ONE* **2016**, *11*, 1–17. [CrossRef] [PubMed]
21. Polikov, V.S.; Tresco, P.A.; Reichert, W.M. Response of brain tissue to chronically implanted neural electrodes. *J. Neurosci. Methods* **2005**, *148*, 1–18. [CrossRef] [PubMed]
22. Pan, Z.; Yan, C.; Peng, R.; Zhao, Y.; He, Y.; Ding, J. Control of cell nucleus shapes via micropillar patterns. *Biomaterials* **2012**, *33*, 1730–1735. [CrossRef] [PubMed]
23. Davidson, P.M.; Özçelik, H.; Hasirci, V.; Reiter, G.; Anselme, K. Microstructured surfaces cause severe but non-detrimental deformation of the cell nucleus. *Adv. Mater.* **2009**, *21*, 3586–3590. [CrossRef]
24. Badique, F.; Stamov, D.R.; Davidson, P.M.; Veuillet, M.; Reiter, G.; Freund, J.; Franz, C.M.; Anselme, K. Biomaterials Directing nuclear deformation on micropillared surfaces by substrate geometry and cytoskeleton organization. *Biomaterials* **2013**, *34*, 2991–3001. [CrossRef] [PubMed]
25. Tan, J.L.; Tien, J.; Pirone, D.M.; Gray, D.S.; Bhadriraju, K.; Chen, C.S. Cells lying on a bed of microneedles: An approach to isolate mechanical force. *Proc. Natl. Acad. Sci. USA* **2003**, *100*, 1484–1489. [CrossRef] [PubMed]
26. Shakoor, A.; Muhammad, R.; Thomas, N.L.; Silberschmidt, V.V. Mechanical and thermal characterisation of poly (L-lactide) composites reinforced with hemp fibres. *J. Phys. Conf. Ser.* **2013**, *451*, 12010. [CrossRef]

27. Petersen, K.E. Silicon as a Mehcanical Material. *Proc. IEEE* **1982**, *70*, 420–455. [CrossRef]
28. Johnson, K.L. *Contact Mechanics*; Cambridge University Press: Cambridge, UK, 1985.
29. Jahed, Z.; Shahsavan, H.; Verma, M.S.; Rogowski, J.L.; Seo, B.B.; Zhao, B.; Tsui, T.Y.; Gu, F.X.; Mofrad, M.R.K. Bacterial networks on hydrophobic micropillars. *ACS Nano* **2017**, *11*, 675–683. [CrossRef] [PubMed]
30. Jahed, Z.; Lin, P.; Seo, B.B.; Verma, M.S.; Gu, F.X.; Tsui, T.Y.; Mofrad, M.R.K. Responses of Staphylococcus aureus bacterial cells to nanocrystalline nickel nanostructures. *Biomaterials* **2014**, *35*, 4249–4254. [CrossRef] [PubMed]
31. World Health Organization; International Agency for Research on Cancer. *World Cancer Report 2014*; IARC Press: Lyon, France, 2014.
32. Laermer, F.; Schilp, A. Method of Anisotropically Etching Silicon. U.S. Patent 5,501,893, 26 March 1996.
33. Özçelik, H.; Padeste, C.; Hasirci, V. Systematically organized nanopillar arrays reveal differences in adhesion and alignment properties of BMSC and Saos-2 cells. *Coll. Surf. B Biointerfaces* **2014**, *119*, 71–81.
34. Li, P.; Wang, D.; Li, H.; Yu, Z.; Chen, X.; Fang, J. Identification of nucleolus-localized PTEN and its function in regulating ribosome biogenesis. *Mol. Biol. Rep.* **2014**, *41*, 6383–6390. [CrossRef] [PubMed]
35. Park, S.; Ang, R.R.; Duffy, S.P.; Bazov, J.; Chi, K.N.; Black, P.C.; Ma, H. Morphological differences between circulating tumor cells from prostate cancer patients and cultured prostate cancer cells. *PLoS ONE* **2014**, *9*, e85264. [CrossRef] [PubMed]
36. Nexcelom Bioscience LLC, NCI-60 Cancer Cell Lines. 2017. [Online]. Available online: http://www.nexcelom.com/Applications/bright-field-analysis-of-nci-60-cancer-cell-lines.php (accessed on 16 June 2017).
37. Albuschies, J.; Vogel, V. The role of filopodia in the recognition of nanotopographies. *Sci. Rep.* **2013**, *3*. [CrossRef] [PubMed]
38. Jahed, Z.; Molladavoodi, S.; Seo, B.B.; Gorbet, M.; Tsui, T.Y.; Mofrad, M.R.K. Cell responses to metallic nanostructure arrays with complex geometries. *Biomaterials* **2014**, *35*, 9363–9371. [CrossRef] [PubMed]
39. Aalipour, A.; Xu, A.M.; Leal-ortiz, S.; Garner, C.C.; Melosh, N.A. Plasma membrane and actin cytoskeleton as synergistic barriers to nanowire cell penetration. *Langmuir* **2014**, *30*, 12362–12367. [CrossRef] [PubMed]
40. Robinson, J.T.; Jorgolli, M.; Shalek, A.K.; Yoon, M.; Gertner, R.S.; Park, H. Vertical nanowire electrode arrays as a scalable platform for intracellular interfacing to neuronal circuits. *Nat. Nanotechnol.* **2012**, *7*, 180–184. [CrossRef] [PubMed]
41. Xie, X.; Aalipour, A.; Gupta, S.V.; Melosh, N.A.; Science, M.; States, U.; Sciences, T.; Francisco, S.; Francisco, S.; States, U. Determining the time window for dynamic nanowire cell penetration. *ACS Nano* **2015**, *8*, 11667–11677. [CrossRef] [PubMed]
42. Berthing, T.; Bonde, S.; Rostgaard, K.R.; Madsen, M.H.; Sørensen, C.B.; Nygård, J.; Martinez, K.L. Cell membrane conformation at vertical nanowire array interface revealed by fluorescence imaging. *Nanotechnology* **2012**, *23*, 415102. [CrossRef] [PubMed]

MDPI

Article

Influence of Microencapsulated Phase Change Material (PCM) Addition on (Micro) Mechanical Properties of Cement Paste

Branko Šavija , Hongzhi Zhang * and Erik Schlangen

Microlab, Delft University of Technology, 2628 CN Delft, The Netherlands; b.savija@tudelft.nl (B.Š.); erik.schlangen@tudelft.nl (E.S.)
* Correspondence: h.zhang-5@tudelft.nl; Tel.: +31-015-278-8986

Received: 10 July 2017; Accepted: 24 July 2017; Published: 27 July 2017

Abstract: Excessive cracking can be a serious durability problem for reinforced concrete structures. In recent years, addition of microencapsulated phase change materials (PCMs) to concrete has been proposed as a possible solution to crack formation related to temperature gradients. However, the addition of PCM microcapsules to cementitious materials can have some drawbacks, mainly related to strength reduction. In this work, a range of experimental techniques has been used to characterize the microcapsules and their effect on properties of composite cement pastes. On the capsule level, it was shown that they are spherical, enabling good distribution in the material during the mixing process. Force needed to break the microcapsules was shown to depend on the capsule diameter and the temperature, i.e., whether it is below or above the phase change temperature. On the cement paste level, a marked drop of compressive strength with increasing PCM inclusion level was observed. The indentation modulus has also shown to decrease, probably due to the capsules themselves, and to a lesser extent due to changes in porosity caused by their inclusion. Finally, a novel micro-cube splitting technique was used to characterize the tensile strength of the material on the micro-meter length scale. It was shown that the strength decreases with increasing PCM inclusion percentage, but this is accompanied by a decrease in measurement variability. This study will contribute to future developments of cementitious composites incorporating phase change materials for a variety of applications.

Keywords: cement paste; nanoindentation; PCM; microcapsules; tensile strength; porosity

1. Introduction

Reinforced concrete is a construction material of choice for structures built in challenging environments. Compared to materials such as steel and timber, it has relatively good properties in aggressive conditions, leading to its high durability. Nevertheless, concrete is a quasi-brittle material susceptible to cracking due to mechanical and environmental loading [1]. Cracks in the concrete can promote deterioration by allowing rapid ingress of chloride [2] or carbon dioxide [3]. This will then lead to a fast corrosion initiation and propagation [4]. It is therefore of practical importance to avoid the occurrence of excessive cracking.

There are different strategies of achieving this, depending on the underlying cause of cracking. For example, if the mechanical loading is the cause, cracking can be limited by using fiber reinforcement, for example polyvinyl alcohol (PVA) fibers [5,6]. On the other hand, if the cracking is caused by thermal variations (such as early age temperature rise due to cement hydration or freeze-thaw damage), controlling the temperature is a good option. To achieve this, incorporation of phase change materials (PCMs) has been proposed in the past few years [7–10]. Phase change materials are combined (sensible-and-latent) thermal storage materials that can store and dissipate energy in the form of heat [8]. PCMs are usually added to the concrete mix as either microencapsulated particles [8], within embedded pipes [11], or as part of lightweight aggregates [12].

When microencapsulated PCMs are added to the mix, they influence the mechanical properties of the cement paste and, consequently, concrete [8,9,13,14]. This is (presumably) because the microcapsules are softer than the matrix material. The decrease of compressive strength seems to be more pronounced than of tensile strength [8,9,13].

In this study, the effect of PCM microcapsule addition on micromechanical properties of cement pastes is studied. Micromechanical testing of both the capsules and the composite paste is performed together with various characterization techniques. The focus is on a newly developed experimental technique based on nanoindentation, namely microcube testing [15]. This technique enables mechanical testing of cement paste on the representative length scale, i.e., the micrometer scale. This study provides insight into causes of changes in mechanical properties and their practical implications.

2. Experimental Program

2.1. Materials

In order to quantify the influence of microcapsule addition on the mechanical properties of cement based materials, studies were performed on the binding phase of concrete, i.e., the cement paste. For the purpose of material characterization, cement paste specimens were prepared. All pastes used ordinary Portland cement (CEM I 42.5 N) as a binder and a water-to-cement ratio of 0.45. Four different mixtures were used, with different levels of PCM addition: a reference mixture and mixtures containing 10%, 20%, and 30% of PCM microcapsules per volume, respectively.

Microcapsules used in this study are composed of a paraffinic phase change core and a melamine formaldehyde (MF) shell, with a core-to-shell ratio (mass based) of around 11.8 [13]. Enthalpy of phase change provided by the manufacturer was 143.5 J/g and the median particle size 22.53 μm.

The pastes were mixed in accordance with EN 196-3:2005+A1:2008 (E) using a Hobart mixer. First, the dry material (cement and PCM powder) was placed in a bowl. Water was added within 10 s. This was followed by mixing for 90 s at low speed. The mixer was then stopped for 30 s during which all paste adhering to the wall and the bottom part of the bowl was scrapped using a metal scraper and added to the mix. The mixing was then resumed for additional 90 s. The total mixer running time was around 3 min. The mix was then cast in plastic cylinders with an inner diameter of 34 mm and height of 58 mm. The cylinders were then sealed and rotated slowly for around 24 h in order to avoid bleeding. The pastes were then cured in sealed conditions until needed.

2.2. Microcapsule Characterization

2.2.1. Differential Scanning Calorimetry (DSC)

Differential scanning calorimetry (DSC) was used to investigate the thermal properties of microencapsulated PCMs. This included the onset and peak temperatures and enthalpy. The thermal program was as follows: the sample was heated from −20 °C to 100 °C and then cooled back to −20 °C in a nitrogen environment. The rate of heating and cooling was set to 5 °C per min.

2.2.2. Scanning Electron Microscopy (SEM)

The microstructure of microencapsulated PCMs was observed using a Philips XL30 Environmental Scanning Electron Microscope (ESEM) (FEI, Eindhoven, The Netherlands). The microcapsules were sprinkled on top of a glass plate which was coated with superglue to ensure bonding, and were subsequently imaged in the secondary electron (SE) mode.

2.2.3. Particle Size Distribution

The particle size distribution and the mean particle size of PCM microcapsules were determined by laser diffraction.

2.2.4. Compression of Microcapsules

In order to examine the influence of temperature on the mechanical properties of microcapsules, compression of individual microcapsules was performed. Microcapsules were sprinkled on top of a stage and individual microcapsules were identified and subjected to loading using a flat tiped indenter with a diameter of 135 µm (Figure 1). This method was initially proposed by the authors of [16]. This was done for room-temperature conditions (>25 °C, above the phase change temperature as determined by DSC), and for 15 °C using a temperature stage (below the phase change temperature). The relationship between maximum load and capsule diameter was determined for both temperatures.

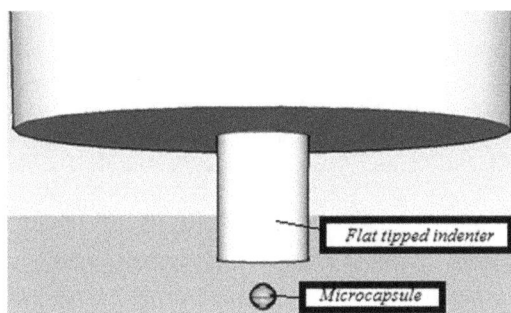

Figure 1. Nanoindentation setup used to measure the force-displacement relationship of phase change material (PCM) microcapsules.

2.3. Cement Paste Characterization

2.3.1. Compressive Strength Development

Compressive strength of cement paste specimens with different levels of PCM addition was measured at 1 day, 3 days, 7 days, 14 days, and 28 days. Cylindrical paste specimens with a diameter of 34 mm were cut to height of 40 mm (by cutting off the ends of the 58 mm high cylinder) and exposed to uniaxial compression. The loading rate of 1 kN/s was applied until specimen failure. Three specimens were tested for each condition.

2.3.2. Porosity and Pore Size Distribution

Porosity of the cement paste samples was determined using Mercury Intrusion Porosimetry (MIP). MIP is a commonly used technique for porosity investigation of cement based materials [17]. Although heavily criticized [18], this technique can be considered appropriate for comparative purposes as used herein.

For testing, a Micrometrics PoroSizer 9320 (Micrometrics, Norcross, GA, USA) device was used with a maximum pressure of 207 MPa. The contact angle and the surface tension of the mercury were set to 139°and 485 mN/m, respectively.

Prior to testing, hydration of the specimens was stopped using solvent exchange by isopropanol [19]. Paste specimens (obtained from cylinders by cutting) were submerged five times and taken out for a period of one min in order to enable a fast exchange of water and the solvent. Afterwards, they were placed in isopropanol for a prolonged period of time. This was followed by crushing and then vacuum drying for at least three months until the specimens were completely dry (determined by monitoring their weight over time).

Porosity and pore size distribution was measured at 3, 7, and 28 days for each cement paste mix.

2.3.3. Nanoindentation Testing

Micromechanical properties of cement paste mixtures with different additions of PCM microcapsules were measured by nanoindentation technique. Nanoindentation enables determination of local mechanical properties of tested volumes from the indentation load/displacement curve [20]. In the past decade, it has been commonly used to investigate the micromechanical properties of cementitious materials [21–25].

The elastic modulus of the indented material can be obtained from the following Equation:

$$\frac{1}{E_r} = \frac{1 - v_s^2}{E_s} + \frac{1 - v_i^2}{E_i} \tag{1}$$

where v_s is the Poisson's ratio of the tested material, v_i the Poisson's ratio of the indenter (0.07), E_s the Young's modulus of the sample and E_i the Young's modulus of the indenter (1141 GPa). It is assumed that during the unloading phase only elastic displacements are recovered, and that the reduced elastic modulus, E_r, can be determined using the slope of the unloading curve:

$$S = \frac{dP}{dh} = \frac{2}{\sqrt{\pi}} E_r \sqrt{A} \tag{2}$$

Here, S is the elastic unloading stiffness defined as the slope of the upper portion of the unloading curve during the initial stages of unloading, P is the load, h the displacement relative to the initial undeformed surface, and A the projected contact area at the peak load.

Nanoindentation tests were performed for specimens after 28 days of hydration. Prior to testing, hydration was stopped as described. Discs cut from the cylindrical pastes were glued onto a glass holder. The specimens were ground using sandpaper, during which ethanol was used as a cooling liquid. After grinding, samples were polished with 6 µm (5 min), 3 µm (5 min), 1 µm (10 min), and 0.25 µm (30 min) diamond paste on a lapping table. After each polishing step, samples were soaked into an ultrasonic bath to remove any residue. Sample preparation was performed just prior to testing to avoid carbonation of the tested surface.

An Agilent Nanoindenter G200 (Keysight, Santa Rosa, CA, USA) equipped with a diamond Berkovich tip was used for nanoindentation. For each specimen, a series of 20 × 20 indents were performed on a tightly spaced grid, with spacing of 20 µm between indents. Indentation depth was set to 700 nm. The Continuous Stiffness Method (CSM) proposed by Oliver and Pharr [20], which provides continuous measurements of elastic modulus as a function of indentation depth, was used to analyze the results. The average E modulus was determined in the loading range between 500–650 nm. For the calculation, Poisson's ratio of the indented material was taken as 0.18.

2.3.4. Microcube Splitting

While nanoindentation can be considered appropriate for measuring the elastic properties of cement paste and its individual phases, more complex procedures are needed for measuring strength properties at the micrometer length scale. This is because no relation between the indentation hardness and strength has been found so far for cement based materials [15,26]. Therefore, more advanced procedures that use e.g., nanoindentation equipment need to be used.

Recently, several authors have proposed measuring the tensile strength of cement paste [27] and it's individual phases [28] using micro-cantilever bending tests. This technique has been previously used for micromechanical testing of other quasi-brittle materials such as e.g., nuclear graphite [29,30]. This technique involves focused ion milling of a cantilever beam in the material, typically in the size range of up to 10 μm. Such cantilever beam is subsequently loaded in bending and tested until failure, providing a measure of the elastic modulus and the flexural strength of the tested microvolume. A major drawback of this approach is the fact that specimen preparation is very time consuming, so a relatively small number of specimens can be prepared and analyzed. Keeping in mind that on the μm length scale high scatter of measured mechanical properties can be expected [29,30] and that a large number of tests need to be performed for the measurements to be statistically reliable, herein a different approach is followed.

In this work, a recently developed method for creating a grid of micro-cubes (100 × 100 × 100 μm), developed by the authors of [15], is used. The method is shortly presented here. Cement paste specimens aged 28 days (with the hydration halted as previously described) were first glued on top of a glass substrate. Then, it is necessary to make the specimen thickness equal to the desired thickness (100 μm in this case), and this was done using a Struers Labopol-5 thin sectioning machine. The specimen is then ready for creation of the micro-cube grid. This is done using a precise diamond saw (MicroAce Series 3, Loadpoint, Swindon, UK) that is commonly employed in the semiconductor industry to create silicon wafers. To prevent chipping of the edges of the micro-cubes during cutting, a thin layer of soluble glue was applied on the surface of the thin section, which was later removed by soaking the specimen for a short time in acetone. In the machine, a 260 μm thick blade was run in two perpendicular directions over the specimen and the glass substrate (Figure 2). The procedure results in a grid of micro-cubes (100 × 100 × 100 ± 4 μm) that are used for micromechanical testing (Figure 3).

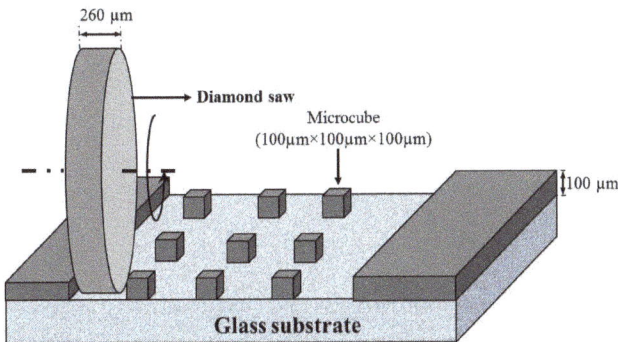

Figure 2. A schematic view of the specimen preparation procedure [15].

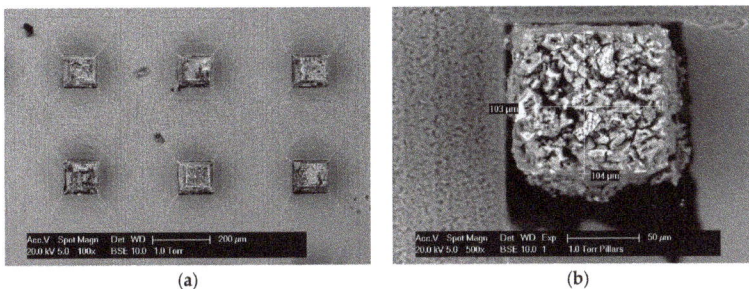

Figure 3. SEM images (**a**) A grid of micro-cubes on a glass substrate; (**b**) A single microcube.

For testing of the micro-cubes, the nanoindenter is employed (Figure 4). For the purpose of this splitting test, a diamond cylindrical wedge tip (radius 9.6 μm, length 200 μm) was used in order to apply the load across the middle axis. The experiments were run using displacement control with a loading rate of 50 nm/s.

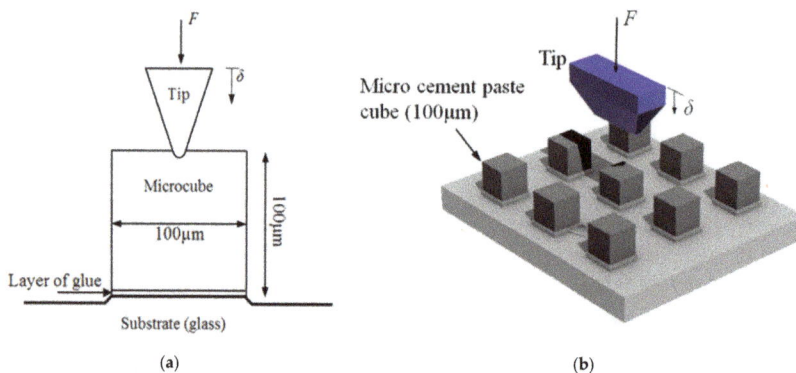

(a)   (b)

Figure 4. Schematic illustration of (**a**) A contact between the indenter tip and a single microcube; (**b**) the knife-tip loading procedure.

3. Results and Discussion

3.1. Microcapsule Characterization Results

Differential scanning calorimetry (DSC) curves of the PCM microcapsules are shown in Figure 5a for both the heating and the cooling regime. The heat of fusion during the phase change was determined as the area under the heat flow curve during the phase transition. Measured heat of fusion was 146.7 J/g, which corresponds well with the value provided by the manufacturer (143.5 J/g). The onset of phase change corresponding to melting is measured at 19.07 °C, and the endothermic peak at 22.07 °C. Considering their phase change temperature, these particles are suitable for applications such as reduction of temperature rise in young concrete for structures cast in moderate climatic conditions, as shown by previous finite element (FE) analyses by the authors [10] and others [31,32].

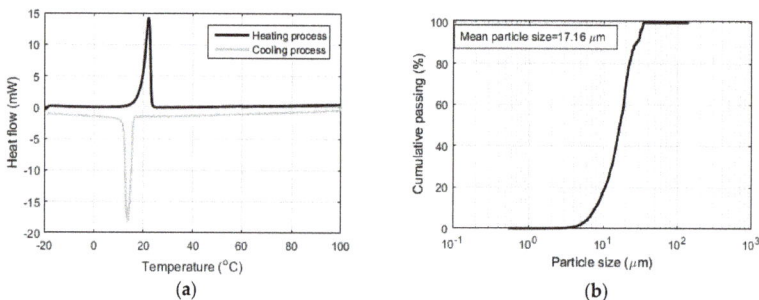

(a)   (b)

Figure 5. (**a**) Differential scanning calorimetry (DSC) thermograph of the PCM microcapsules; (**b**) Particle size distribution of microencapsulated PCM. (Adapted from study [9]).

In order to observe individual PCM microcapsules, they were sprinkled on a superglue layer on top of a glass substrate and placed inside the ESEM chamber. Imaging was performed using the secondary electron mode, acceleration voltage of 7 kV and 200× magnification. A micrograph of microencapsulated PCMs is shown in Figure 6. It can be seen that microcapsules are spherical in shape with a range of different diameters, which is beneficial for proper dispersion inside the cementitious matrix, as shown by [13] using micro-computed X-ray tomography.

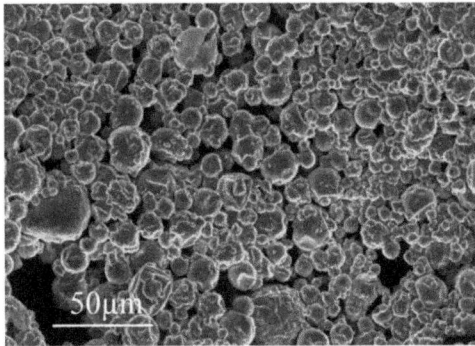

Figure 6. A micrograph of dispersed PCM microcapsules (adapted from [9]).

Particle size distribution of the microcapsules is shown in Figure 5b, with a mean particle size of 17.16 µm, which is somewhat smaller than reported by the manufacturer (22.53 µm).

Compression testing of individual microcapsules was performed as described previously. An example of a microcapsule before and after compression testing is shown in Figure 7. As the capsule ruptures, the encapsulated content (paraffin wax that acts as a phase change material in this case) is squeezed out, as seen in Figure 7b.

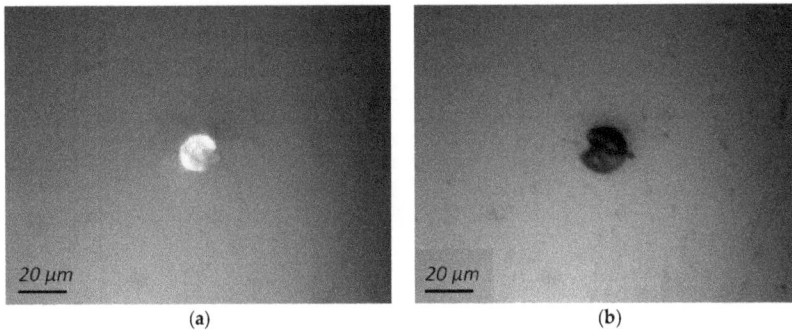

(a) (b)

Figure 7. An individual microcapsule imaged in the nanoindenter (**a**) before compression testing; (**b**) after compression testing.

While it is possible to relate the rupture force with the capsule diameter when punching with a Berkovich indenter is used, as shown by [33], herein the approach using a flat indenter tip was used. The method was initially described by [16] who tested brittle microcapsules (microballons) that show a distinct plateau in the load-displacement curve, indicating capsule failure. Since the capsules tested herein are not brittle, the identification of "failure" load was not as simple. A typical force-displacement curve of a microcapsule is shown in Figure 8a. However, the bump in the curve was not always visible, so it was not possible to determine the rupture force for all tested microcapsules. In the analysis provided, only capsules with a clear rupture point are included. Diameters of individual microcapsules are measured using microscopic images taken in the nanoindenter before the testing, such as the one shown in Figure 7a.

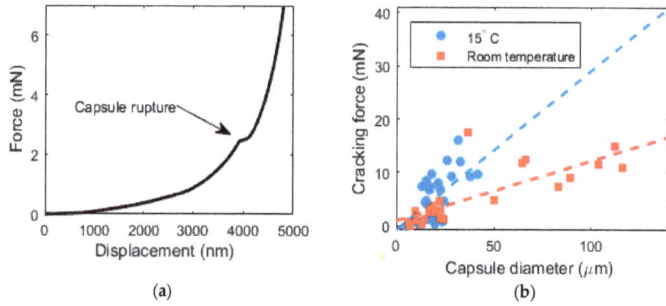

(a)   (b)

Figure 8. (a) A typical load vs. displacement curve measured in the capsule compression test; (b) A relationship between capsule diameter and cracking force for capsules below and above the phase change temperature (dashed lines indicate a linear fit).

It is clear that the rupture force of the microcapsules exhibits a size dependence, as capsules with larger diameters clearly require more force to rupture Figure 8b. It is also interesting that the capsule strength exhibit temperature dependence, as the capsules tested below the phase change temperature (at 15 °C) need a higher rupture force compared to those tested at room temperature (above 25 °C). This is because the encapsulated material seems to contribute to the load bearing capacity when it is in the solid phase, but not when it is in the liquid phase. Although the influence of temperature on the mechanical properties of the cement paste with microencapsulated PCM addition was not tested here, it will be a part of further research.

3.2. Cement Paste Characterization Results

3.2.1. Compressive Strength Results

The development of compressive strengths as a function of time for cement pastes with different percentages of microencapsulated PCM additions is given in Figure 9.

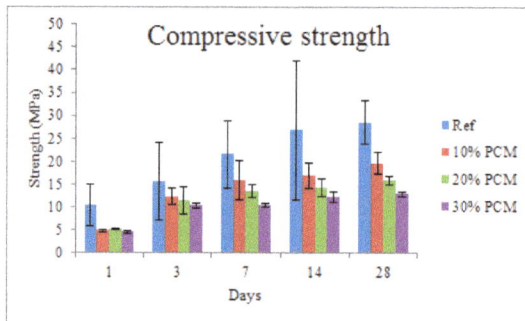

Figure 9. Development of paste compressive strength as a function of PCM addition percentage (error bars indicate standard deviation).

After 1 day, there is a marked difference between the strength of plain cement paste and the pastes with incorporation of PCM microcapsules. However, at this age, there is no significant difference between specimens with different amounts of PCM microcapsules. This changes already after 3 days, when a clear decrease of compressive strength with PCM addition percentage is observed. This trend remains valid until 28 days, when the compressive strength decreases by 31.2%, 44.5%, and 54.8% for the 10%, 20%, and 30% volumetric PCM inclusion, respectively.

3.2.2. Porosity Measurements

In Figure 10, pore size distributions for pastes aged 3, 7, and 28 days with varying PCM inclusion percentages are given. From these curves, critical pore diameters are extracted as follows: a peak in the differential PSD curve is defined as the critical pore diameter [17,34]. If two peaks are observed (such is the case with 30% PCM sample after 3 days of hydration), then the highest peak is defined as the critical pore diameter. MIP measurements provide, in addition, a measure of the total percolated pore volume. The percolated pore volume and critical pore diameter for all the pastes are given in Figure 11.

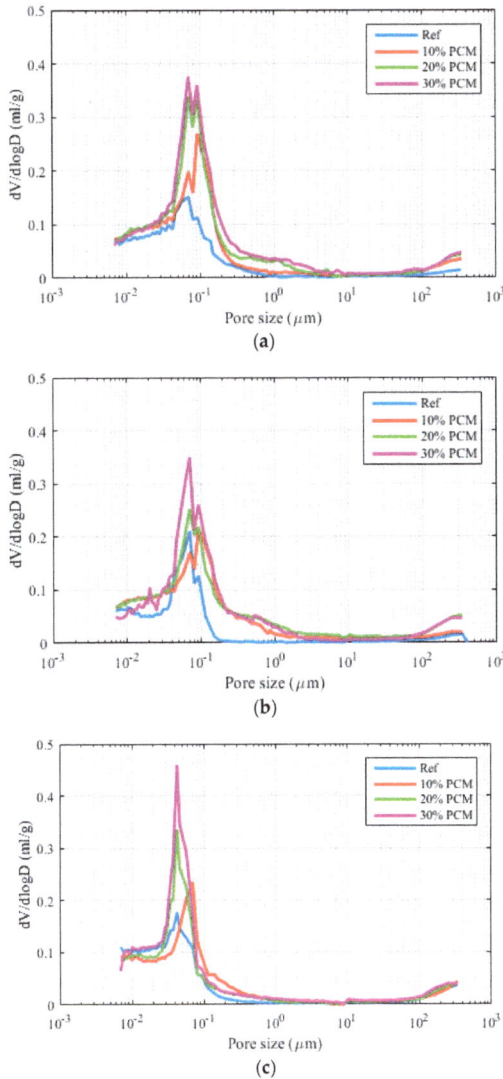

Figure 10. The effect of PCM microcapsule addition on the pore size distribution in cement paste samples after (**a**) 3 days; (**b**) 7 days and (**c**) 28 days of hydration.

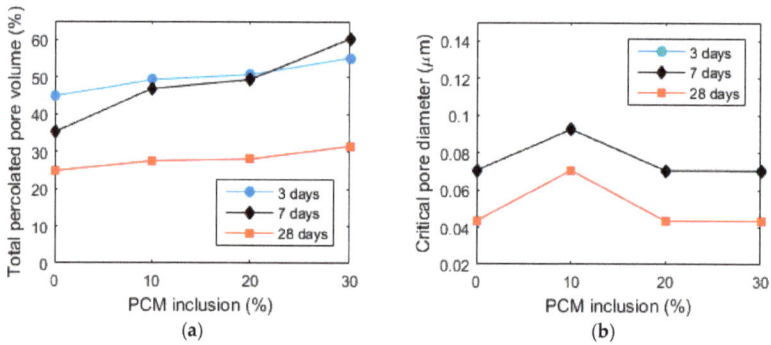

Figure 11. (**a**) Total percolated pore volume and (**b**) Critical pore diameter for cement pastes with various PCM inclusion percentages at different ages.

It can be observed that the total percolated pore volume increases with the increase in PCM dosage, more so for early hydration ages (3 and 7 days), and significantly less for the age of 28 days. Therefore, it is unlikely that the strength of the composite pastes is influenced by the increase in porosity caused by microencapsulated PCM addition. It is probable that the major cause of the strength drop is the addition of weak inclusions in the form of PCM microcapsules.

The critical pore diameter, on the other hand, remains constant for different paste formulations at the same age (apart from the 10% specimen, which shows a somewhat larger critical pore diameter). For the 3 and 7 day old pastes, the critical pore diameters do not change for any of the paste formulations. The critical pore diameters do decrease for the 28 day formulations. It needs to be noted that the critical pore diameter is a controlling parameter for durability of concrete. For example, chloride diffusion coefficient shows a linear relationship with the critical pore diameter, while the permeability shows a power relationship [35]. Furthermore, a recent study has shown that the addition of microencapsulated PCMs has very little influence on water absorption [36]. It is important that the long term durability of the cement paste with incorporated microencapsulated PCMs will not be affected.

3.2.3. Nanoindentation Results

In Figure 12, histograms of measured elastic modulus for 28-day old pastes with various percentages of PCM inclusions are given. As the percentage of PCM microcapsules increases it can be seen that histograms shift towards lower elastic modulus values. Although PCM microcapsules at the specimen surface are likely to be damaged during the specimen preparation procedure, the moduli reduced with PCM inclusion percentage increase. This is because nanoindentation is a volumetric measurement: a test will sample the material under the indenter up to a certain depth depending on different factors, as shown by [26]. Since MIP measurement showed no significant increase in porosity of the paste phase, this is most probably the reason.

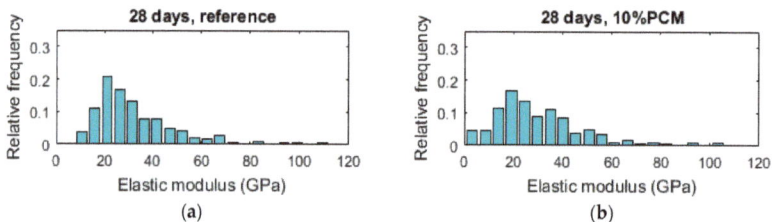

Figure 12. *Cont.*

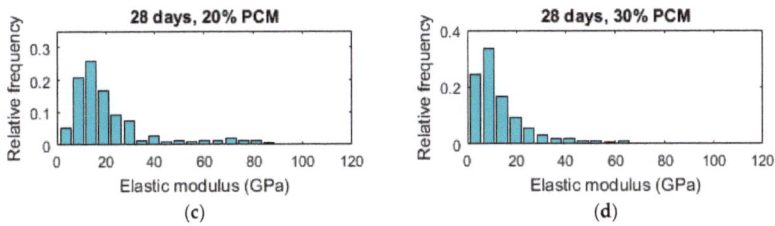

Figure 12. Histograms of elastic moduli for pastes with various percentage of PCM microcapsules measured by nanoindentation. (**a**) 28 days, reference paste; (**b**) 28 days, 10% PCM paste; (**c**) 28 days, 20% PCM paste; (**d**) 28 days, 30% paste.

In Figure 13, mean values of elastic modulus for pastes with increasing percentages of PCM inclusions are given. It is clear that the mean elastic modulus decreases with the increased amount of PCM inclusions in the mix. This is expected, as the addition of relatively large inclusions was previously shown to linearly decrease the elastic modulus with increasing inclusion volume in model quasi-brittle materials [37]. The mean elastic modulus measured by nanoindentation decreases by 6.1%, 33.9%, and 58.8% for the 10%, 20%, and 30% volumetric PCM inclusion, respectively, compared to the reference. While the decrease in strength may be considered detrimental for structural use of cementitious materials, a decrease of the elastic modulus may be beneficial for certain applications of crack control, since it will lead to lower stress build up in e.g., restrained deformation condition [8,10].

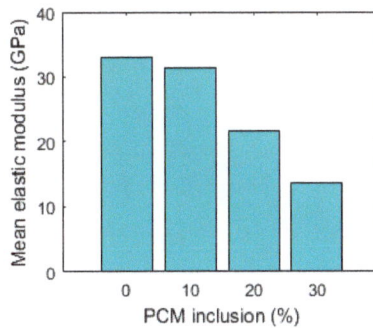

Figure 13. Mean elastic modulus of cement paste as a function of PCM volume fraction, measured by nanoindentation.

3.2.4. Microcube Splitting Results

The micro-cube splitting test performed in this work results in a load vs. displacement curve for a tested micro-cube. A typical load-displacement curve is shown in Figure 14. The curve shows two distinct regimes. Regime 1 signifies a nearly linear load-displacement curve until the peak is reached. After the peak load, the system enters an unstable regime (regime 2), which signifies a rapid crack propagation and failure of the micro-cube. Due to limitations in speed of the displacement control, the post-peak behaviour cannot be measured at present.

The setup of the micro-cube splitting test is similar to the Brazilian test (NEN-EN 12390-6 Standard) for splitting tensile strength assessment of cement based materials. The difference is in the boundary condition at the bottom: in the standard Brazilian test, a linear support is used. In the micro-cube splitting test, the specimen is clamped (glued) to the bottom (Figure 15).

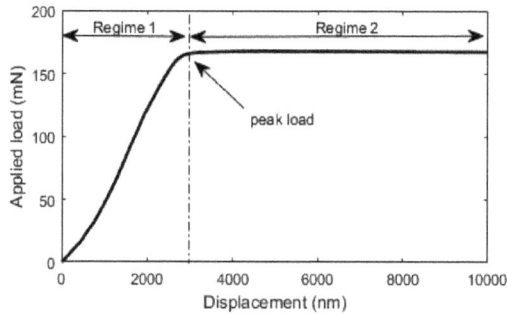

Figure 14. A typical load vs. displacement curve measured in the microcube splitting test.

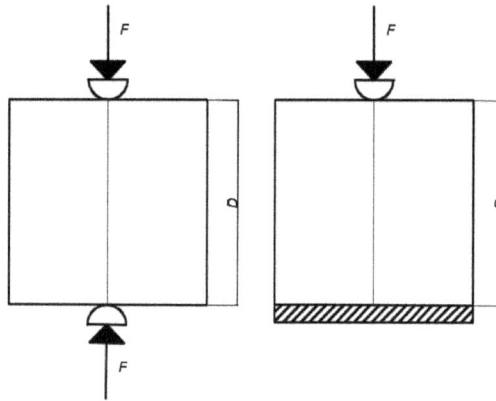

Figure 15. Schematics of the Brazilian splitting test (**left**) and the microcube splitting test (**right**). *D* is the specimen height, and *F* is the applied force.

In the Brazilian splitting test, a line load is applied on the top and the bottom surface of the specimen, leading to an almost uniform distribution of horizontal splitting stresses in the middle of the specimen. The magnitude of failure splitting stress can be determined using linear elastic theory as [38]:

$$f_{st} = \frac{2P}{\pi D^2} \tag{3}$$

In Equation (3), P is the maximum load, and D the specimen height. In the micro-cube splitting test, the bottom side is glued to the glass plate, leading to a somewhat different stress distribution. As shown by Zhang et al. [39], a modification of Equation (3) can be used to calculate the splitting stress in this case as:

$$f_{st} = 0.73 \cdot \frac{2P}{\pi D^2} \tag{4}$$

For the tests performed herein, Equation (4) was used to calculate the splitting strength of micro-cubes based on the peak load P, as shown in Figure 14.

For each mixture, a large number of micro-cubes were fabricated and tested as previously described. The results are summarized in Table 1. Histograms of splitting tensile strengths of measured micro-cubes are given in Figure 16. From Table 1 it can be seen that mean splitting tensile strength of cement micro-cubes decreases with the increasing PCM inclusion percentage. This is shown in Figure 17.

Table 1. Summary of micro-cube splitting results.

Mixture	Number of Micro-Cubes Tested	Mean Splitting Strength (MPa)	Standard Deviation (MPa)
Reference	118	19.39	5.68
10% PCM	105	14.98	4.19
20% PCM	166	10.35	3.41
30% PCM	98	10.03	2.92

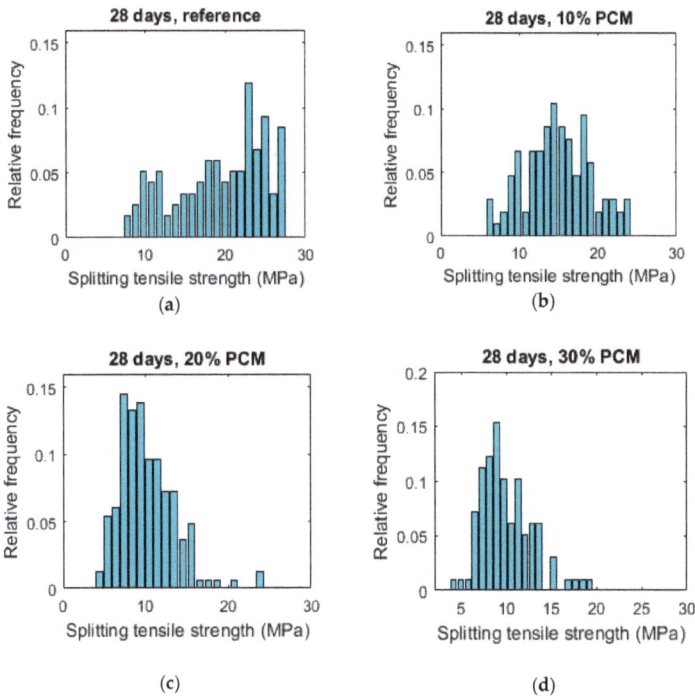

Figure 16. Histograms of splitting tensile strengths of micro-cubes made of pastes with various percentage of PCM microcapsules. (**a**) 28 days, reference paste; (**b**) 28 days, 10% PCM paste; (**c**) 28 days, 20% PCM paste; (**d**) 28 days, 30% PCM paste.

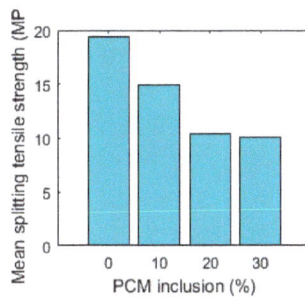

Figure 17. Mean splitting tensile strength of micro-cubes made of cement paste as a function of PCM volume fraction, measured by micro-cube splitting.

Histograms of splitting tensile strengths shift toward lower values with increasing PCM percentages. This is accompanied by a narrower distribution, signified in a lower standard deviation (Table 1). This means that paste specimens with lower PCM contents are stronger on average but also have more weak spots (represented by weak micro-cubes) compared to the specimens with higher PCM inclusion percentages. This probably explains a difference between micro-scale results obtained in this work and tests on larger (mortar) specimens employing similar materials: for example, Ref. [13] found that flexural strength of mortar specimens incorporating (similar) PCM microcapsules is only marginally affected by PCM addition, while the compressive strength is markedly lower. Unlike the elastic modulus of composite materials which is influenced by the properties of material components and their relative amounts, the (fracture) strength is also governed by the weakest link in the system. It is therefore desirable to analyse the obtained results using Weibull statistics. The probability of failure can then be written as [40]:

$$P_f = 1 - \exp\left[-\left(\frac{\sigma}{\sigma_0}\right)^m\right] \tag{5}$$

Here, P_f is the probability of failure, m the Weibull modulus, and σ_0 the scaling parameter (i.e., the stress corresponding to 63% probability of failure). Figure 18 shows the micro-cube splitting tensile strength tests for the tested cement pastes in a Weibull coordinate system.

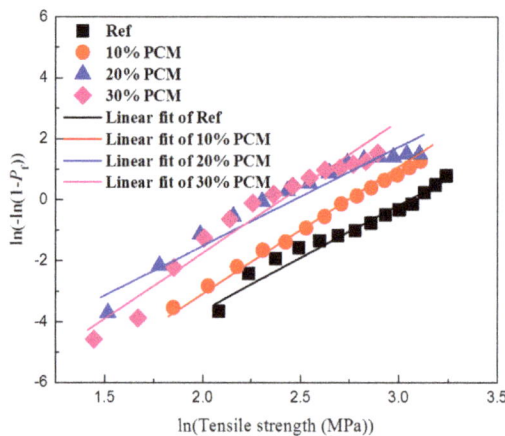

Figure 18. Weibull plot for measured splitting tensile strength of cement paste micro-cubes with different PCM inclusion percentages.

Table 2. Weibull parameters for the measured micro-cube splitting tensile strength.

Mixture	Number of Micro-Cubes Tested	Weibull Modulus, m	Scaling Parameter, σ_0 (MPa)
Reference	118	3.26	21.16
10% PCM	105	4.17	15.50
20% PCM	166	3.24	11.85
30% PCM	98	4.31	11.12

All tested mixtures show a good linear fit, with a coefficient of determination (R^2) higher than 0.95. The Weibull modulus and the scaling parameter for the tested pastes were fitted using the least squares method and are given in Table 2. The Weibull modulus increases with the increase in PCM inclusion percentage, with the exception of the 20% PCM specimen which does not follow this trend. This signifies a decrease in variability in measured micro-cube strength values for pastes with increasing inclusion percentages. This is also evident in a lower standard deviation, as given in Table 1. The scaling parameter obtained from the analysis (Table 2) shows a decrease with the increase in PCM inclusion percentage, similar to the previously shown trend for the mean splitting strength (Table 1). This analysis indicates that, although on average there is a large decrease of micro-cube splitting tensile strengths with increasing PCM inclusion percentage, the macroscopic tensile strength is not that different because it is governed by the weakest link in the system [41]. Furthermore, previous studies [8] have shown that the addition of compliant PCM microcapsules increases the toughness of the matrix by crack deflection and the microcapsule deformation ability. Since it was also shown recently that the inclusion of PCM microcapsules does not negatively affect the volume stability of cement-based composites [36], it is unlikely that it will increase shrinkage induced cracking either.

4. Summary and Conclusions

In this work, a detailed micromechanical characterization of cement pastes incorporating microencapsulated phase change materials (PCMs) has been performed. It was shown that the microcapsules used were spherical with a relatively fine particle size, enabling good dispersion in the cementitious matrix during the mixing process. Compression testing of individual microcapsules showed a linear relationship between the rupture force and the capsule diameter. Furthermore, it showed that there is a temperature dependence of the rupture force: capsules tested below the phase change temperature (when the core is solid) needed a higher force to rupture compared to capsules tested above the phase change temperature (when the core is liquid).

Then, cement pastes with varying PCM inclusion percentages (0–30% per volume) were prepared and characterized. As expected, compressive strength of cement pastes showed a reduction with increasing PCM inclusion percentages for pastes aged up to 28 days. Porosity of cement pastes was characterized by MIP, showing an increase in total percolated porosity with increasing PCM addition level. This increase was much more pronounced for early ages (3 and 7 days), and relatively minor for 28 day old paste specimens. Therefore, it was concluded that the change in porosity is probably only a minor factor causing decrease in strength with increasing PCM inclusion percentages. Furthermore, the critical pore diameter, which is an important parameter governing transport properties and durability of cement based materials, was shown to be independent of the PCM inclusion percentage but dependent on hydration age. This is consistent with recent studies showing that PCM microcapsule addition does not have a detrimental effect on durability of cementitious composites [36]. Nanoindentation of 28 day old cement pastes has shown a decrease in elastic modulus with increasing PCM percentages, consistent with previous studies [8]. This was attributed mainly to the addition of compliant inclusions in the form of PCM microcapsules. Furthermore, a new micro-cube splitting technique was used to characterize splitting strength of cement pastes with varying percentages of PCM inclusions on the micro-metre length scale, which is an appropriate length scale for testing the complex micromechanical properties of concrete's binding phase. It was found that, although pastes with higher PCM inclusion percentages showed a significantly lower average micro-cube splitting strength, the scatter in the measurements (i.e., standard deviation) was also lower. Consequently, pastes with lower PCM percentages have a relatively higher percentage of weak spots (in this case a percentage of micro-cubes weaker than the average), leading to their lower macroscopic tensile strength. This is considered to be a reason that the macroscopic tensile or flexural strength was found to be much less affected by the PCM addition compared to the compressive strength [8,9]. It should be noted, however, that due to the size of micro-cubes ($100 \times 100 \times 100$ µm), the size of microcapsules contained in these specimens was limited.

This study focused on small-scale characterization of cement pastes with PCM inclusions. Cementitious composites with PCM inclusions can be used as smart materials in a variety of applications: to promote thermal comfort in building applications [42], to melt ice and snow [11], or mitigate early and late age cracking [10,31]. Each of these applications can be achieved by adjusting the phase change temperature, the amount of phase change microcapsules and their latent heat. This study provides a basis for future developments of cementitious composites incorporating phase change materials for a variety of applications.

Acknowledgments: The first author gratefully acknowledges funding from European Union's Seventh Framework Programme for research, technological development and demonstration under the ERA-NET Plus Infravation programme, grant agreement No. 31109806.0001. The authors would like to thank Encapsys, LLC, for providing the encapsulated PCMs. The contribution of Johan Bijleveld for performing DSC measurements, Natalie Carr for particle size distribution measurements, and Arjan Thijssen for MIP measurements is gratefully acknowledged.

Author Contributions: Branko Šavija and Erik Schlangen devised the experimental program. Branko Šavija and Hongzhi Zhang performed experiments and analyzed the data. All authors wrote the manuscript.

Conflicts of Interest: The authors declare no conflict of interest.

References

1. Van Mier, J.G. *Concrete Fracture: A Multiscale Approach*; CRC Press: Boca Raton, FL, USA, 2012.
2. Šavija, B. Experimental and Numerical Investigation of Chloride Ingress in Cracked Concrete. Ph.D. Thesis, Delft University of Technology, Delft, The Netherlands, 2014.
3. De Schutter, G. Quantification of the influence of cracks in concrete structures on carbonation and chloride penetration. *Mag. Concr. Res.* **1999**, *51*, 427–435. [CrossRef]
4. Blagojević, A. The Influence of Cracks on the Durability and Service Life of Reinforced Concrete Structures in Relation to Chloride-Induced Corrosion. Ph.D. Thesis, Delft University of Technology, Delft, The Netherlands, 2016.
5. Qian, S.; Zhou, J.; De Rooij, M.; Schlangen, E.; Ye, G.; Van Breugel, K. Self-healing behavior of strain hardening cementitious composites incorporating local waste materials. *Cem. Concr. Compos.* **2009**, *31*, 613–621. [CrossRef]
6. Šavija, B.; Luković, M.; Schlangen, E. Influence of Cracking on Moisture Uptake in Strain-Hardening Cementitious Composites. *J. Nanomech. Micromech.* **2016**, *7*, 04016010. [CrossRef]
7. Bentz, D.P.; Turpin, R. Potential applications of phase change materials in concrete technology. *Cem. Concr. Compos.* **2007**, *29*, 527–532. [CrossRef]
8. Fernandes, F.; Manari, S.; Aguayo, M.; Santos, K.; Oey, T.; Wei, Z.; Falzone, G.; Neithalath, N.; Sant, G. On the feasibility of using phase change materials (PCMs) to mitigate thermal cracking in cementitious materials. *Cem. Concr. Compos.* **2014**, *51*, 14–26. [CrossRef]
9. Šavija, B.; Luković, M.; Kotteman, G.M.; Figuieredo, S.C.; de Mendoça Filho, F.F.; Schlangen, E. Development of ductile cementitious composites incorporating microencapsulated phase change materials. *Int. J. Adv. Eng. Sci. Appl. Math.* **2017**, *2017*, 1–12. [CrossRef]
10. Šavija, B.; Schlangen, E. Use of phase change materials (PCMs) to mitigate early age thermal cracking in concrete: Theoretical considerations. *Constr. Build. Mater.* **2016**, *126*, 332–344. [CrossRef]
11. Farnam, Y.; Krafcik, M.; Liston, L.; Washington, T.; Erk, K.; Tao, B.; Weiss, J. Evaluating the use of phase change materials in concrete pavement to melt ice and snow. *J. Mater. Civ. Eng.* **2015**, *28*, 1–10. [CrossRef]
12. Aguayo, M.; Das, S.; Castro, C.; Kabay, N.; Sant, G.; Neithalath, N. Porous inclusions as hosts for phase change materials in cementitious composites: Characterization, thermal performance, and analytical models. *Constr. Build. Mater.* **2017**, *134*, 574–584. [CrossRef]
13. Aguayo, M.; Das, S.; Maroli, A.; Kabay, N.; Mertens, J.C.; Rajan, S.D.; Sant, G.; Chawla, N.; Neithalath, N. The influence of microencapsulated phase change material (PCM) characteristics on the microstructure and strength of cementitious composites: Experiments and finite element simulations. *Cem. Concr. Compos.* **2016**, *73*, 29–41. [CrossRef]
14. Hunger, M.; Entrop, A.; Mandilaras, I.; Brouwers, H.; Founti, M. The behavior of self-compacting concrete containing micro-encapsulated phase change materials. *Cem. Concr. Compos.* **2009**, *31*, 731–743. [CrossRef]
15. Zhang, H.; Šavija, B.; Chaves Figueiredo, S.; Lukovic, M.; Schlangen, E. Microscale Testing and Modelling of Cement Paste as Basis for Multi-Scale Modelling. *Materials* **2016**, *9*, 907. [CrossRef]
16. Koopman, M.; Gouadec, G.; Carlisle, K.; Chawla, K.; Gladysz, G. Compression testing of hollow microspheres (microballoons) to obtain mechanical properties. *Scr. Mater.* **2004**, *50*, 593–596. [CrossRef]

17. Scrivener, K.; Snellings, R.; Lothenbach, B. *A Practical Guide to Microstructural Analysis of Cementitious Materials*; Crc Press: Boca Raton, FL, USA, 2015.
18. Diamond, S. Mercury porosimetry: An inappropriate method for the measurement of pore size distributions in cement-based materials. *Cem. Concr. Res.* **2000**, *30*, 1517–1525. [CrossRef]
19. Zhang, J.; Scherer, G.W. Comparison of methods for arresting hydration of cement. *Cem. Concr. Res.* **2011**, *41*, 1024–1036. [CrossRef]
20. Oliver, W.C.; Pharr, G.M. Measurement of hardness and elastic modulus by instrumented indentation: Advances in understanding and refinements to methodology. *J. Mater. Res.* **2004**, *19*, 3–20. [CrossRef]
21. Constantinides, G.; Ulm, F.-J.; Van Vliet, K. On the use of nanoindentation for cementitious materials. *Mater. Struct.* **2003**, *36*, 191–196. [CrossRef]
22. Luković, M.; Šavija, B.; Dong, H.; Schlangen, E.; Ye, G. Micromechanical study of the interface properties in concrete repair systems. *J. Adv. Concr. Technol.* **2014**, *12*, 320–339. [CrossRef]
23. Šavija, B.; Luković, M.; Hosseini, S.A.S.; Pacheco, J.; Schlangen, E. Corrosion induced cover cracking studied by X-ray computed tomography, nanoindentation, and energy dispersive X-ray spectrometry (EDS). *Mater. Struct.* **2015**, *48*, 2043–2062. [CrossRef]
24. Ulm, F.-J.; Vandamme, M.; Jennings, H.M.; Vanzo, J.; Bentivegna, M.; Krakowiak, K.J.; Constantinides, G.; Bobko, C.P.; Van Vliet, K.J. Does microstructure matter for statistical nanoindentation techniques? *Cem. Concr. Compos.* **2010**, *32*, 92–99. [CrossRef]
25. Zheng, K.; Lukovic, M.; De Schutter, G.; Ye, G.; Taerwe, L. Elastic Modulus of the Alkali-Silica Reaction Rim in a Simplified Calcium-Alkali-Silicate System Determined by Nano-Indentation. *Materials* **2016**, *9*, 787. [CrossRef]
26. Luković, M.; Schlangen, E.; Ye, G. Combined experimental and numerical study of fracture behaviour of cement paste at the microlevel. *Cem. Concr. Res.* **2015**, *73*, 123–135. [CrossRef]
27. Chen, S.J.; Duan, W.H.; Li, Z.J.; Sui, T.B. New approach for characterisation of mechanical properties of cement paste at micrometre scale. *Mater. Des.* **2015**, *87*, 992–995. [CrossRef]
28. Němeček, J.; Králík, V.; Šmilauer, V.; Polívka, L.; Jäger, A. Tensile strength of hydrated cement paste phases assessed by micro-bending tests and nanoindentation. *Cem. Concr. Compos.* **2016**, *73*, 164–173. [CrossRef]
29. Liu, D.; Flewitt, P.E. Deformation and fracture of carbonaceous materials using in situ micro-mechanical testing. *Carbon* **2017**, *114*, 261–274. [CrossRef]
30. Šavija, B.; Liu, D.; Smith, G.; Hallam, K.R.; Schlangen, E.; Flewitt, P.E. Experimentally informed multi-scale modelling of mechanical properties of quasi-brittle nuclear graphite. *Eng. Fract. Mech.* **2016**, *153*, 360–377. [CrossRef]
31. Arora, A.; Sant, G.; Neithalath, N. Numerical simulations to quantify the influence of phase change materials (PCMs) on the early-and later-age thermal response of concrete pavements. *Cem. Concr. Compos.* **2017**, *81*, 11–24. [CrossRef]
32. Young, B.A.; Falzone, G.; Zhenye, S.; Thiele, A.; Wei, Z.; Neithalath, N.; Sant, G.; Pilon, L. Early-age temperature evolutions in concrete pavements containing microencapsulated phase change materials. *Constr. Build. Mater.* **2017**, *147*, 466–477. [CrossRef]
33. Lv, L.; Schlangen, E.; Yang, Z.; Xing, F. Micromechanical Properties of a New Polymeric Microcapsule for Self-Healing Cementitious Materials. *Materials* **2016**, *9*, 1025. [CrossRef]
34. Yu, Z. Microstructure Development and Transport Properties of Portland Cement-fly Ash Binary Systems: In View of Service Life Predictions. Ph.D. Thesis, Delft University of Technology, Delft, The Netherlands, 2015.
35. Halamickova, P.; Detwiler, R.J.; Bentz, D.P.; Garboczi, E.J. Water permeability and chloride ion diffusion in Portland cement mortars: Relationship to sand content and critical pore diameter. *Cem. Concr. Res.* **1995**, *25*, 790–802. [CrossRef]
36. Wei, Z.; Falzone, G.; Wang, B.; Thiele, A.; Puerta-Falla, G.; Pilon, L.; Neithalath, N.; Sant, G. The durability of cementitious composites containing microencapsulated phase change materials. *Cem. Concr. Compos.* **2017**, *81*, 66–76. [CrossRef]
37. Liu, D.; Šavija, B.; Smith, G.E.; Flewitt, P.E.J.; Lowe, T.; Schlangen, E. Towards understanding the influence of porosity on mechanical and fracture behaviour of quasi-brittle materials: Experiments and modelling. *Int. J. Fract.* **2017**, *205*, 57–72. [CrossRef]
38. Rocco, C.; Guinea, G.V.; Planas, J.; Elices, M. Size effect and boundary conditions in the Brazilian test: Experimental verification. *Mater. Struct.* **1999**, *32*, 210–217. [CrossRef]

39. Zhang, H.; Šavija, B.; Schlangen, E. Micro-cube splitting test for tensile strength characterisation of cement paste at micro scale. *Constr. Build. Mater.* **2017**, *140*, 10–15.

40. Pizette, P.; Martin, C.; Delette, G.; Sornay, P.; Sans, F. Compaction of aggregated ceramic powders: From contact laws to fracture and yield surfaces. *Powder Technol.* **2010**, *198*, 240–250. [CrossRef]

41. Bažant, Z.P.; Planas, J. *Fracture and Size Effect in Concrete and Other Quasibrittle Materials*; CRC Press: Boca Raton, FL, USA, 1997; Volume 16.

42. Snoeck, D.; Priem, B.; Dubruel, P.; De Belie, N. Encapsulated Phase-Change Materials as additives in cementitious materials to promote thermal comfort in concrete constructions. *Mater. Struct.* **2016**, *49*, 225–239. [CrossRef]

materials

MDPI

Article

Characterisation of Asphalt Concrete Using Nanoindentation

Salim Barbhuiya * and Benjamin Caracciolo

Department of Civil Engineering, Curtin University, Perth 6845, Australia;
benjamin.caracciolo@graduate.curtin.edu.au
*   Correspondence: salim.barbhuiya@curtin.edu.au; Tel.: +61-892-662-392

Received: 30 June 2017; Accepted: 17 July 2017; Published: 18 July 2017

Abstract: In this study, nanoindentation was conducted to extract the load-displacement behaviour and the nanomechanical properties of asphalt concrete across the mastic, matrix, and aggregate phases. Further, the performance of hydrated lime as an additive was assessed across the three phases. The hydrated lime containing samples have greater resistance to deformation in the mastic and matrix phases, in particular, the mastic. There is strong evidence suggesting that hydrated lime has the most potent effect on the mastic phase, with significant increase in hardness and stiffness.

Keywords: asphalt concrete; hydrated lime; nanoindentation; Young's modulus; hardness

1. Introduction

Asphalt concrete (AC) is widely used around the world due to its primary function in road and pavement construction. AC is a composite material composed of a heterogeneous mix of asphalt binder and mineral aggregates. It is generally agreed that AC consists of three distinct phases: mastic, matrix, and aggregate phases [1]. The mastic phase is defined by asphalt binder interacting with fines (particles passing the 0.075 mm sieve) to form a thin film that coats and binds the aggregate particles. The matrix phase is defined as the mix of asphalt binder and fillers passing the 4.75 mm sieve.

The capacity to study the mechanical behaviour of materials is expanding with the introduction of new technologies that can probe locally at nanoscale resolutions and with great control over experimental variables. Using nanoindentation, a multi-structured, multi-phased material like AC can be examined under a variety of loading conditions. Whilst nanoindentation has been used as a means of testing material properties for decades [2–4], studies exploring its use on asphalt binder and AC have until recently been limited [5–7]. There are difficulties in performing the test on asphalt binder due to the soft and highly viscoelastic nature of the material [8]. It was found that the use of a spherical indenter was better suited to the binder, but the Berkovich indenter tip was able to produce successful results on concrete [9].

In order to modify the properties of asphalt binder and asphalt concrete mixes various fillers (aggregate particles fine than 75 μm) are added. Some of the commonly used fillers include Portland cement, hydrated lime, fly ash, limestone dust and clay particles [10]. It is well documented that filler exerts a significant effect on the characteristics and performance of AC mixes [11,12]. It is believed that good packing of coarse, fine, and filler aggregates provides a strong backbone for the AC mix [13]. The presence of filler in AC mixes is even more important because of its possible interaction with asphalt. Owing to the larger surface area, fillers absorb more asphalt and its interaction with asphalt leads to better performance of AC mixes. Fillers are also found to improve the temperature susceptibility and durability of asphalt binder and AC mixes [14].

Calcium hydroxide, known more commonly as hydrated lime, is one of the fillers frequently used in AC to improve the asphalt-aggregate bond and the binder's resistance to water-induced damage [15–19]. However, limited research is available on its effects on load-displacement behaviour and nanomechanical properties of individual AC phases. In this study, nanoindentation is used to assess the load-displacement characteristics of AC with and without hydrated lime and obtain the nanomechanical properties in terms of Young's modulus and hardness in the mastic, matrix, and aggregate phases.

2. Experimental Methods

2.1. Materials

Two hot mix AC were selected for testing in this study (one with hydrated lime and one without). Each mix contained a heterogeneous blend of crushed granite aggregates (10 mm) and asphalt binder. A 10 mm size was chosen as this is a common fine aggregate size used on roads, pathways and car parks. The aggregates were sourced from Boral and BGC and dried in the Geomechanics Lab of Curtin University Australia. The standard addition of hydrated lime in AC mixes in Australia is 1.5% of the total weight of the dry mix [20]. Therefore, in this study the hydrated lime was added at 1.5%. The mix details are summarised in Table 1. The term "base" in Table 1 refers to a mix not containing hydrated lime. Other properties of the mixes are given in Table 2.

Table 1. Mix details.

Materials (%)	Base (without Hydrated Lime)	With Hydrated Lime
10 mm aggregates	43.7	43.7
5 mm aggregates	11.4	11.4
Dust	28.5	28.1
Washed dust	11.4	10.4
Hydrated lime	0	1.5

Table 2. Properties of the mixes.

Properties	Values
Binder type	C320
Air voids	3.0–7.0%
VMA	$\geq$16.0%
Stability	$\geq$8.0 kN
Flow	2–4 mm
Compaction	75 blows

2.2. Sample Preparation

The samples were mixed and compacted as per AS 2150–2005 [21]. The compacted cylindrical samples were then cut down to smaller size in order for them to fit the maximum thickness requirements of the indenter. The surfaces of the samples were sliced off using a concrete cutter and then divided into 10 mm cubic sections using a precision saw for the recommended nanoindentation size. Four samples from each mix were selected for resin casting, a process that suspends the sample in resin, creating total stability for grinding, polishing, and nanoindentation. Once the resin solidified 24 h later, the samples were ground using a grinder-polisher machine. Before polishing, a step called impregnation was used. The process of impregnation helped to fill the valleys in the structure of the sample, creating an even surface. The samples were covered with a thin coat of EpoThin and placed in a vacuum chamber, eradicating all air bubbles. Following this, the samples were grinded once again in order to expose the surface. Preparation of the samples is complete after the final stage of polishing. Four decreasing sizes of Polycrystalline Diamond Suspension (9 μm, 6 μm, 3 μm, and 1 μm) were used to lubricate a polishing cloth, which rotates under the samples. The diamond particles consist of single grains that have sharp edges, working to remove excess particles and enabling a smooth surface finish.

2.3. Nanoindentation

Nanoindentation tests were carried out using Agilent Nano Indenter G200 (Keysight Technologies, Inc., Santa Rosa, CA, USA). Poisson's ratio for both AC and indenter tip were assumed based on average values from previous literature [22]. In order to appropriately represent the diversities that exist within AC, it was important to perform a sufficient number of indents over a variety of areas for each sample. The following five sets of tests were conducted using nanoindentation:

Set 1: Two indents on a randomly selected area of mastic and two indents on a randomly selected area of aggregate (Figure 1). This was the initial test to assess the loading conditions and whether the 20 s dwell time was sufficient to minimise the effects of creep. It provides a basic reference for the mastic/aggregate characteristics of each sample, although four indents are not representative of the overall sample.

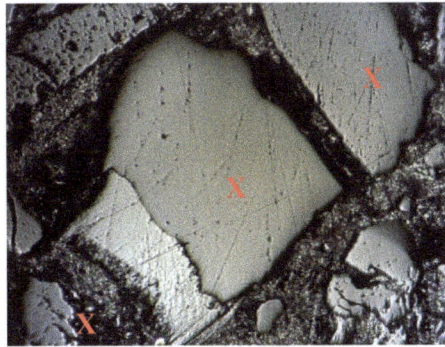

Figure 1. Indentation sites for Set 1 (marked with X).

Set 2: A 10 × 10 grid (100 indents) on a manually selected area of each sample containing mastic, matrix, and aggregate (Figure 2). The grid was organised with each indent spaced at 40 μm apart in order to avoid residual impressions from previous indents. Areas were chosen such that there was a relatively even mix of each phase present in order to gather a good representation of data.

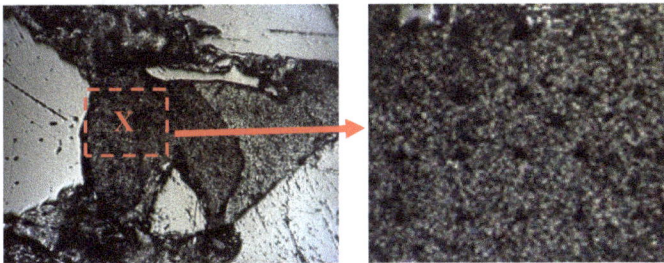

Figure 2. Indentation sites for Set 2 (marked with X).

3. Results and Discussion

3.1. Load-Displacement Characteristics

Figure 3 shows the load-displacement characteristics of two separate indents on the mastic region. First, it is evident that both curves in Figure 3 have an almost identical shape, with one displacing more during loading. It is also evident that there is a period of creep during maximum load, which was expected given the viscoelastic nature of the mastic. However, the unloading curve appears to elicit an elastic response, which indicates the dwell period of 20 s was enough to offset the effects of the creep. An elastic unloading curve is necessary in order to use the Oliver-Pharr method [23] of analysis. Both curves are set to 95% unloading. This means when the indenter unloads to approximately 0.5 mN, a further dwell period of 75 s is employed in order to account for thermal drift. The horizontal line at the end of the unloading curve is a measure of the displacement during this period and accounted for automatically as drift correction. This is the case for all indentations performed. Figure 3 also shows a difference of approximately 200 nm in indentation depth between the two tests. The diversity within the mastic means it is highly unlikely that two indentation points will be exactly the same, even within the same region.

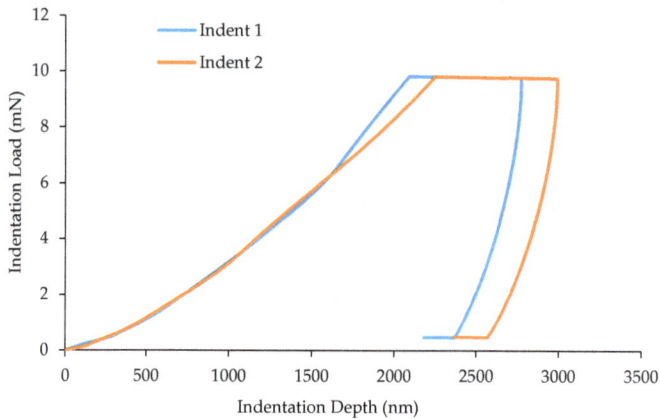

Figure 3. Load-displacement curve for two separate indents on mastic phase.

Figure 4 illustrates the diversity within the mastic by showing 10 indents. It is notable that the shape of the unloading curves for the tests in Figures 5 and 6 are consistent in having an elastic response. There is a noticeable amount of hysteresis at the beginning of most curves before 1000 nm depth. After this, the data normalises and follows a reliable shape. It is possible that this is due to imperfections on the surface before the indenter reaches material with relative consistency.

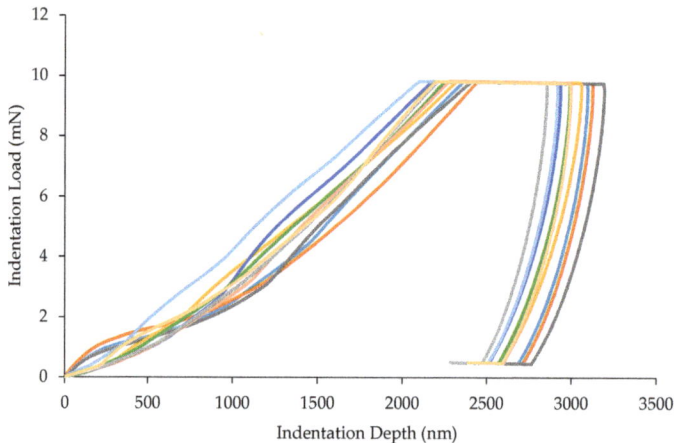

Figure 4. Load-displacement curve for 10 indents on mastic phase.

Figure 5 shows two separate indents performed on the aggregate phase of the sample. The characteristics of the aggregate curves compared to those on the mastic are evident in the reduction of indentation depth and lower creep response during the dwell period. Indentation depth at the peak load of 10 mN is in the range of 200–300 nm indicating a much harder material when contrasted with the 2000–3000 nm range of the mastic. Figure 6 displays 10 tests performed on a section of aggregate, spaced 40 µm apart. The overall variance between indents on a particular region of aggregate is also less apparent as can be seen in Figure 6. Over the 10 tests, the change in indentation depth at max load was just 72 nm, which is indicative of a fairly consistent aggregate material. This can also be seen in the "smoothness" of the loading curve compared to those on the mastic.

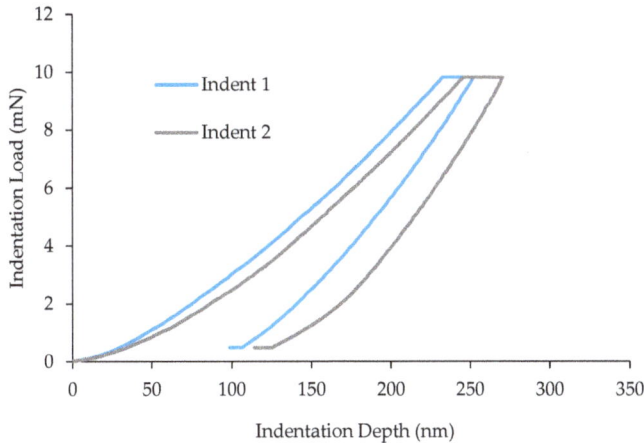

Figure 5. Load-displacement curve for two separate indents on aggregate phase.

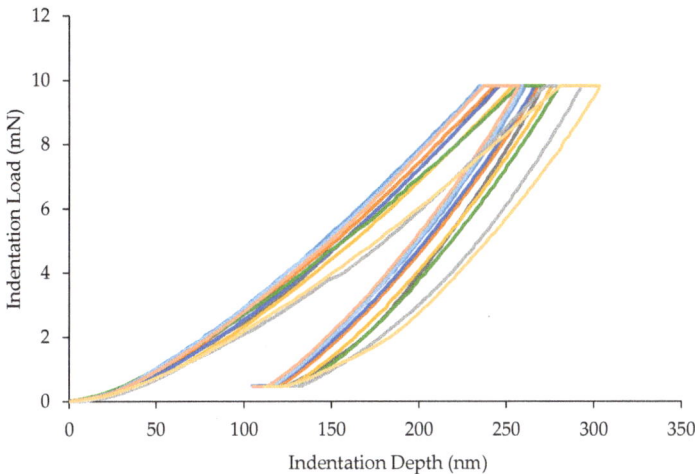

Figure 6. Load-displacement curve for 10 indents on aggregate phase.

The relative difference of each of the phases in AC without hydrated lime is more apparent when displayed on the same scale as shown in Figure 7. There is a clear distinction between the aggregate and mastic/matrix phases when looking at the indentation depth and creep during dwell time. The loading curves for both the mastic and matrix phases show a much greater elastic-plastic response than the aggregate due to the softness of the regions. It can also be seen that the displacement during dwell time for the mastic phase lessens during the matrix phase and is relatively small during the aggregate phase. The unloading portion of the load-displacement curves for each region displays elastic flow.

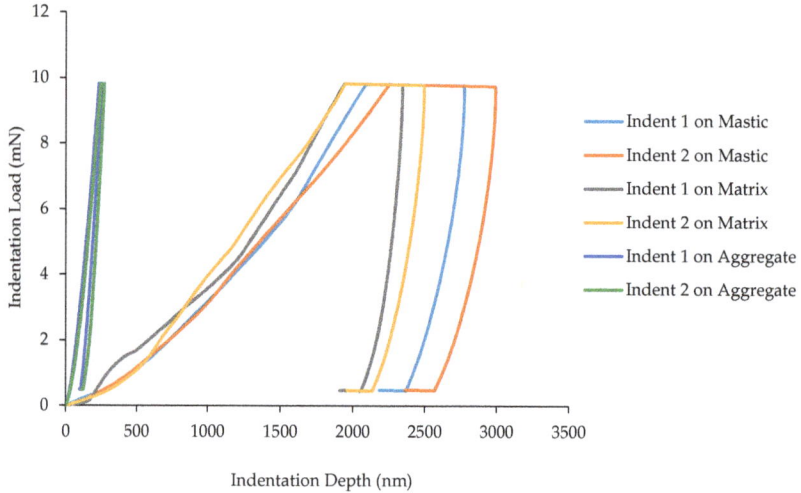

Figure 7. Load-displacement curve for all phases in AC without hydrated lime.

Figure 8 shows a series of indentations taken on the sample containing hydrated lime. It appears that the addition of hydrated lime to the mix does not directly affect the elastic-plastic properties of the loading curve or the elastic response of the unloading curve. What is immediately apparent, however, is the reduction in displacement compared to the sample without the lime. Considering the documented effects of hydrated lime as a stiffening agent, this is an expected result. The hydrated lime addition appears to have a greater effect on displacement in the mastic and matrix phases. This is logical considering the additive bonds itself with asphalt binder, which is mostly present within these phases.

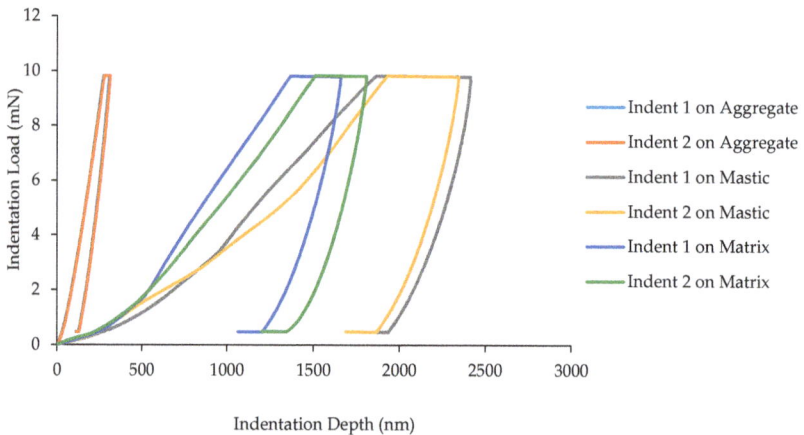

Figure 8. Load-displacement curve for all phases in AC with hydrated lime.

Figure 9 displays nanoindentation tests performed on the mastic phase of samples with hydrated lime. It can be seen from the figure that the difference in indentation depth between samples with and without hydrated lime is obvious. However, comparing a small number of indents does not necessarily provide an accurate representation of a material with the complexities of asphalt concrete. Figure 10 presents the results of two 10 × 10 grids of indentations for both the samples. The result is an average of 186 successful indents for the hydrated lime containing sample and 167 successful indents for the sample without hydrated lime.

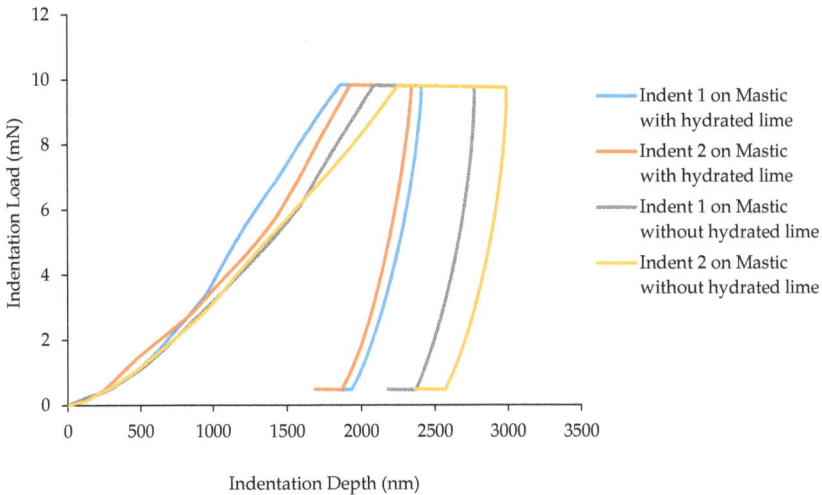

Figure 9. Load-displacement curve for mastic phases with hydrated lime.

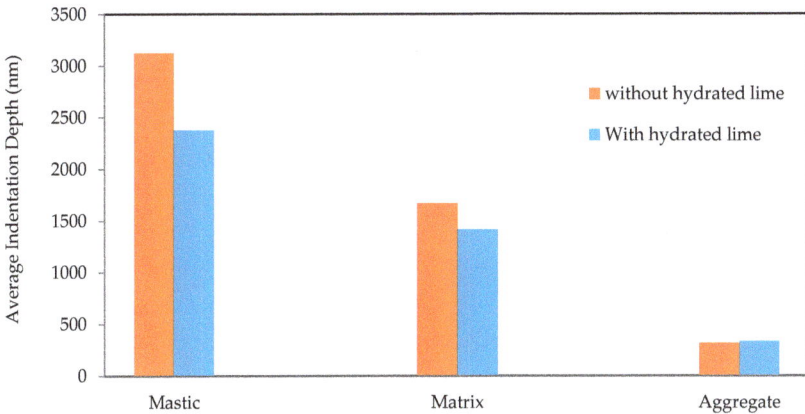

Figure 10. Average indentation depth for all phases.

Overall, there is reasonable confirmation that the hydrated lime containing samples have greater resistance to deformation in the mastic and matrix phases, in particular, the mastic. Again, there appears to be a negative differential in the aggregate phase, although 16 nm could be accounted for as natural variance within the material. There is certainly no indication from the results that hydrated lime acts as a stiffener in the aggregate phase.

3.2. Nanomechanical Properties

The nanomechanical properties (Young's modulus and hardness) in sample without hydrated lime are shown in Figure 11. It can be seen that there is a trend of increasing hardness as Young's modulus increases. The Young's modulus range seems to be under 4 GPa for mastic, 4–12 GPa for matrix, and 12–100 GPa for aggregate. The aggregate phase in particular has a wide range of Young's modulus and hardness values compared to the other phases. Figure 12 takes a closer a look at the Young's modulus and hardness values of the mastic region. The hardness still appears to increase slightly with the Young's modulus, although it is less clear at this scale.

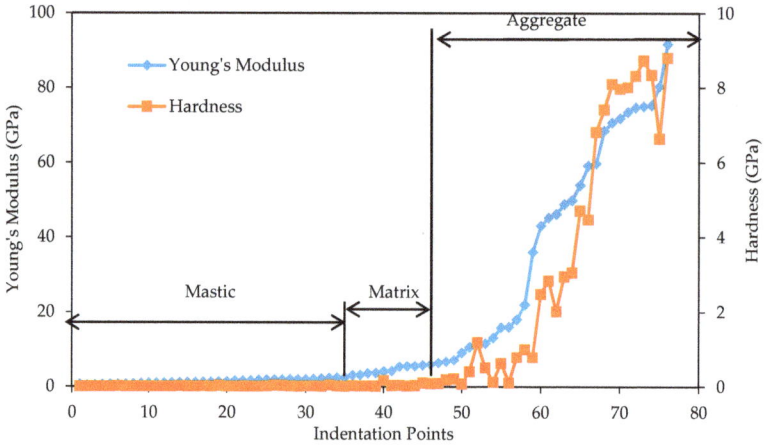

Figure 11. Young's modulus and hardness of all phases in sample without hydrated lime.

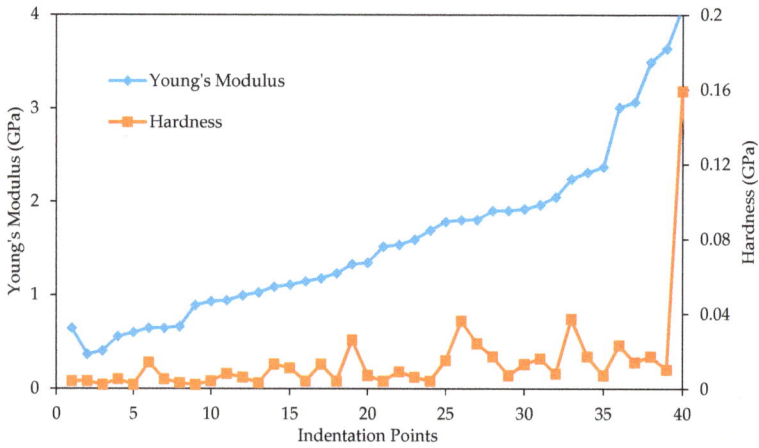

Figure 12. Young's modulus and hardness in mastic phase of sample without hydrated lime.

Figure 13 compares the Young's modulus and hardness on mastic and matrix phase for samples with and without hydrated lime. There is a clear observable trend for the sample containing hydrated lime having higher Young's modulus throughout both phases. Similarly, the majority of hardness values are higher for the hydrated lime containing sample, with the exception of point 40 which appears to be an outlier, and point 48. The difference between the samples decreases from point 35, which is close to when the indents begin appearing in the matrix phase. There is some indication that the hydrated lime has less of an effect on Young's modulus and hardness in the matrix region. It should be clarified that since this data was processed from the 10 × 10 grid of indentations, the number of indents for mastic/matrix/aggregate was not the same for both samples. In fact, the mastic region tested on the sample containing hydrated was larger and had a greater quantity of indents. In order to directly compare the different phases, some of the mastic data for samples with hydrated lime was omitted. The omitted indents display similar hardness/modulus characteristics to those included and do not contradict observations from Figure 13.

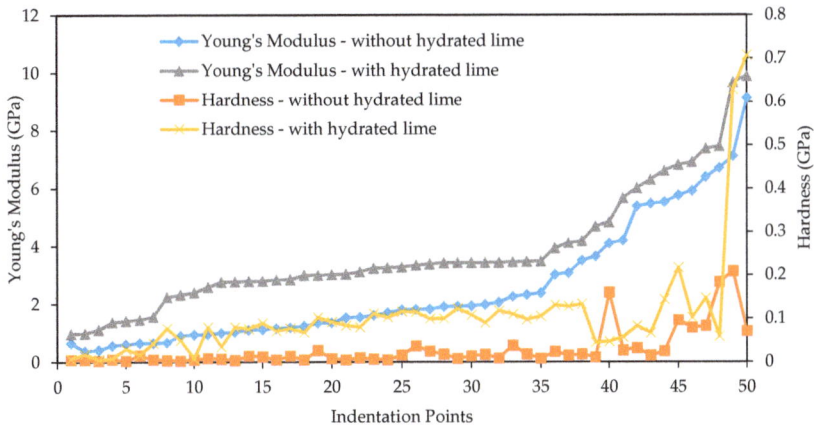

Figure 13. Young's modulus and hardness on mastic and matrix phase for samples with and without hydrated lime.

Figures 14 and 15 shows the average Young's modulus and hardness values respectively obtained from the indentation grids. The Young's modulus and hardness for samples with hydrated lime are higher than those of the samples without them. In mastic phase there is an increase of 31% in the Young's modulus and 153% in the hardness due to the addition of hydrated lime. In matrix phase there is an increase of 6% in the Young's modulus and 114% in the hardness due to the addition of hydrated lime.

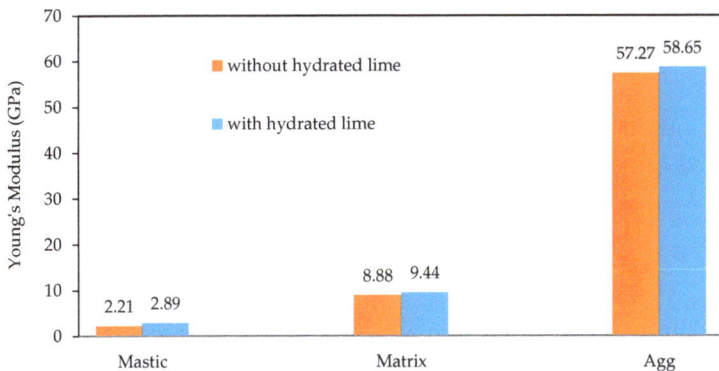

Figure 14. Average Young's modulus values obtained from the indentation grids.

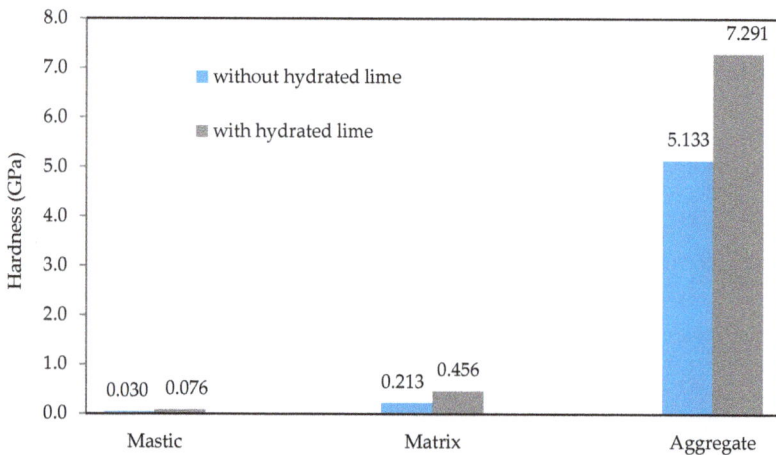

Figure 15. Average hardness values obtained from the indentation grids.

4. Conclusions

Using nanoindentation with a Berkovich indenter, it was possible to successfully capture the heterogeneity of asphalt concrete, with distinctly different properties for mastic, matrix, and aggregate phases. A dwell time of 20 s was sufficient in limiting creep for the majority of indents, except those on very soft mastic. The effects of creep were present on the unloading curve for indents with a dwell time under 20 s. The Young's modulus and hardness for samples with hydrated lime are higher than the samples without them. In the mastic phase, there is an increase of 31% in the Young's modulus and 153% in the hardness due to the addition of hydrated lime. In the matrix phase, there is an increase of 6% in the Young's modulus and 114% in the hardness due to the addition of hydrated lime. Future study on correlating the nanoscale testing with the macroscale testing presents exciting possibilities to improve current understanding of asphalt concrete.

Author Contributions: Salim Barbhuiya conceived and designed the experiments and wrote the paper. Benjamin Caracciolo performed the experiments and analyzed the data.

Conflicts of Interest: The authors declare no conflict of interest.

References

1. Tarefder, R.A.; Faisal, H.M. Nanoindentation characterization of asphalt concrete aging. *J. Nanomech. Micromech.* **2014**, *4*, 1–28. [CrossRef]
2. Briscoe, B.J.; Fiori, L.; Pelillo, E. Nano-indentation of polymeric surfaces. *J. Phys. D Appl. Phys.* **1998**, *31*, 2395–2405. [CrossRef]
3. Gregory, J.R.; Spearing, S.M. Nanoindentation of neat and in situ polymers in polymer-matrix composites. *Compos. Sci. Technol.* **2005**, *65*, 595–607. [CrossRef]
4. Chen, X.; Vlassak, J.J. Numerical study on the measurement of thin film mechanical properties by means of nanoindentation. *J. Mater. Res.* **2001**, *16*, 2974–2982. [CrossRef]
5. Allen, R.G.; Little, D.N.; Bhasin, A.; Lytton, R.L. Identification of the composite relaxation modulus of asphalt binder using AFM nanoindentation. *J. Mater. Civ. Eng.* **2013**, *25*, 530–539. [CrossRef]
6. Faisal, H.M.; Tarefder, R.A.; Weldegiorgis, M. Nanoindentation characterization of moisture damage in different phases of asphalt concrete. *Adv. Civ. Eng. Mater.* **2015**, *4*, 31–46. [CrossRef]
7. Hossain, M.I.; Faisal, H.M.; Tarefder, R.A. Determining effects of moisture in mastic materials using nanoindentation. *Mater. Struct.* **2016**, *49*, 1079–1092. [CrossRef]

8. Tarefder, R.A.; Zaman, A.M.; Uddin, W. Determining hardness and elastic modulus of asphalt by nanoindentation. *Int. J. Geomech.* **2010**, *10*, 106–116. [CrossRef]

9. Tarefder, R.A.; Faisal, H.M. Effects of dwell time and loading rate on the nanoindentation behavior of asphaltic materials. *J. Nanomech. Micromech.* **2012**, *3*, 17–23. [CrossRef]

10. Zulkati, A.; Diew, Y.; Delai, D.S. Effects of fillers on properties of asphalt–concrete mixture. *J. Transp. Eng.* **2012**, *138*, 902–910. [CrossRef]

11. Taha, R.; Al-Rawas, A.; Al-Harthy, A.; Qatan, A. Use of cement bypass dust as filler in asphalt concrete mixtures. *J. Mater. Civ. Eng.* **2002**, *14*, 338–343. [CrossRef]

12. Aljassar, A.H.; Metwali, S.; Ali, M.A. Effect of filler types on Marshall stability and retained strength of asphalt concrete. *Int. J. Pavement Eng.* **2004**, *5*, 47–51. [CrossRef]

13. Qiu, Y. Design and Performance of Stone Mastic Asphalt in Singapore Conditions. Ph.D. Thesis, Nanyang Technological University, Singapore, 2006.

14. Gorkem, C.; Sengoz, B. Predicting stripping and moisture induced damage of asphalt concrete prepared with polymer modified bitumen and hydrated lime. *Constr. Build. Mater.* **2009**, *23*, 2227–2236. [CrossRef]

15. Huang, S.C.; Robertson, R.E.; Branthaver, J.F.; Petersen, J.C. Impact of lime modification of asphalt and freeze-thaw cycling on the asphalt-aggregate interaction and moisture resistance to moisture damage. *J. Mater. Civ. Eng.* **2005**, *17*, 711–718. [CrossRef]

16. Little, D.N.; Petersen, J.C. Unique effects of hydrated lime filler on the performance-related properties of asphalt cements: Physical and chemical interactions revisited. *J. Mater. Civ. Eng.* **2005**, *17*, 207–218. [CrossRef]

17. Lesueur, D.; Little, D.N. Effect of hydrated lime on rheology, fracture and aging of bitumen. *Transp. Res. Rec.* **1999**, 93–105. [CrossRef]

18. Lesueur, D.; Petit, J.; Ritter, H.J. The mechanisms of hydrated lime modification of asphalt mixtures: A state-of-the-art review. *Road. Mater. Pavement Res.* **2013**, *14*, 1–16. [CrossRef]

19. Veytskin, Y.; Bobko, C.; Castorena, C.; Kim, Y.R. Nanoindentation investigation of asphalt binder and mastic cohesion. *Constr. Build. Mater.* **2015**, *100*, 163–171. [CrossRef]

20. *Transport and Main Roads Specifications MRTS30 Asphalt Pavements*; Technical Specification; Queensland Government: Brisbane, Queensland, Australia, 2016; p. 56.

21. AS 2150-2005. *Hot Mix Asphalt—A Guide to Good Practice*; Standards Australia: Sydney, Australia.

22. Fischer-Cripps, A.C. Critical review of analysis and interpretation of nanoindentation test data. *Surf. Coat. Technol.* **2006**, *200*, 4153–4165. [CrossRef]

23. Oliver, W.C.; Pharr, G.M. An improved technique for determining hardness and elastic-modulus using load and displacement sensing indentation experiments. *J. Mater. Res.* **1992**, *7*, 1564–1583. [CrossRef]

materials

MDPI

Article

Full-Field Indentation Damage Measurement Using Digital Image Correlation

Elías López-Alba * and Francisco A. Díaz-Garrido

Departamento de Ingeniería Mecánica y Minera, Campus las Lagunillas, Universidad de Jaén, 23071 Jaén, Spain; fdiaz@ujaen.es
* Correspondence: elalba@ujaen.es; Tel.: +34-953-21-28-62

Received: 2 June 2017; Accepted: 6 July 2017; Published: 10 July 2017

Abstract: A novel approach based on full-field indentation measurements to characterize and quantify the effect of contact in thin plates is presented. The proposed method has been employed to evaluate the indentation damage generated in the presence of bending deformation, resulting from the contact between a thin plate and a rigid sphere. For this purpose, the 3D Digital Image Correlation (3D-DIC) technique has been adopted to quantify the out of plane displacements at the back face of the plate. Tests were conducted using aluminum thin plates and a rigid bearing sphere to evaluate the influence of the thickness and the material behavior during contact. Information provided by the 3D-DIC technique has been employed to perform an indirect measurement of the contact area during the loading and unloading path of the test. A symmetrical distribution in the contact damage region due to the symmetry of the indenter was always observed. In the case of aluminum plates, the presence of a high level of plasticity caused shearing deformation as the load increased. Results show the full-field contact damage area for different plates' thicknesses at different loads. The contact damage region was bigger when the thickness of the specimen increased, and therefore, bending deformation was reduced. With the proposed approach, the elastic recovery at the contact location was quantified during the unloading, as well as the remaining permanent indentation damage after releasing the load. Results show the information obtained by full-field measurements at the contact location during the test, which implies a substantial improvement compared with pointwise techniques.

Keywords: contact; indentation; damage; 3D digital image correlation

---

1. Introduction

The mechanical contact and the indentation damage experimented between two bodies under loading have been extensively studied and investigated in the past. The first study to develop a theory of the behavior of two elements in contact was provided by Hertz [1]. However, in many situations, the limits of Hertz's theory are exceeded when a permanent indentation, once the yield strength of the material is exceeded, occurs during the experiment. Some efforts have been made to consider the effect of permanent indentation [2], even for the unloading path [3]. Early studies were focused on the contact analysis of the elastic/elastoplastic behavior of isotropic materials [4]. Decades after, the contact analysis in anisotropic and orthotropic materials was an important issue in the analysis of new materials [5–7]. However, when half space conditions are not achieved, bending stresses due to the indenter displacement are superimposed on the contact stress problem [8]. For a flexible target, the surface under contact will experience indentation and a force-deflection relationship due to the deformation of the target [9].

Many works have been conducted involving experiments using plates and a spherical indenter to validate analytical and numerical models [10–12]. In most of the studies, bending was avoided by assuming that the plate was rigidly supported or half-space conditions were achieved.

In a real situation when the plate is loaded, the contact behavior depends on the plate thickness and its deflection [13]; therefore, the size of the contact damage will depend on this effect. Recently, Chen et al. [14] developed an analytical model that incorporated the influence of the specimen thickness to explain the effect of bending during contact. None of the reported methodologies were able to differentiate the displacements that were caused by bending and contacting [6,7,15]. Swanson and Rezaee [16] emphasized the importance of the depth of penetration and the size of the residual crater after unloading as a result of the material softening, concluding that the assumption of a half-space gave an underestimated stiffness value. Other authors [17] highlighted the importance of the contact area investigating the effect of impact velocity on the indentation produced. Thus, they employed a marker paint to measure, after the experiment, the maximum contact area during impact, finding some differences in the indentation measurement depending on the velocity test.

The mentioned research used a pointwise technique, which obtained the maximum indentation, but does not offer information regarding the complete region of interest. This paper presents a full-field experimental methodology to characterize the contact damage size evolution and indentation depth of thin plates in the presence of bending deformation. A better understanding of the region of interest could be achieved using this methodology, compared with the traditional pointwise techniques. A quasi-static contact experiment, using a rigid bearing sphere on aluminum plates, has been adopted to quantify the contact damage size and the maximum indentation depth occurring during the tests. In a conventional contact study, the indenter hides the area of interest presenting a limitation for its measurement during the experiment. In the present work, a full-field, optical technique, namely 3D Digital Image Correlation (3D-DIC), has been adopted to obtain information about the contact zone. Using the recorded information and knowing the indenter geometry, it was possible to apply a geometrical relationship obtaining information related to the hidden contact area. Preliminary studies using DIC to analyze the contact phenomenon were reported for other applications [18]; however, no research was found related to the determination of the evolution of contact damage area in real time.

Experiments were performed using aluminum plates with thicknesses of 2, 3, 4, 5 and 6 mm. The contact damages and the bending deformation experienced variation depending on the stiffness of the specimen. The evolution of the contact damage for the loading and unloading path was analyzed under bending deformation. In the current paper, an alternative methodology based on the 3D-DIC technique has been developed and implemented to evaluate using the evolution of contact during indentation experiments. The proposed methodology provides full-field information from the rear face of the specimen where contact occurred, making it possible to evaluate the evolution of contact damage during the tests. By implementing this methodology, it has been possible to observe and evaluate the material elastic recovery and the generation of permanent damage of the specimens during experiments, showing the ability and potential of the proposed methodology.

2. Experimental Methodology

In a mechanical contact between a rigid sphere and a plate, the contact damage is the crater generated by the sphere in the plate under loading conditions. Considering the sphere fixed and displacing the half space plate against the sphere, the indentation would be the z displacement experienced by the plate. We remark that the indentation α is defined as the thickness reduction experienced by the plate, considering a rigid sphere. Figure 1a shows a theoretical half space case when bending does not exists. The measured Δz displacement is the indentation α that occurred during contact. In this case, Δz is equal to the displacement of the plate d as a result of the movement of the plate against the indenter. When bending occurred (Figure 1b), measuring on the upper face of the plate, opposite of the contact area, will allocate the point with the minimum out of plane displacement that provides the indentation α for each loading step. From Figure 1b, it can be observed that maximum out of plane displacement, d, will occur at the edge of the plate, and it will provide the bending displacement. Thus, the total bending experienced by the plate will be the difference between d and α. In Figure 1b, an alternative measurement system would be necessary to quantify the indentation measurement.

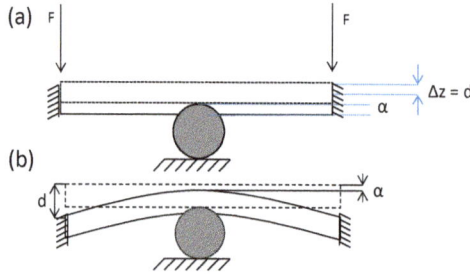

Figure 1. Mechanical contact between a plate and a sphere. (**a**) Schematic illustration showing the indentation and no bending produced in the plate; (**b**) indentation measurement principle during a contact experiment with the presence of bending.

One major problem in a contact experiment is that the contact area is hidden. The proposed methodology provides an indirect measurement of the contact area during the test. Therefore, only information is extracted from the upper face of the specimen when contact occurs. Full-field information is provided by 3D-DIC positioning the cameras focusing on the upper face of the specimen. By measuring the out of plane displacements provided by the 3D-DIC technique, it is possible to extract the contact area during the loading and unloading path of the test. Figure 2 shows a scheme of the original position of the plate and the bearing ball when contact starts. Every point P(x,y) experiences an out of plane displacement measured by 3D-DIC. It must be remarked that for the present analysis, it is assumed that the bearing ball has a very high stiffness, and consequently, it does not suffer any deformation during contact.

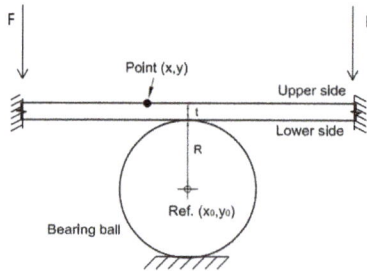

Figure 2. Original schematic of the bearing ball and the specimen.

Figure 3 shows the contact behavior between a plate and a sphere. When the load starts increasing, part of the sphere penetrates into the specimen, generating a contact damage area. The limit of this contact area is defined by the position of the sphere contour, where no contact with the plate exists. As the load starts increasing, the specimen thickness decreases at the contact region. This thickness reduction is not uniform, presenting deeper indentation at the center of the contact area (t'). The thickness (t'') reduces towards the center of the contact area (t'), as is shown in Figure 3. Outside of the contact region, the thickness should be constant and equal to the original thickness of the specimen (t). Thus, every point P on the upper surface of the specimen experiences a displacement to a new position, P'', on the deformed geometry, as shown in Figure 3.

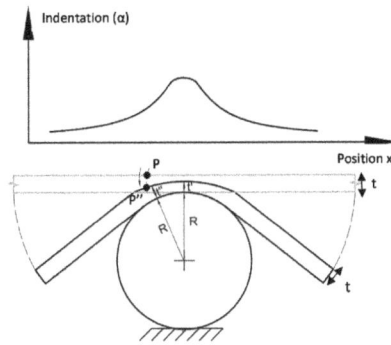

Figure 3. Schematic showing the contact behavior between a plate and a sphere. Detail of the damage generated due to contact.

An experiment was performed to validate the thickness reduction evolution, as was indicated in Figure 3. For that, the contact region of a sphere with a 2 mm-thick specimen under loading condition was studied, observing the contact damage through the thickness. Results shown in Figure 4 represent the thickness reduction in the contact region and the thickness constant value outside this area.

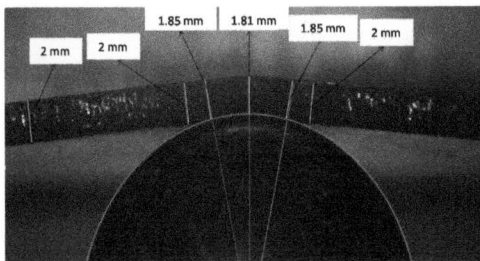

Figure 4. Experiment showing the contact area and the thickness reduction during the loading for a 2 mm-thick specimen and a sphere through the thickness.

Figure 5 illustrates the coordinate position on the plate. The position of any point at the upper surface of the specimen can be defined as a vector T with coordinates x and y referring to the point of maximum indentation (x_0, y_0), as shown in Figure 5. The components of the vector T are obtained from 3D-DIC results.

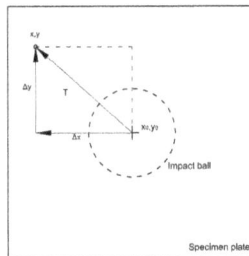

Figure 5. Definition of the point coordinates (x,y) at the upper specimen surface referring to the minimum out of plane displacement point coordinates (x_0, y_0).

Figure 6 shows how the position of a point P changes to P″ when contact damage occurs. Point O represents the center of the sphere. If the radial distance from the center of the sphere OP″ or R + t″, is smaller than R + t, it means that the specimen has a thickness reduction. Therefore, it represents the contact damage region. At the limit of the contact region, the radial distance OP″ is R + t (where R is the radius of the sphere).

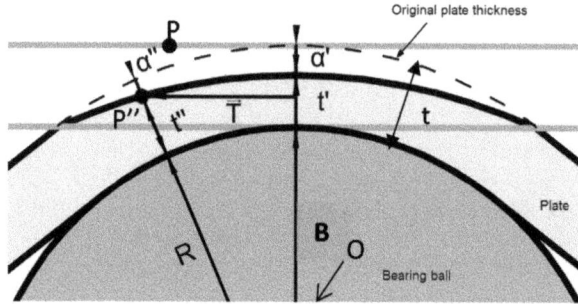

Figure 6. Sketch showing the adopted criteria to identify if a surface point is affected by thickness reduction due to contact.

Therefore, the geometrical position of OP″ is defined as the sum of vectors T and B. The vector T is the projected distance on the reference plane xy shown in Figure 5, and B is P″ distance in the thickness direction with respect to the sphere center and measured with 3D-DIC. Thus, from Figure 6, the following equations could be obtained:

$$B = R + t - \alpha'' \tag{1}$$

$$T = \sqrt{(x^2 + y^2)} \tag{2}$$

$$OP'' = \sqrt{(B^2 + T^2)} \tag{3}$$

where R is the radius of the sphere, t the original plate thickness, α'' the out of plane displacement experienced by point P after the load application and x and y are the coordinate of the point P″. Thus, the condition to have contact damage would be:

$$OP'' \leq R + t \tag{4}$$

where t is the original thickness of the plate and t″ the specimen thickness on the contact region at the position of P″.

Reorganizing Equations (4) and (5), it is achieved that where the sphere is in contact with the plate, contact damage will be present during the test.

$$OP'' - R \leq t \tag{5}$$

If during the loading path, shear stress is predominant because higher loads are applied, it produces a higher deformation with a bigger thickness reduction in the region surrounding the center of the contact area, having a thickness distribution profile similar to that shown in Figure 7 [19].

Based on this methodology, a script was programmed into Matlab to post-process the images captured during the test. The geometry of the contact damage was obtained during the test to evaluate the thickness reduction.

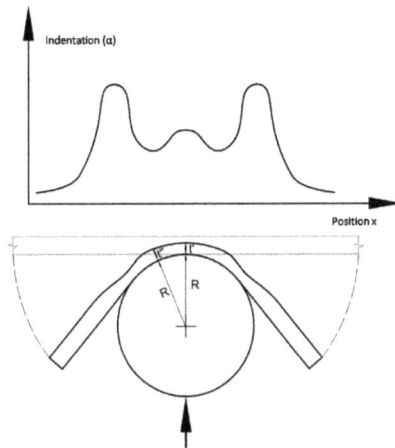

Figure 7. Scheme of the damage contact region when shear is predominant increasing the load applied.

3. Specimen Preparation and Experimental Setup

Specimens were manufactured from commercial AL-1050 H-14 (Dissa, Jaén, Spain, Table 1). The geometry adopted for experiments was square plates of 150 mm × 150 mm. The plates were flat, and no warps were observed over the surface.

Table 1. AL-1050 H-14 properties.

Density (g/cm^3)	Hardness Brinell (HB)	E (GPa)	ν	Yield Stress (MPa)
2.7	34	69	0.33	105

The specimen surface was treated with white paint and subsequently a random speckle pattern over the white surface spraying a matt black paint. The random distribution is needs to apply 3D-DIC. The thickness of the applied coating was around 1 μm. Therefore, it was negligible compared with the magnitude of the measured displacements. The specimen was clamped 15 mm at each edge using a loading frame, leaving a free area of 120 mm^2.

A calibrated stereoscopic system using two monochromatic CCD 5 Mpixels cameras (brand Allied Vision Technologies, model Stingray F-504B/C (Allied Vision Technologies GmbH, Stadtroda, Germany) was employed with two 23-mm focal length lenses (brand Schneider). Both cameras were synchronized and monitored from a laptop connected to a Data Acquisition system DAQ module. The specimen surface was properly illuminated. Finally, the displacements fields occurring at the specimen surface during loading and unloading were measured using a commercial software package (Vic-3D by Correlated Solutions Inc., West Columbia, SC, USA) [20].

Tests were conducted using a MTS 370.02 servohydraulic machine (MTS Systems Corporation, Eden Prairie, MN, USA) with a maximum load capacity of 25 kN where a loading frame designed specifically for these test was clamped. The loading frame moved in the vertical direction (using the hydraulic actuator) through four guides to ensure a normal application of monotonic loading during experiments (Figure 8). Specimens were clamped into the bottom part of the loading frame. A 20-mm spherical indenter (steel ball bearing) was screwed to the load cell using a specially-designed adaptor, recording the load magnitude during experiments. The plate was previously lubricated, decreasing friction between the bodies.

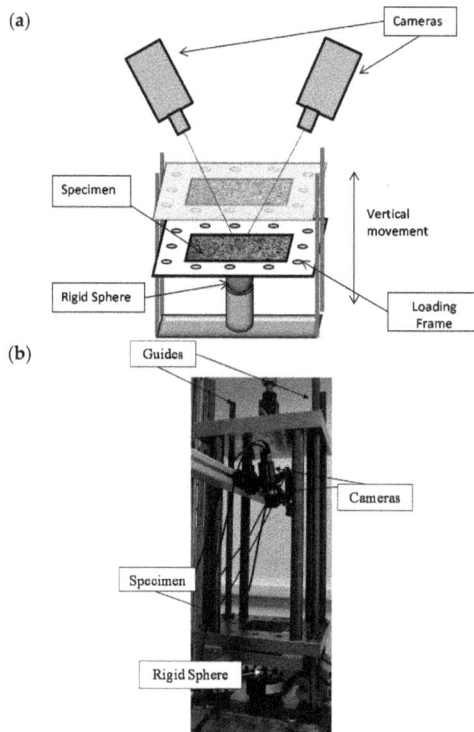

Figure 8. Experimental setup adopted. (**a**) Schematic illustration showing the setup to perform the test; (**b**) setup adopted for the experimental methodology proposed.

Before starting the test, a 25-N preload was applied to guarantee the contact between the plate and the sphere (indenter). Subsequently, the test started, and a transistor-transistor logic pulse (TTL) was automatically generated to command the image acquisition from the DIC system. The hydraulic machine was generating +5 V TTL pulses to trigger the cameras in a synchronized way governing the optical system. Experiments were controlled using the displacement movement experienced by the hydraulic cylinder. For each displacement step, the load value, the cylinder displacement value and the image acquisition time were recorded. The damage area of the plate was indirectly quantified from displacements measured at the upper side of the contact surface.

The cylinder was moved with a displacement speed of 0.1 mm/s during loading and unloading. The maximum cylinder displacement was limited to 2 mm for aluminum specimens. Images were captured and post-processed to obtain the out of plane displacement field to infer the indentation and the contact damage area experienced by the specimen during each deformation step.

4. Validation Methodology

An initial experiment was conducted to evaluate the measurement uncertainty of the 3D-DIC technique for the analysis of contact problems using the proposed setup. In this case, the experiment was conducted without a specimen. Movement of the hydraulic cylinder was transferred to the movable part of the loading frame, and only a rigid body motion was measured without any load application. The actuator was moved 3 mm in six displacement steps of 0.5 mm. Hence, displacement readings provided by the hydraulic machine LVDT sensor were compared with those measured by 3D-DIC to obtain the measurement uncertainty of the technique. Thus, a statistical analysis was performed, and the mean, μ, and standard deviation, s, of the difference between 3D-DIC and the LVDT measurements were evaluated for each displacement step [21]. In general terms, the trend of the evaluated experimental results matched with the LVDT readings with a very low scatter. The average mean for the different displacements step was 0.0011 mm and the average standard deviation 0.0004 mm. Statistical calculations clearly show the high level of concordance.

In order to ensure that the setup, used in this work, satisfies the quasi-static contact law, two experiments were conducted to quantify its accuracy and repeatability. Indentation measurements were performed 10 times with the aid of a dial indicator and 10 times using 3D-DIC for a 2 mm-thick specimen. During the experiment, the hydraulic cylinder was moved 2 mm down (once the indenter touched the specimen) in steps of 0.1 mm at a speed of 0.1 mm/s. A dwell was programmed between load steps to trigger the cameras using a TTL pulse generated by the servohydraulic machine. The dial indicator, with an accuracy of 0.001 mm, was placed at the center of the sphere over the specimen.

Figure 9 shows a comparison of the results using both techniques. In all of the cases, differences were smaller than 6% and used as a reference to calculate the 6% dispersion bands of the 3D-DIC results. These differences in the indentation values can be attributed to the position of the dial indicator or to a small loss of perpendicularity of the dial indicator needle when the specimen deforms due to bending.

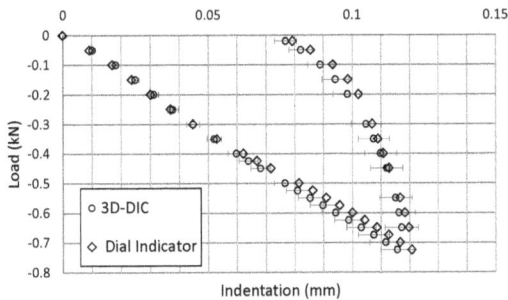

Figure 9. Results comparison using 3D Digital Image Correlation (3D-DIC) and the dial indicator.

As has been presented, the repeatability of the maximum thickness reduction measurements (maximum indentation) has been achieved. Therefore, an agreement using two different experimental techniques has been demonstrated. The 3D-DIC technique provides the indentation measurement and the information related to the contact area size between the indenter and the specimen. In addition, the results obtained measuring with the full-field technique offer the information related to the thickness reduction in the contact damage region where the sphere was in contact with the plate. With the dial indicator, it is possible to measure only at one point, and the difficulties to measure are high in the presence of any gap in the setup. Moreover, it has been observed that results obtained using 3D-DIC showed less scattering than those obtained using a dial indicator. It can be concluded that the adopted setup provides a robust and repeatable procedure to satisfy the quasi-static contact law of the material in the contact damage region.

5. Results and Discussion

To determine the contact damage, images were captured during the test and subsequently post-processed using the 3D-DIC technique. A facet size of 25 × 25 pixels with two pixels of overlap was defined. Figure 10 shows an image captured by one of the cameras. The out of plane displacement profile along a line AA' centered at the specimen is measured. In addition, Figure 10a, a detail of the full-field area where the sphere is in contact with the plate is shown. In Figure 10b–e, the out of plane displacement for different displacements steps is shown.

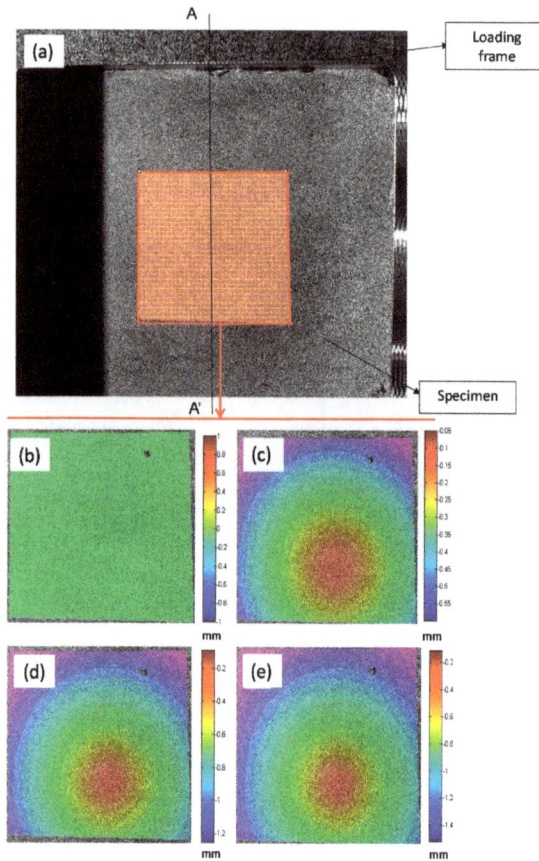

Figure 10. Region of interest for contact analysis. (**a**) Specimen image before starting a quasi-static test. Out of plane displacements measured in a 2 mm-thick specimen at the region of interest for different loads; (**b**) 0 mm to 0 N; (**c**) −0.05 mm to 320 N; (**d**) −0.096 mm to 611 N; (**e**) −0.111 mm to 731 N.

With the information extracted from the profile, the total movement of the hydraulic cylinder and the evolution of the out of plane displacements experienced by the specimen can be obtained. Figure 11 shows the out of plane deformation for a 3 mm-thick aluminum specimen at 20%, 40%, 60%, 80% and 100% of the cylinder displacement, respectively, corresponding to 100% to −2 mm. For the maximum cylinder displacement, the specimen showed a thickness reduction of 0.19 mm in the region where the sphere is contacting the plate. This minimum out of plane displacement evaluated represents the indentation generated by the sphere on the plate. To perform a profile comparison for the different deformation steps, Figure 12 shows the normalized out of plane displacement by −2 mm (maximum displacement of the cylinder) versus the specimen length for all of the thickness of the aluminum specimens tested. A zoom of the region of interest was shown. As for high thicknesses, the sphere penetrates deeply into the plate, resulting in greater contact damage.

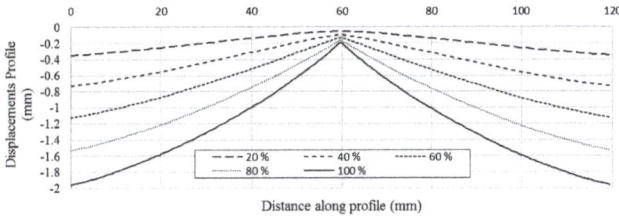

Figure 11. Evolution of the out of plane displacements along the profiles AA′ (defined in Figure 10a) at different displacement steps (in percentage) for a 3 mm-thick specimen, corresponding 100% to −2 mm cylinder displacement.

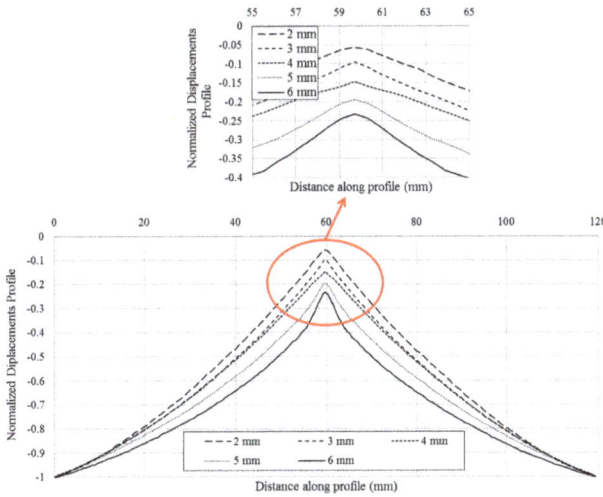

Figure 12. Normalized out of plane displacements profile by −2 mm cylinder movement versus distance along profile for different specimen thickness (2, 3, 4, 5 and 6 mm).

From the presented results, it is concluded that there was a thickness reduction in the contact area. This reduction was largest at the center of the contact area (top of the sphere in contact with the plate). Thus, the indentation increased with the specimen thickness for the same cylinder displacement. This is attributed to the influence of the specimen thickness and supported by the fact that bending decreases when the specimen thickness increases. Thus, a stiffer specimen will experience more contact damage and less bending deformation than a lower stiffness specimen (thinner specimen). An example of this effect is presented in Figure 12 with a zoom where the minimum out of plane displacement occurred.

During the unloading path of the experiments, a recovery deformation was observed due to the elastic behavior of the material. Figure 13 shows a comparison between the loading steps and recovery deformation for the 6 mm-thick specimen. The absence of contact in the unloading path was identified as the moment at which the sphere lost contact with the plate, and consequently, no load was detected by the load cell. Thus, the measured out of plane displacements were associated with a permanent indentation in the contact region, as is shown in blue color. Therefore, it was possible to obtain a complete contact law for loading and unloading, as well as quantification of the bending deformation of the specimen. Similar results were obtained for different specimen thicknesses. Different recovery deformations were observed depending on the specimen stiffness.

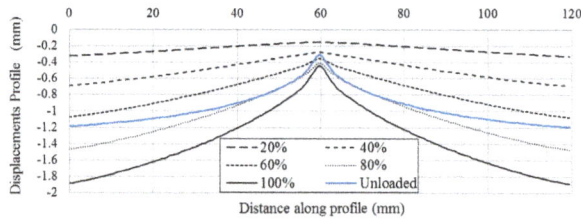

Figure 13. Evolution of the out of plane displacements along the profile AA′ (defined in Figure 10a) at different displacement steps (in percentage) during the loading and unloading final step for a 6 mm-thick specimen.

Figure 14 shows the results for a 2 mm-thick specimen. The geometry of the contact damage is shown at different displacement steps of 20%, 40%, 60%, 80% and 100% of the cylinder displacement. For each displacement, the applied load is shown, with a maximum load of 731 N measured. A symmetrical distribution of damage was experienced due to the symmetry of the indenter. At the maximum load, the initiation of shear deformation was observed. This effect happened because a flattening is present where maximum thickness reduction should occur. This event implies a higher thickness reduction surrounding the top point of the sphere in contact with the plate. However, this effect should be highlighted for other tests with higher contact damage produced. Figure 14a–e represents the thickness reduction and therefore the contact damage produced by the sphere in contact with the plate; a full-field view of the contact damage is shown. In Figure 14f, the contact damage profiles along line AA′, defined in Figure 10a, are shown for each displacement of the cylinder.

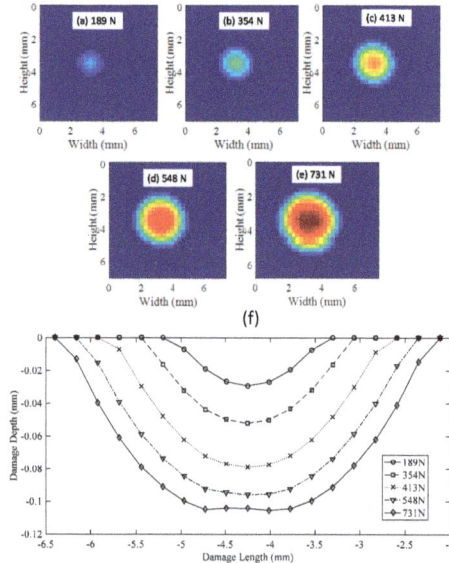

Figure 14. Damage geometry in the contact region during the unloading path for a 2 mm-thick specimen. (a–e) Geometry of the contact damage region for different displacement steps (20%, 40%, 60%, 80% and 100% cylinder displacement with a maximum of −2 mm) during the loading path for a 2 mm-thick specimen. (f) Profiles of the contact damage depth for the different displacements steps.

For a 3 mm-thick aluminum specimen, the results are shown in Figure 15. The maximum load applied to move the hydraulic cylinder at −2 mm was 1758 N. In this case, no shear nor flattening effects were observed in the contact region reaching a maximum indentation depth of −0.19 mm. Figure 15a–e represents the thickness reduction and therefore the contact damage size produced by the sphere in contact with the plate. Figure 14f illustrates the AA' profiles (defined in Figure 10a) of the contact damage showing the depth damage for every step plotted. Similar results were obtained for thickness of 4 and 5 mm.

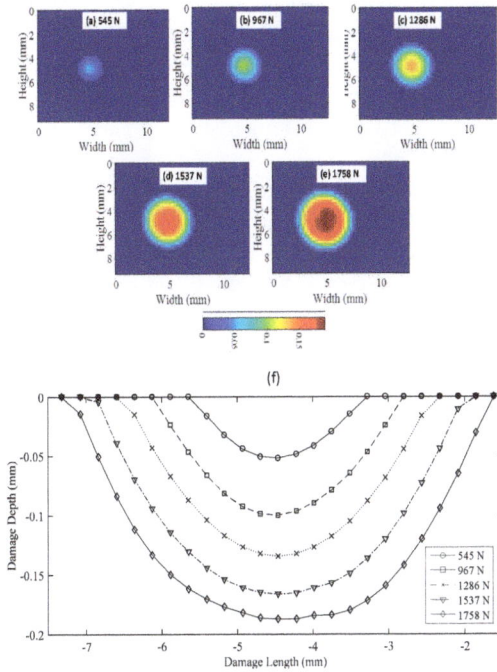

Figure 15. Damage geometry in the contact region during the unloading path for 3 mm thick specimen. (**a–e**) Geometry of the contact damage region for different displacement steps (20%, 40%, 60%, 80% and 100% cylinder displacement with a maximum of −2 mm) during the loading path for 3 mm thick specimen; (**f**) Profiles of the contact damage depth for the different displacements steps.

Figure 16 shows the geometry of the contact damage for a 6 mm-thick specimen. The maximum applied load for this test was 6274 N at −2 mm cylinder displacement, reaching a maximum indentation in the peak point of the bearing ball of −0.447 (blue color). It is observed that the specimen experienced more thickness reduction surrounding the peak point of the sphere in the contact area. This is attributed to the shear effect explained previously. Results of the geometrical contact damage were plotted for the maximum load state and for the unloading path. Once the maximum displacement was reached (blue color), the plate was unloaded, showing the elastic behavior of the contact damage. From this figure, the recovery deformation experienced by the specimen is observed when the load decreases, until non-contact between both bodies at 0 N (red color). This instant resulted in permanent contact damage. Similar results were obtained for different thicknesses. Figure 15a shows the maximum contact damage at maximum displacement of −2 mm and a load of 6274 N. Figure 15b–e illustrates the evolution of the contact damage during the unloading path until permanent contact damage was present at 0 N. Figure 16f shows the profiles AA' (defined in Figure 10a) at the different steps indicated previously.

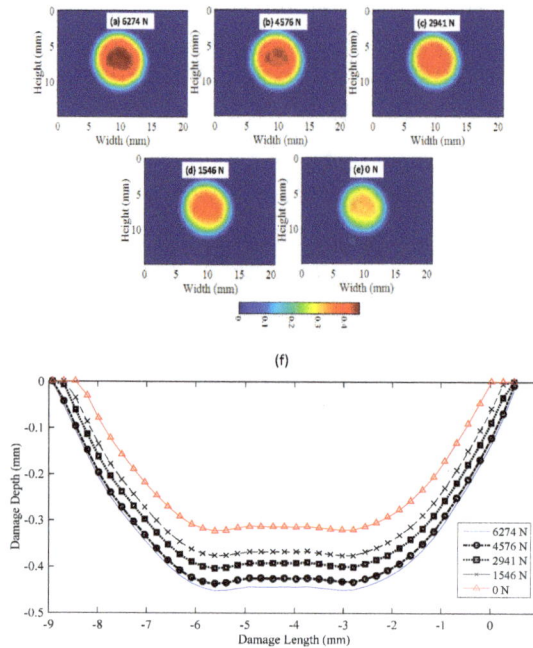

Figure 16. Damage geometry in the contact region during the unloading path for a 6 mm-thick specimen. (a) Maximum load reached to 100% of the cylinder displacement with a maximum of −2 mm; (b–d) Stages representing the elastic recovery during the unloading path; (e) Permanent damage in the plate; (f) Profiles of the contact damage depth for the different stages during the unloading path.

With a full-field view provided by 3D-DIC, it was possible to quantify the size of the contact damage. As was observed in Figures 14 and 15 when the plate is thicker, the contact damage area is bigger and deeper. For 2 mm thick (Figure 14e), the maximum contact damage had a diameter of 3.83 mm. For 3 mm thick (Figure 15e), the maximum contact damage had a diameter of 5.25 mm. Figure 16a shows the contact damage at maximum load (−2 mm cylinder displacement) for a 6 mm-thick specimen; in this case, the contact damage had a diameter of 9.3 mm. Monitoring the test during the loading and unloading path, it was possible to evaluate the elastic recovery experienced by the material. Figure 16e shows the permanent contact damage of 8.12 mm in diameter. Thus, 12.69% of elastic recovery was experienced compared with the maximum load state.

In this study, a robust methodology has been presented that uses full-field displacement measurements to observe the evolution of contact damage with bending deformation. This offers a better understanding of the behavior of the plate during the loading and unloading path.

6. Conclusions

A full-field experimental methodology based on 3D-DIC has been presented to determine the real contact damage on metallic specimens. The adopted setup takes into account the influence of bending during contact experiments. Using a geometrical evaluation, it was possible to provide an indirect measurement of the contact region. The proposed methodology has been validated for the indentation depth using a pointwise conventional indentation measuring techniques. The differences between the indentation depths obtained with both techniques were less than 6%. To illustrate the proposed methodology, contact experiments using aluminum plates with 2, 3, 4, 5 and 6 mm thicknesses were conducted. Results from experiments made it possible to quantify the experimental contact damage geometry and the maximum indentation depth for different specimen's thicknesses with the presence of bending deformation. It can be concluded that the contact damage area increases when the specimen thickness increases. When the bending deformation decreases, it was observed that the contact damage area was higher. The recovered elastic indentation during the unloading and the permanent contact damage created have been also successfully quantified.

The adopted experimental methodology could potentially be used in future work to evaluate the specimen contact behavior at different loading rates, material stiffnesses, bending deformations of the specimens and the extrapolation of the methodology to dynamic events, such as impact.

Author Contributions: Elías López-Alba performed the experiments, analysis of results and the writing of the paper. Francisco A. Díaz-Garrido performed the analysis of results and the paper revision.

Conflicts of Interest: The authors declare no conflicts of interest.

References

1. Hertz, H. Ueber die Berührung fester elastischer Körper. *J. Reine Angew. Math.* **1882**, *1882*, 156–171.
2. Goldsmith, W. *Impact*; Dover Publications Inc.: New York, NY, USA, 2003.
3. Crook, A.W. A Study of some impacts between metal bodies by a piezo-electric method. In *Proceedings of the Royal Society of London A: Mathematical, Physical and Engineering Sciences, London, UK, May 1952*; The Royal Society: London, UK, 1952; pp. 377–390.
4. Johnson, K.L. *Contact Mechanics*; Cambridge University Press: Cambridge, UK, 1985.
5. Swanson, S.R. Hertzian contact of orthotropic materials. *Int. J. Solids Struct.* **2004**, *41*, 1945–1959. [CrossRef]
6. Tan, T.M.; Sun, C.T. Use of statical indentation laws in the impact analysis of laminated composite plates. *J. Appl. Mech.* **1985**, *52*, 6–12. [CrossRef]
7. Yang, S.H.; Sun, C.T. Indentation Law for Composite Laminates. In *Composite Materials: Testing and Design (6th Conference)*; ASTM International: West Conshohocken, PA, USA, 1982; pp. 425–449.
8. Timoshenko, S.; Goodier, J.N. *Theory of Elasticity*, 3rd ed.; McGraw-Hill: New York, NY, USA, 1951.
9. Zukas, J.A.; Theodore, N. *Impact Dynamics*; Wiley: Hoboken, NJ, USA; New York, NY, USA, 1982.
10. Abrate, S. *Impact on Composite Structures*; Cambridge University Press: Cambridge, UK, 1998.
11. Christoforou, A.P. On the contact of a spherical indenter and a thin composite laminate. *Compos. Struct.* **1993**, *26*, 77–82. [CrossRef]
12. Yigit, A.S.; Christoforou, A.P. On the impact of a spherical indenter and an elastic-plastic transversely isotropic half-space. *Compos. Eng.* **1994**, *4*, 1143–1152. [CrossRef]
13. Wu, E.; Shyu, K. Response of composite laminates to contact loads and relationship to low-velocity impact. *J. Compos. Mater.* **1993**, *27*, 1443–1464. [CrossRef]
14. Chen, P.; Xiong, J.; Shen, Z. Thickness effect on the contact behavior of a composite laminate indented by a rigid sphere. *Mech. Mater.* **2008**, *40*, 183–194. [CrossRef]
15. Wu, E.; Yen, C.-S. The contact behavior between laminated composite plates and rigid spheres. *J. Appl. Mech.* **1994**, *61*, 60–66. [CrossRef]
16. Swanson, S.R.; Rezaee, H.G. Strength loss in composites from lateral contact loads. *Compos. Sci. Technol.* **1990**, *38*, 43–54. [CrossRef]
17. Lee, C.-Y.; Liu, D. Effect of impact velocity on the indentation of thick composite laminate. *Exp. Tech.* **2009**, *33*, 59–64. [CrossRef]
18. Brake, M.R.; Aragon, D.S.; Reu, P.L.; Van Goethem, D.J.; Bejarano, M.V.; Sumali, H.; Volk, C. Experimental Measurements of Rebound for the Validation of an Elastic-Plastic Contact Model. In Proceedings of the ASME 2012 International Mechanical Engineering Congress and Exposition, Houston, TX, USA, 12–15 November 2012; SciTech Connect: Houston, TX, USA, 2012.
19. Rodríguez-Millán, M.; Vaz-Romero, A.; Rusinek, A.; Rodríguez-Martínez, J.A.; Arias, A. Experimental study on the perforation process of 5754-H111 and 6082-T6 aluminium plates subjected to normal impact by conical, hemispherical and blunt projectiles. *Exp. Mech.* **2014**, *54*, 729–742. [CrossRef]

20. Sutton, M.A.; Orteu, J.J.; Schreier, H.W. *Image Correlation for Shape, Motion and Deformation Measurements: Basic Concepts, Theory and Applications*; Springer: New York, NY, USA, 2009.

21. Nurse, A.D.; Patterson, E.A. Determination of predominantly mode II stress intensity factors from isochromatic data. *Fatigue Fract. Eng. Mater. Struct.* **1993**, *16*, 1339–1354. [CrossRef]

![materials logo] *materials*

MDPI

Article

A Validation Approach for Quasistatic Numerical/Experimental Indentation Analysis in Soft Materials Using 3D Digital Image Correlation

Luis Felipe-Sesé [1],*, Elías López-Alba [1], Benedikt Hannemann [2], Sebastian Schmeer [2] and Francisco A. Diaz [1]

[1] Departamento de Ingeniería Mecánica y Minera, Campus las Lagunillas, Universidad de Jaén, 23071 Jaén, Spain; elalba@ujaen.es (E.L.-A.); fdiaz@ujaen.es (F.A.D.)

[2] Institute for Composite Materials (IVW), Kaiserslautern University of Technology, 67663 Kaiserslautern, Germany; benedikt.hannemann@ivw.uni-kl.de (B.H.); sebastian.schmeer@ivw.uni-kl.de (S.S.)

* Correspondence: lfelipe@ujaen.es; Tel.: +34-953-21-23-53

Received: 29 May 2017; Accepted: 23 June 2017; Published: 28 June 2017

Abstract: A quasistatic indentation numerical analysis in a round section specimen made of soft material has been performed and validated with a full field experimental technique, i.e., Digital Image Correlation 3D. The contact experiment specifically consisted of loading a 25 mm diameter rubber cylinder of up to a 5 mm indentation and then unloading. Experimental strains fields measured at the surface of the specimen during the experiment were compared with those obtained by performing two numerical analyses employing two different hyperplastic material models. The comparison was performed using an Image Decomposition new methodology that makes a direct comparison of full-field data independently of their scale or orientation possible. Numerical results show a good level of agreement with those measured during the experiments. However, since image decomposition allows for the differences to be quantified, it was observed that one of the adopted material models reproduces lower differences compared to experimental results.

Keywords: Digital Image Correlation; instrumented indentation; numerical validation; Image Decomposition; strain analysis

1. Introduction

Today, the investigation of engineering problems involving the use of large deformation materials such as silicon, rubber, or biological materials has focused the attention of many researchers [1]. One typical field of research for such materials is the analysis of their contact behaviour during indentation experiments. These materials exhibit large displacements under contact loading due to their physical behaviour. One major difficulty in the analysis of contact problems in hyperplastic materials is that their mechanical behaviour cannot always be described by theoretical models [2]. Some analytical [3–9] and numerical [10–12] studies can be found in the literature that have contributed to a better knowledge of such a problem, but they often consider different assumption for their analysis, such as assuming small strains below the elastic limit [6,8,11], considering half-plane elastic for both the indenter and the soft body material [7,9,11,12], assuming that the contact area is much smaller than the radius of the indenter and the elements are frictionless [7,10,11], or that the external angle of the indenter is very small [8]. Nevertheless, some of these assumptions are not always well supported by lab experiments. Although experimental techniques constitutes an important alternative to validate analytical models and numerical studies, no substantial work can be found in the literature [2,13–16] for the analysis of large deformation materials.

In this paper, one of the best established optical techniques for full field displacement measurement is applied to the analysis of large strains due to indentation in a hyperplastic material. The adopted technique is 3D digital image correlation (3D-DIC) [17], which employs at least two cameras for a stereoscopic visualization of the surface of the specimen with a random distribution of light intensity (speckle). The element surface can be unambiguously distinguished by its random neighbourhood pattern. In this way, a group of pixels corresponding to a specific location at the element surface is identified on a sequence of images captured during deformation with the aid of an image correlation algorithm [17]. Displacement fields of the studied surface are obtained, and from them, strain maps can be also calculated.

3D-DIC is employed for the analysis of contact experiments using a wedge-shape indenter on a soft material cylinder. The experiment was performed to obtain displacements and strains fields. Two analytical studies employing Finite Element Method representing the experimentation were also performed. Those analyses employed two different mathematical formulation to determine the material behaviour.

During the experiments, it was not possible to guarantee avoiding solid rigid displacement, which could affect to the displacement measurements with 3D-DIC. For that purpose, it is believed that strain maps more adequately represent the mechanical behaviour of the material during the test rather than displacement maps. In order to evaluate the accuracy of the numerical simulation performed using FEM models, strain maps from the experiments and FEM were compared. The compassion of strain maps obtained by employing different methodologies is a challenging task. In this paper, a novel methodology based on the decomposition of the strain maps from both experiment and simulation was evaluated to perform quantitative comparisons of the indentation process at different displacement steps. The adopted comparison methodology is based on an Image Decomposition algorithm [18]. It consists of the decomposition of each data field into a feature vector that is independent from scale or orientation of the original data map. It is necessary to decompose each of the displacement maps captured along the test in order to study an event through time. Nevertheless, this paper presents a novel procedure to encode the full experimentation in a single strain map (based on Tchebichef shape descriptors [19]). This procedure decreases the dimensionality of the comparison process compared to some previous work [2,20].

2. Methodology

2.1. Digital Image Correlation

The full field technique employed for the measurement of strains at the surface of the specimen was Digital Image Correlation 3D (3D-DIC). This technique requires a stereoscopic digital camera system acquiring images from two different points of view of the area of interest in order to measure the 3D shape and displacements occurring at that surface as observed in Figure 1. Additionally, the surface should present a randomly distributed speckle pattern to ease the tracking of the points that compose that region of interest. This is usually composed of white speckles over a black background (as observed in the detail of Figure 1). DIC divides each of the images into small subsets [21], as observed in Figure 1. Every subset has a unique intensity pattern within the subset due to the randomness of the speckle pattern.

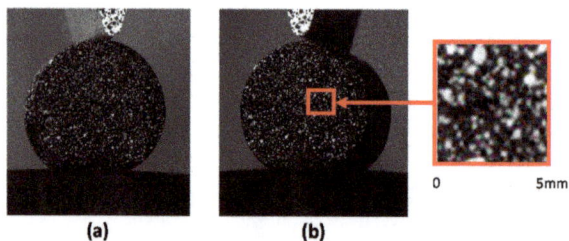

(a) **(b)**

Figure 1. Image captured by left camera (**a**) and right camera (**b**) of a stereoscopic DIC system.

The most similar intensity pattern of this subset in the first speckle image is then searched for in a second speckle image in an area around the same pixel position. Once the intensity pattern of the subsets closely coincides, the displaced pixel is found. Since two cameras are observing the same area of interest, a calibration procedure is required to accurately determine the three-dimensional position of the subsets. This calibration procedure consists of the calculation of intrinsic (focal length, image size, aberration of lenses) and extrinsic (related to the position of the cameras and the specimen) parameters of the optical set-up [22,23].

2.2. Experimental Set-Up for Contact Experiments

The experimental work presented consisted of the indentation of a cylinder made of a black rubber material employed in automotive applications. The material cannot be disclosed for confidentially reasons, although this does not affect the aim of the paper, which is the validation of the procedure. The shape of the specimen was 15 mm in height and 25 mm in diameter, and the contact area was lubricated with oil.

The experimental set up is shown in Figure 2. For indentation, a 2024 aluminum wedge was employed as illustrated in Figure 3. To control the displacement of the indenter, an Instron testing machine (model 5967 with 30 kN load capacity) was employed. The indentation was achieved by performing a displacement of the indenter from 0 mm to −5 mm and then unloading the indenter back to 0mm. The speed of the displacement was set to 10 mm/min.

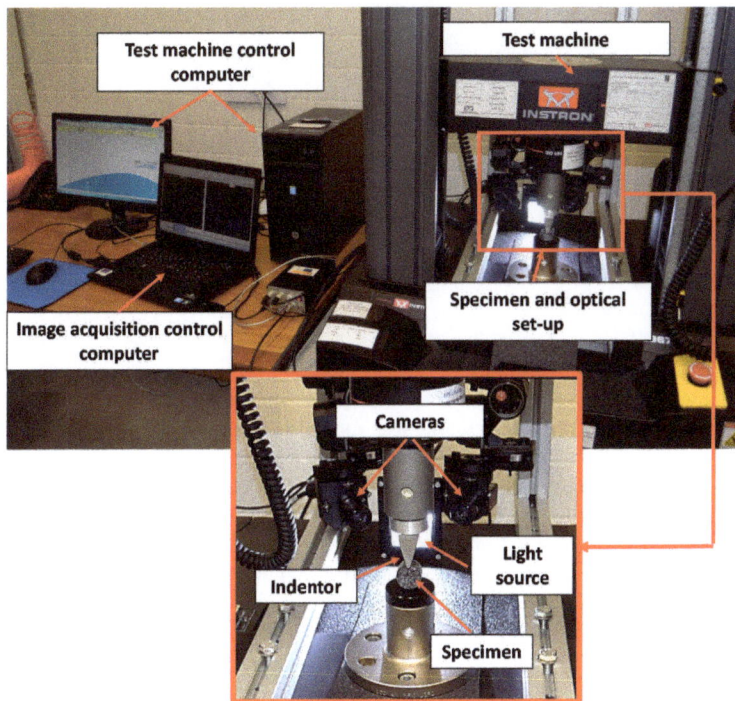

Figure 2. Illustration of the experimental set-up with detail of the optical arrangement.

The full-field response of the specimen during the indentation was captured by a stereoscopic system composed of two CCD monocromatic cameras (brand Allied model Stingray) with 2452 × 2056 pixels resolution and a 35 mm f 1.6 focal length lenses (brand Goyo Optical Inc., Saitama,

Japan). Since the rubber material was completely black, the speckle pattern was obtained by spraying white paint over the surface of the specimen. During the tests, two images were synchronously captured by the stereoscopic camera system at different indentation steps, from 0 mm to −5 mm and back to 0 every 0.5 mm increment (a total of 21 steps and 42 images). The trigger signal was commanded by the testing machine and was registered by VicSnap commercial software from Correlated Solutions.

The calibration was performed employing a 12 × 9 targets with 5 mm spacing calibration target supplied with the commercial software VIC 3D (Correlated Solutions).

Since DIC only measure 3D displacements maps, strains maps require a postprocesing analysis. In this case, strain maps were calculated employing a Lagrange Tensor and using a subset size of 15 pixels to obtain horizontal (ε_{xx}) and vertical (ε_{yy}) strains.

2.3. Numerical Modelling

An explicit numerical analysis was conducted using Abaqus 6.14 (commercially available software) for a 25 mm diameter rubber cylinder 15 mm high under lateral indentation at 10 mm/min using a 20 mm thick aluminum wedge with a tip radius of 1.68 mm. An included angle of 73.45° was also employed, as shown in Figure 3.

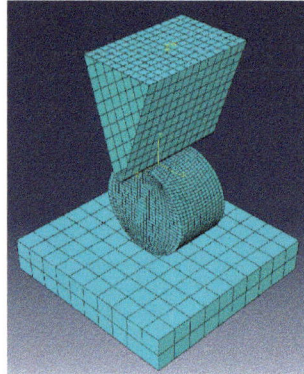

Figure 3. Detail of the finite element mesh employed for numerical modeling.

The indenter was modelled as a non-deformable body using 520 bidimensional rigid solid elements (type R3D4). The rubber cylinder was modelled using 9800 solid reduced integration elements (type C3D8R). The size of these elements varied according to their position in the cylinder from 0.5 mm to 1 mm. The minimum element size controlled the minimum time-step in the explicit solution. The values were chosen to give an appropriate temporal and spatial resolution. Additionally, friction coefficient was estimated to be 0.5, according to previous studies [24,25].

The rubber is known to exhibit a non-linear elastic behavior that can be modelled using different models [26]. In this paper, two different materials models are evaluated. The first material model (FEM Model A) is Neo-Hookean [27]: the expression of the material's strain energy function is described by Equation (1). The second (FEM Model B) is Van der Waals [28], which follows Equation (2).

$$W = C_1(I_1 - 3), \tag{1}$$

$$W = \mu c_e \ln\left(1 - \sqrt{\frac{I_1 - 3}{c_e}}\right) - \mu c_e \sqrt{\frac{I_1 - 3}{c_e}}, \tag{2}$$

where W is the strain-energy density, μ is the shear modulus, C_1 is a material constant, I_1 is the first invariant of the right Cauchy-Green deformation tensor, and c_e is the chain extensibility dependent of the material.

The material model parameters were obtained by providing empirical data from the compression tests that were used by the Abaqus subroutine to evaluate the response of the material. The uniaxial compression tests were performed on a similar cylinder on which the experimental indentation where performed but with metallic cups on the extremes of the cylinder to homogenize the compression force. The material test specimen was loaded between platens at 10 mm/min up to a compression of 12 mm using a Zwick 1474 machine and performing four cycles of pre-conditioning up to 13 mm compression. An illustration of the compression test and the results from these tests are shown in Figure 4.

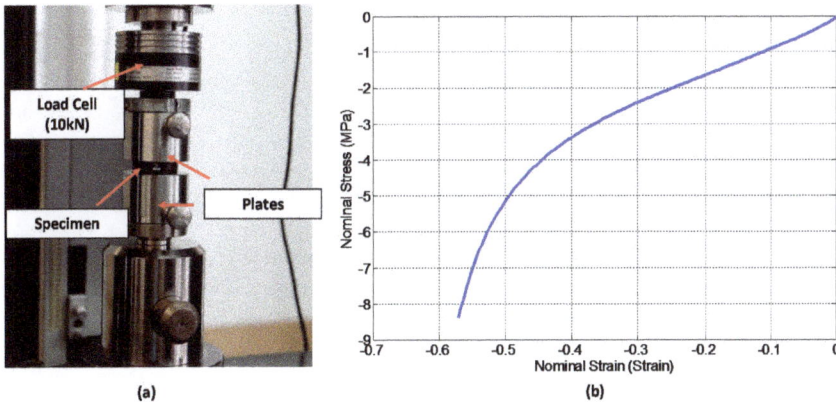

Figure 4. Illustration of compression test. (**a**) Image of the specimen during compression test; (**b**) Stress-Strain response of the rubber.

The applied load was obtained from the load cell of the test machine and used to compute engineering stress by dividing the applied load by the original cross-section area of the specimen (1.963×10^{-3}), while the displacement of the cross-head of the test machine was used to evaluate engineering strain by dividing the measured displacement by the original height of the specimen. Data in Figure 4b were employed to evaluate the material parameters to satisfy Equations (1) and (2) using the EDIT MATERIAL tool from Abaqus/CAE 6.14-2. The movement of the wedge into the cylinder and the release of the wedge after indentation were both modelled using wedge speeds obtained from the experiments measured by applying speckle pattern on the indenter as illustrated in Figures 1 and 2.

2.4. Validation Procedure

The amount of experimental and numerical results for the 21 indentation steps—a total of 126 strain maps considering 21 studied steps employing three different methods (two numerical and one experimental) and two different studied strain maps (ε_{xx} and ε_{yy})—was difficult to manage. Additionally, experimental and numerical results were not directly comparable due to the different angle of view of the cameras and differences in the scale of the data exportation files. However, a comparison between both data sets was required, so an image decomposition method was employed [18]. This method was employed for the validation of the analytical and theoretical models, and its fundamentals consisted of converting the information from strain maps into a feature vector that can be directly compared for each indentation step. The more similar the vectors, the more similar are the results. Using this approach, it is possible to eliminate the influence of the images' size and camera view angles, making it possible to perform a direct comparison of the results.

The adopted image decomposition method was based on Tchebichef polynomials $T(i,j)$ [19] to decompose displacements images $I(i,j)$ into shape descriptors that have the same units of the decomposed strain map (in this case was units of strain mm/mm).

$$I(i,j) = \sum_{k=0}^{N} s_k T_k(i,j),$$ (3)

where the coefficients s_k are called as feature or shape vector for the displacements map $I(i,j)$ and they are determined by:

$$s_k = \sum_{k=0}^{N} I(i,j) T_k(i,j),$$ (4)

Polynomials are dimensionless, and N is the number of moments or shape descriptor. The applicability of this image decomposition method is reduced to rectangular data fields. The required number of shape descriptors (SD) is given by a correlation coefficient between the original and the reconstructed image using N shape descriptors. Thus, a study of the adequate number of descriptors is required. As shown at Figure 5a, as the number of descriptors increases in the horizontal axis, the correlation coefficient (on the right ordinate axis) increases. Figure 5b indicates that as the number of moments increases, the calculated uncertainty (defined as the squared root of the squared difference between the original and the reconstructed images) decreases on the ordinate axis. Finally, for vectors comparison, 50 was considered as an appropriate number of shape descriptors, which is the quantity that raises the asymptotic maximum of the correlation coefficients in Figure 5.

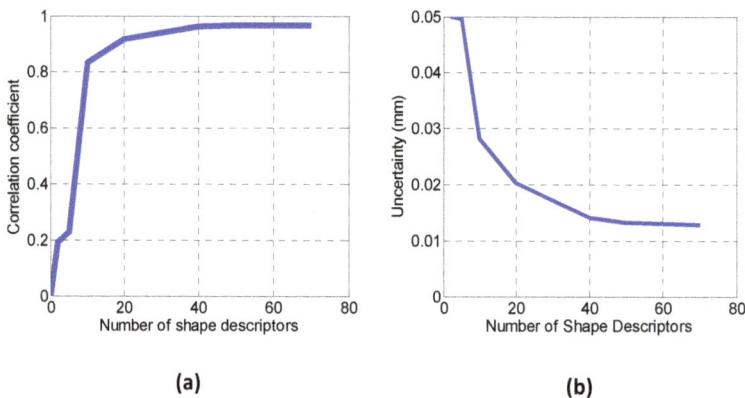

(a)

(b)

Figure 5. (a) Evolution of the correlation coefficient between original and reconstructed strain map and (b) Evolution of the Uncertainty of the reconstruction against the number of shape descriptors employed in the image decomposition.

Nonetheless, it is necessary to compare feature vectors from ε_{xx} and ε_{yy} strains maps from 21 steps, and from experiments and a similar number of them for numerical results that results in 126 feature vectors. In this paper, an alternative comparison process was performed to minimize the quantity of data to compare and to make it possible to employ Tchebichef descriptors with circular strain maps. Hence, the evolution of a vertical and horizontal profile of horizontal (ε_{xx}) and vertical (ε_{yy}) strains maps along time in order was studied in order to encode the evolution in time in a single data field instead of 21 images. Data fields (strains-step maps) composed of 21 columns that correspond to vertical profiles (in the indentation direction) of the strain map along the mid-plane of the specimen at each of the 21 indentation steps were generated, as illustrated in Figure 6. This methodology was applied for ε_{xx} and ε_{yy} strains. Thus, one single image encodes ε_{xx} information and another image encodes ε_{yy} information over the whole test. Additionally, in order to compare the behavior in the normal direction to the load, the same procedures were followed, but in this case, the columns of the image consisted on horizontal profiles of the strain maps along the horizontal mid-plane of the specimen. This procedure makes it possible to compare only 12 strain-time or strain-step maps, six of them representing the vertical profiles of ε_{xx} and ε_{yy} for each of the three set of results obtained, and another six representing the horizontal profiles of those results. This leads to 12 feature vectors instead of 126.

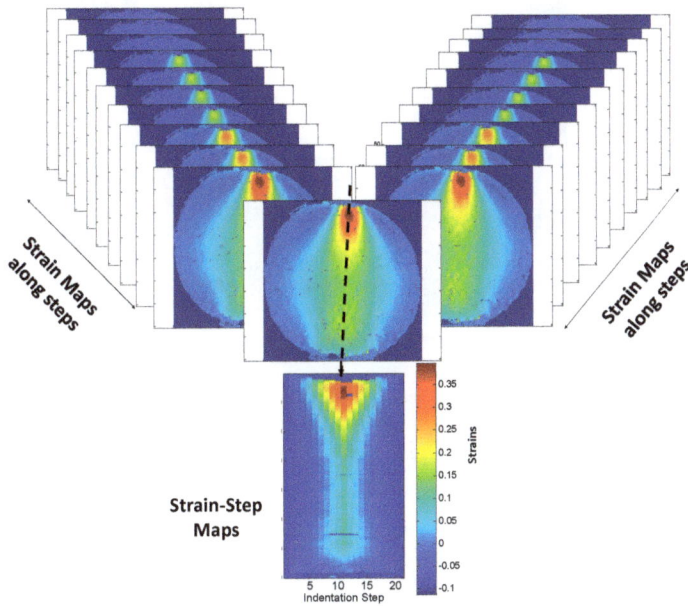

Figure 6. Illustration of the sampling of the strain profiles.

3. Results

The experimental and numerical results of the indentation performed on the rubber cylinder are presented here together with the results of the comparative employing image decomposition. As mentioned above, the large amount of displacement and strain maps makes it necessary to reduce the presented data in order to accelerate the their direct comparison.

For illustration purposes, Figure 7 shows the displacement maps obtained at three different indentation steps (i.e., 2.5, 3.5, and 5 mm) in the three spatial directions. As observed, there are some areas where the results were not successfully monitored due to the large level of deformation achieved. That is especially notorious in the upper area, where the contact with the indenter occurred, and in the lower area, where the specimen was supported.

In addition, some tilting on the displacement maps was also observed. This orientation was due to the angles required in the stereoscopic camera system to perform 3D-DIC. Nonetheless, the orientation of the axis coordinate system was selected to have a Y-axis in the direction of the indention and a perpendicular X-direction in order to have the same coordinates systems for both the experimental and FEM results.

As previously mentioned, the focus was placed on the strain fields instead of the displacement fields to avoid possible solid rigid displacements and to analyze the material mechanical behavior. Hence, Figure 8 presents the experimental and numerical strain maps in X and Y directions (ε_{xx} and ε_{yy} respectively) obtained at maximum indentation of 5 mm on the rubber specimen.

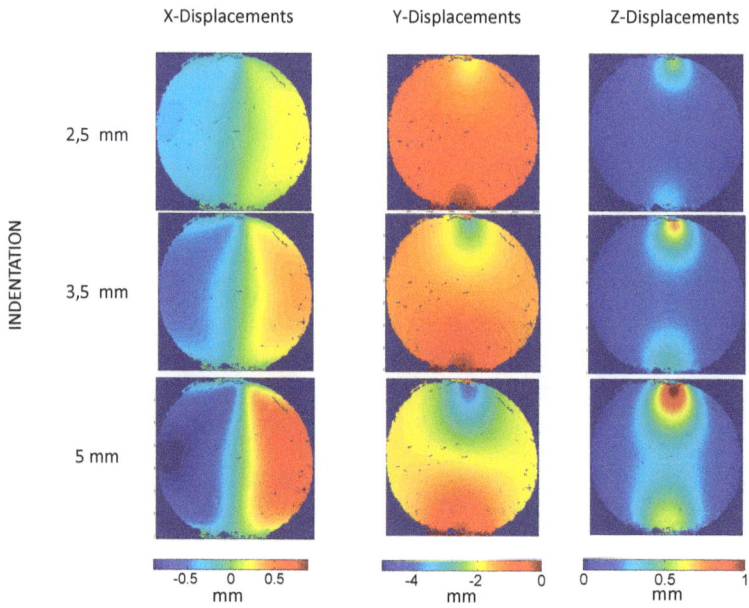

Figure 7. Displacement maps obtained experimentally at 2.5 mm (first row), 3.5 mm (second row) and 5 mm (third row) indentation obtained in Xcdirection (first column), Y direction (second column), and Z direction (third column).

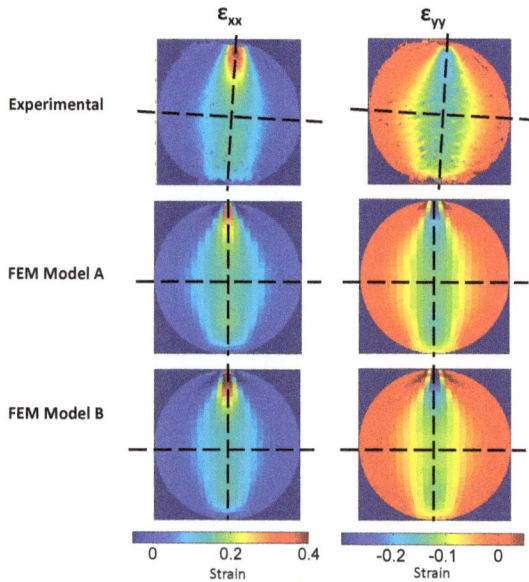

Figure 8. Strain maps in X direction (considered normal to indentation) and Y direction (considered in the direction of the indentation) Obtained experimentally (upper row), and numerically with FEM A and B (central and lower rows).

As observed, strain maps obtained with FEM are very close to those obtained during the experiments. However, it is observed that no direct quantitative full-field comparison is possible due to the differences in size of the strain map and the orientation of the experimental strain maps.

To overcome these drawbacks, the vertical and horizontal profiles are presented in Figure 8 as black dashed lines. The data of 126 strain maps considered in this study are simplified into 12 data fields encoding the strain data (strains-step maps) along the 21 steps, as presented in Figure 9.

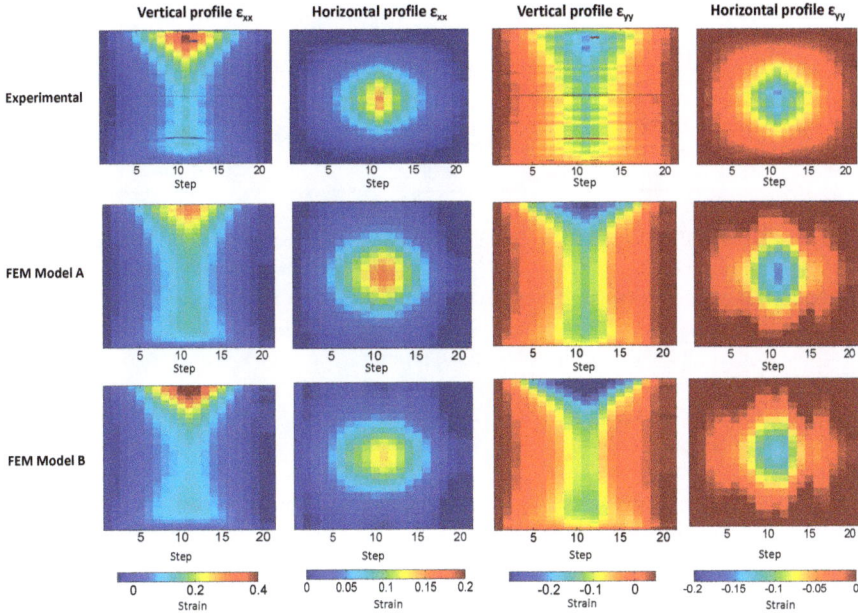

Figure 9. Strain-step maps obtained for ε_{xx} along vertical profile (first column); ε_{xx} along horizontal profile (second column); ε_{yy} along vertical profile (third column) and ε_{yy} along horizontal profile (fourth column) from experimental (first row), FEM A (second row) and FEM B (third row) results.

A good correlation between experimental and numerical results is observed in Figure 9. However, some slight differences are present, and some quantification of them is required to define which numerical model represents the experimentation more adequately.

Feature vectors make it possible to quantify differences by calculating the Euclidean distance between them. The Euclidean distance is simply the straight-line distance between the locations represented by the vectors in a multi-dimensional space, so that two coincident vectors have a Euclidean distance of zero. Table 1 shows the differences obtained between the strain-step maps presented in Figure 9.

Table 1. Euclidean distances between feature vectors representing strain-step maps.

-	FEM A (Strain)	FEM B (Strain)
ε_{xx} Vertical profile strain map	0.0211	0.0236
ε_{xx} Horizontal profile strain map	0.0137	0.0096
ε_{yy} Vertical profile strain map	0.0252	0.0356
ε_{yy} Horizontal profile strain map	0.0138	0.0152

The differences are in a lower order of magnitude of one tenth of the maximum strain value. Since feature vectors have the same length, a correlation coefficient between experimental and both numerical results was also calculated. This concordance correlation coefficient provides an indication of the extent to which the components of the feature vector fall on a straight line of gradient unity when plotted against one another. The concordance coefficients for the shape descriptors are presented in Table 2 as a percentage in order to represent the similitude of the strain-step fields.

Table 2. Correlation coefficient between feature vectors representing strain-step maps.

-	FEM A (Correlation Coef. in %)	FEM B (Correlation Coef. in %)
ε_{xx} Vertical profile strain map	98.1%	97.7%
ε_{xx} Horizontal profile strain map	98.3%	98.4%
ε_{yy} Vertical profile strain map	96.7%	93.3%
ε_{yy} Horizontal profile strain map	95.3%	94.8%

Finally, it is interesting to plot the residuals of the differences between feature vectors in order to observe if any of the shape descriptors show larger differences, which could give information about the evolution of the differences along the studied steps.

Figure 10 presents bar graphics of the residuals corresponding to the four strain-step maps (i.e., the vertical and horizontal profiles for ε_{xx} and ε_{yy} strains). The residuals are calculated by directly subtracting experimental and FEM feature vectors. The focus is placed on the two shape descriptors that offer maximum differences (red circles in Figure 10).

It is clearly observed that differences in the shape descriptors values are very small. Figure 11 shows the kernels of the Tchebichef shape descriptors with higher differences. These kernels illustrate the distribution of the normalized residuals along the strain-step map in a two-dimensional representation. Thus, they also provide information about the evolution of the differences along the indentation.

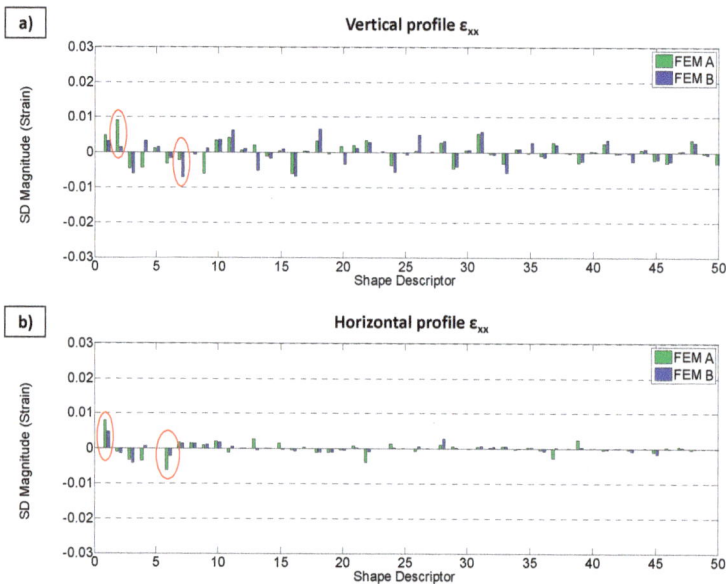

Figure 10. *Cont.*

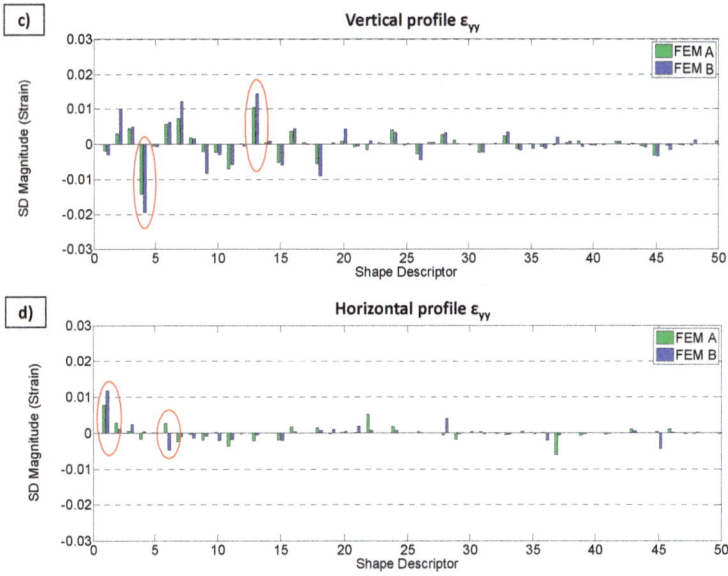

Figure 10. Residuals of the difference between experimental and numerical results (FEM A in green and FEM B in blue) for strain-step maps representing ε_{xx} along vertical (**a**) and horizontal (**b**) profile; and ε_{yy} along vertical (**c**) and horizontal (**d**) profile.

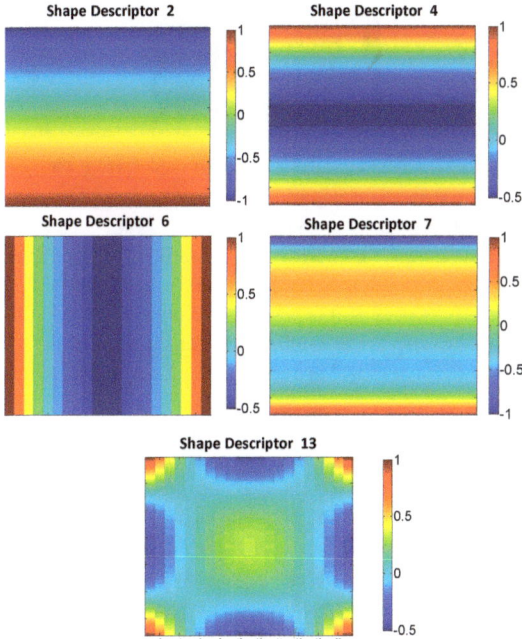

Figure 11. Illustration of the normalized kernels of the shape descriptors with bigger differences between experimental and numerical results.

Bigger differences in ε_{xx} and ε_{yy} strain-step maps for horizontal profiles are observed in SD#1 and #6. In the case of ε_{xx} strain-step maps for the vertical profile, larger differences are observed in SD#2 and #7. Finally, in the case of ε_{yy} strain-step maps for vertical profile, larger differences are observed for SD#4 and #13. The kernels of those shape descriptors are presented in Figure 11. However, the kernel of SD#1 has a homogenous value indicating a full field offset in the strain value along the 21 studied steps, and it has not been considered in Figure 11.

4. Discussion

It is important to emphasize that the reduce area of interest and the high deformation level achieved during deformation made the calculation of the displacements maps using DIC complex. Additionally, as observed in Figures 7 and 8, the indenter and the support of the specimen created some shadows and blind areas that made the visualization of some areas at the instant of maximum indentation difficult. To maximize the visualization of the whole area of interest during the full indentation and release, a specific optic set-up was required, as observed in Figure 2. With this optical arrangement, an uncertainty in the measurement of 31.7 με was achieved when considering the similar set up employed by Tan et al. [2].

Displacement maps obtained by this experimental procedure is presented in Figure 7. As observed, some negligible non-correlated areas are present (dark spots). It particular, some regions close to the indenter and at the lower area of the specimen were not able to be processed due to the high distortion, shadows, and the obstruction of the indenter in the cameras view, as shown in Figure 12.

Figure 12. Image captured at maximum indentation instant.

Those non-data areas are also present in experimental strain maps shown in the first row in Figure 8, where strains at maximum indentation are presented. Moreover, no large differences are observed by comparing those experimental and numerical results from Finite Element model A (and from Finite Element model B). However, it is observed that the FEM A model shows slightly lower maximum values in ε_{xx} (0.32 ε) and lower minimum values in ε_{yy} (0.30 ε), while the FEM B model presents slightly higher values in ε_{xx} (0.44 ε) and ε_{yy} (0.34 ε) compare to experimental ε_{xx} and ε_{yy} values (0.39 and −0.25 strain, respectively).

Moreover, the focus is placed not only in point differences but in performing a full field comparison along the complete experiment. For this purpose, the strain-step images presented in Figure 9 show how the vertical and horizontal profiles of ε_{xx} and ε_{yy} varied along the studied step. Moreover, the creation of these strain-step maps allowed image decomposition to be performed through Tchebichef formulation, which is not applicable in non-rectangular data fields [19]. Moreover, it is observed that these strain-step maps agree between them. To quantify these differences, strain-step maps were decomposed into 50 feature vectors composed (Table 1). FEM model A obtained lower differences that model 2 in all the strain step maps, except for ε_{xx} in the horizontal profile strain map. Larger differences are observed for the ε_{yy} vertical profile strain map for the FEM B model, which represents a difference of 14.2% respect the maximum ε_{yy} experimental value compare to the 10% of difference obtained employing the FEM A model.

The correlation coefficients obtained in Table 2 show a value close to 100%, but again the results from FEM model A show higher concordance with experimental results.

An interesting analysis can be also performed observing the differences in the kernels of the SD as illustrated in Figures 10 and 11.

Looking at the differences in ε_{xx} (Figure 10a), there are more important differences in SD#2 and SD#7 between experimental results and those obtained by the FEM A and FEM B models, respectively. From the representation of kernel of SD#2 in Figure 11, it is observed that the value of experimental strain in the upper area of the strain-step map is slightly bigger than the value of the FEM A model along all the steps of the indentation and, on the contrary, the value of experimental strain in the lower area of the strain-step map is slightly lower than the value of that FEM model. On the other hand, from the representation of kernel of SD#7, it is observed that differences between experimental and FEM B models are also present along the horizontal axis, which means that they are present along all indentation steps. It is also observed that in the upper area close to the edge of the strain-step map, the experimental value is bigger that the simulated value, while in the lower area of the image, the opposite occurs. This indicates that experimental deformation is slightly more focused in the contact area compare to the numerical results. Moreover, the FEM B simulation values are bigger than experimental results just above the half of the strain-step map and are lower just under it.

Differences in ε_{xx} (Figure 10b) suggest that the larger differences between FEM A and experimental results are associated on SD#1 and SD#6. As previously mentioned, SD#1 refers to a full field offset. In this case, the general experimental value of ε_{xx} horizontal profile is lower than the value of FEM A. Additionally, from the kernel of SD#6 in Figure 11, it is observed that an evolution of the difference exists, reaching it maximum values at the central area of the map, which implies the maximum indentation instant. That difference indicates that the values of experimental ε_{xx} are lower than FEM A results mainly at that instant.

With respect to the differences in ε_{yy} (Figure 10c), the differences between shape descriptors are slightly bigger than in the rest of the cases, as illustrated in Figure 10. These differences are more focused in SD#4 and #13 and between the experimental and FEM B results. Attending to the kernels of SD#4, a discordance along the whole indentation is observed where the vertical edges of the vertical profiles are bigger in the FEM B model rather than in the experimental results. Additionally, the values of experimental data in the central area are bigger than those of FEM model B.

Finally, bigger differences between both FEM models and experimental data are focused in SD#1, which means that both FEM models predicted a slightly bigger value in ε_{yy} along the horizontal profile than that which was measured experimentally (Figure 10d). However, this difference is larger for FEM B. SD#13 also presents some discrepancy to the central area of the strain-step map. This discrepancy is different for FEM A and FEM B models. In the case of FEM A, ε_{yy} was slightly larger than the experimental results, and the opposite happened for FEM B.

5. Conclusions

This paper presents a procedure to validate finite element models of indentations in soft materials. The validation was demonstrated by performing an experiment where a rubber cylinder (25 mm diameter and 15 mm high) was subjected to a lateral quasistatic indentation and subsequently released by a rounded edged rigid indenter. Three-Dimensional Digital Image Correlation was employed to calculate three-dimensional displacement fields from which strain maps were calculated. Two finite element models were developed to simulate the experimentation according to two different material models, i.e., Neo Hooken (FEM A) and Van der Walls (FEM B). The validation of those models was performed by comparing strain fields along the complete indentation process by employing image decomposition.

The adopted validation procedure is based on a quantitative comparison of the strain fields in X and Y directions. ε_{xx} and ε_{yy} strain maps for each of the studied steps (21 ε_{xx} strain map and 21 ε_{yy} strain map for each model and experimentation procedure) were combined into two strain-step maps for each procedure. Those strain-step maps where quantitatively compared by employing an Image Decomposition method based on Tchebichef shape descriptors. This methodology quantified differences between experimental results and models that were lower than 10% in case of FEM A. Correlation values between experimental and predicted values were always above 93%, and in the case of FEM A, the correlation factor achieved 97.1%. In addition, differences between strain fields were evaluated along the evolution of the indentation through the study of the kernels of the Shape Descriptors, which offered more important differences.

The validation procedure concluded that the FEM A model (Neo Hookean) better predicts the experimental data than the FEM B model (Van der Walls), which agrees with previous estimations [26].

The quantitative validation approach to comparing strain fields presented herein could provide an intervention methodology to easily validate dynamic (even at high speed) indentation models, enormously reducing the data sets. Additionally it is suitable in cases where the specimen studied is not rectangular, avoiding certain restrictions of some Image Decomposition procedures.

Author Contributions: Luis Felipe-Sesé, Francisco Díaz and Elías López-Alba conceived and designed the experiments and the numerical analysis; Luis Felipe-Sesé and Elías López-Alba performed the experiments and numerical analysis; Benedikt Hannemann and Sebastian Schmeer performed the experiments for material characterization and contributed in the analysis tools; Luis Felipe-Sesé, Francisco Díaz and Elías López-Alba analyzed the data and performed the writing of the manuscript.

Conflicts of Interest: The authors declare no conflict of interest.

References

1. Hamley, I.W. *Introduction to Soft Matter: Synthetic and Biological Self-Assembling Materials*; Wiley: Hoboken, NJ, USA, 2007.
2. Tan, X.B.; Kang, Y.; Patterson, E. An experimental study of the contact of a rounded rigid indenter with a soft material block. *J. Strain Anal. Eng. Des.* **2014**, *49*, 112–121. [CrossRef]
3. Hertz, H. Ueber die Berührung fester elastischer Körper. *J. Die Reine Angew. Math.* **1882**, *1882*, 156–171.
4. Barber, J.; Ciavarella, M. Contact mechanics. *Int. J. Solids Struct.* **2000**, *37*, 29–43. [CrossRef]
5. Johnson, K.L. Motion and Forces at a Point of Contact. In *Contact Mechanics*; Cambridge University Press: Cambridge, UK, 1985.
6. Malits, P. Contact with stick zone between an indenter and a thin incompressible layer. *Eur. J. Mech. A Solids* **2011**, *30*, 884–892. [CrossRef]
7. Dini, D.; Barber, J.R.; Churchman, C.M.; Sackfield, A.; Hills, D.A. The application of asymptotic solutions to contact problems characterised by logarithmic singularities. *Eur. J. Mech. A Solids* **2008**, *27*, 847–858. [CrossRef]
8. Truman, C.E.; Sackfield, A.; Hills, D.A. Contact mechanics of wedge and cone indenters. *Int. J. Mech. Sci.* **1995**, *37*, 261–275. [CrossRef]
9. Ciavarella, M.; Hills, D.A.; Monno, G. Contact problems for a wedge with rounded apex. *Int. J. Mech. Sci.* **1998**, *40*, 977–988. [CrossRef]
10. Jayadevan, K.R.; Narasimhan, R. Finite element simulation of wedge indentation. *Comput. Struct.* **1995**, *57*, 915–927. [CrossRef]
11. Jaffar, M.J. Frictionless contact between an elastic layer on a rigid base and a circular flat-ended punch with rounded edge or a conical punch with rounded tip. *Int. J. Mech. Sci.* **2002**, *44*, 545–560. [CrossRef]
12. Jaffar, M.J. Computation of stresses and deformations for a two-dimensional sliding contact between an elastic layer and a rigid indenter with a rounded profile. *J. Strain Anal. Eng. Des.* **2003**, *38*, 161–168. [CrossRef]
13. Xiao, X.; Kang, Y.; Hou, Z.; Qiu, W.; Li, X.; Li, X. Displacement and strain measurement by circular and radial gratings moiré method. *Exp. Mech.* **2010**, *50*, 239–244. [CrossRef]
14. Burguete, R.L.; Patterson, E.A. A photoelastic study of contact between a cylinder and a half-space. *Exp. Mech.* **1997**, *37*, 314–323. [CrossRef]
15. Han, Y.; Rogalsky, A.D.; Zhao, B.; Kwon, H.J. The application of digital image techniques to determine the large stress–strain behaviors of soft materials. *Polym. Eng. Sci.* **2012**, *52*, 826–834. [CrossRef]

16. Felipe-Sesé, L.; Diaz-Garrido, F.A.; Patterson, E.A. Exploiting measurement-based validation for a high-fidelity model of dynamic indentation of a hyperelastic material. *Int. J. Solids Struct.* **2016**, *97–98*, 520–529.

17. Schreier, H.; Orteu, J.-J.; Sutton, M.A. *Image Correlation for Shape, Motion and Deformation Measurements*; Springer: Boston, MA, USA, 2009.

18. Sebastian, C.; Hack, E.; Patterson, E. An approach to the validation of computational solid mechanics models for strain analysis. *J. Strain Anal. Eng. Des.* **2012**, *48*, 36–47. [CrossRef]

19. Mukundan, R.; Ong, S.H.; Lee, P.A. Image analysis by Tchebichef moments. *IEEE Trans. Image Process.* **2001**, *10*, 1357–1364. [CrossRef] [PubMed]

20. Felipe-Sesé, L.; Siegmann, P.; Díaz, F.A.; Patterson, E.A. Integrating fringe projection and digital image correlation for high-quality measurements of shape changes. *Opt. Eng.* **2014**, *53*, 044106. [CrossRef]

21. Pan, B.; Qian, K.; Xie, H.; Asundi, A. Two-dimensional digital image correlation for in-plane displacement and strain measurement: A review. *Meas. Sci. Technol.* **2009**, *20*, 1–17. [CrossRef]

22. Sutton, M.A.; Orteu, J.-J.; Schreider, H. *Image Correlation for Shape, Motion and Deformation Measurements Basic Concepst, Theory and Applications*; Springer: Boston, MA, USA, 2009; Volume XXXIII.

23. Sutton, M.; Wolters, W.; Peters, W.; Ranson, W.; McNeill, S. Determination of displacements using an improved digital correlation method. *Image Vis. Comput.* **1983**, *1*, 133–139. [CrossRef]

24. Greenwood, J.A.; Tabor, D. The friction of hard sliders on lubricated rubber: The importance of deformation losses. *Proc. Phys. Soc.* **1958**, *71*, 989–1001. [CrossRef]

25. Mofidi, M.; Prakash, B.; Persson, B.N.J.; Albohl, O. Rubber friction on (apparently) smooth lubricated surfaces. *J. Phys. Condens. Matter* **2007**, *20*, 085223. [CrossRef]

26. Reppel, T.; Dally, T.; Weinberg, K. On the elastic modeling of highly extensible polyurea. *Tech. Mech.* **2013**, *33*, 19–33.

27. Treloar, L.R.G. Stress-strain data for vulcanized rubber under various types of deformation. *Rubber Chem. Technol.* **1944**, *17*, 813–825. [CrossRef]

28. Kilian, H.-G. A molecular interpretation of the parameters of the van der Waals equation of state for real networks. *Polym. Bull.* **1980**, *3*, 151–158. [CrossRef]

![materials logo] *materials*

MDPI

Article

Effect of Applied Stress on the Mechanical Properties of a Zr-Cu-Ag-Al Bulk Metallic Glass with Two Different Structure States

Heng Chen, Taihua Zhang and Yi Ma *

Institution of Micro/Nano-Mechanical Testing Technology & Application, College of Mechanical Engineering, Zhejiang University of Technology, Hangzhou 310014, China; hengchen@zjut.edu.cn (H.C.); zhangth@zjut.edu.cn (T.Z.)
* Correspondence: may@zjut.edu.cn; Tel.: +86-571-88320132

Received: 5 June 2017; Accepted: 21 June 2017; Published: 27 June 2017

Abstract: In order to investigate the effect of applied stress on mechanical properties in metallic glasses, nanoindentation tests were conducted on elastically bent Zr-Cu-Ag-Al metallic glasses with two different structure states. From spherical *P-h* curves, elastic modulus was found to be independent on applied stress. Hardness decreased by ~8% and ~14% with the application of 1.5% tensile strain for as-cast and 650 K annealed specimens, while it was slightly increased at the compressive side. Yield stress could be obtained from the contact pressure at first pop-in position with a conversion coefficient. The experimental result showed a symmetrical effect of applied stress on strengthening and a reduction of the contact pressure at compressive and tensile sides. It was observed that the applied stress plays a negligible effect on creep deformation in as-cast specimen. While for the annealed specimen, creep deformation was facilitated by applied tensile stress and suppressed by applied compressive stress. Strain rate sensitivities (SRS) were calculated from steady-state creep, which were constant for as-cast specimen and strongly correlated with applied stress for the annealed one. The more pronounced effect of applied stress in the 650 K annealed metallic glass could be qualitatively explained through the variation of the shear transformation zone (STZ) size.

Keywords: metallic glass; nanoindentation; applied stress; hardness; pop-in; creep

---

1. Introduction

Metallic glass is scientifically defined as amorphous alloy which has a non-crystalline, but short-range order structure [1]. Due to its unique atomic configuration, metallic glass is one of the important parts of condensed matter physics. This new-structure material is promising for use in engineering fields because of its excellent mechanical properties, such as high strength, large elastic limit and good wear resistance [2–4]. However, the localized shear banding is dominating in plastic deformation of bulk metallic glass, causing catastrophic failure and the limited ductility severely hinders its practical application [5,6]. In order to overcome the above problems, numerous efforts have been focused on exploring new compositions in the search for ductile "perfect production", without sacrificing high strength in the last two decades [7–9]. Importantly, plastic behaviors were widely studied to reveal the intrinsic deformation mechanism and to establish structural-properties correlation in metallic glasses [10–12]. In the last decade, a strong size effect on deformation behavior was validated in metallic glasses [13–15]. The plasticity could be improved remarkably without sacrificing high strength at the micro/nano scale, even combined with the transition of deformation modes (localized to homogeneous) [16]. The free volume mode and shear transformation zone (STZ) mode have been successfully applied to analyze the low-temperature deformation of metallic glasses [17,18]. Several effective methods, such as introducing crystalline secondary phase and increasing free volume content, have been developed to promote plasticity of metallic glasses [19,20]. Effects of surface treatments e.g., rolling and shot peening are also validated on metallic glasses [21,22]. Essentially, it is the modulation of shear banding events rather than changing deformation mode (localized to non-localized), which apparently increases plastic strain and delays fracture. In accordance with blocking effects of the secondary phase, residual strain/stress can be introduced into metallic glasses by surface treatments, hence suppressing both the nucleation and propagation of shear bands in metallic glasses.

The applied strain/stress effect on mechanical properties in metallic glasses has attracted many investigations that used pre-straining method [23–27]. From the viewpoint of engineering, it is necessary to assess the merits of surface pretreating in structural materials or anticipate material reliability under complex-stress situation. It is expected that the structure configuration would be disturbed by stress fluctuation and in turn cause alterations in the deformation mechanism. Both experiment and simulation results have reported that hardness [23], yield stress [24], creep flow [25] and shear banding morphology [26] were closely related to the type and magnitude of applied strain/stress. Using atomistic modeling, free volume evolution was speculated under applied strain/stress and expected to lead to the difference of mechanical properties [23,24,27]. It has been revealed that more excess free volume can be created in metallic glasses which own higher atomic packing densities (lower initial free volume fraction) under elasto-static stress [28]. As a consequence, the initial structure state would play an important role on the effect of applied strain/stress on mechanical properties. To the author's best knowledge, there has been no report hitherto that investigated applied strain/stress effect concerns with different structure states. With this in mind, a Zr-Cu-Ag-Al bulk metallic glass which has high forming ability, high yield strength and large plasticity was prepared [29]. High temperature annealing was performed to attain structure relaxation. A home-made apparatus is used to elastically bend the specimen for introducing applied strain/stress. Relying on nanoindentation technology, mechanical properties can be studied at small regions which are subjected to various applied strain/stress. Due to its high accuracy, the variation of mechanical properties on the applied stress can be obtained correctly in instrumented nanoindentation and in turn the residual stress can be extracted [30,31]. In the present work, we aim to study the effect of applied strain/stress on mechanical properties and their correlation with structure states in metallic glasses.

2. Materials and Methods

$Zr_{46}Cu_{37.6}Ag_{8.4}Al_8$ alloy ingots were prepared from high pure elements (99.99%) by arc mixing in a Ti-gettered argon atmosphere. Alloy sheets with a rectangular cross-section of 2 mm × 10 mm were obtained by injecting alloy melt into the copper mold. The as-cast specimens with a dimension of 1 mm × 2 mm × 10 mm were cut for structure characterization and mechanical testing. The annealing was performed at the chamber of magnetron sputtering with an ultra-low base vacuum and argon protective atmosphere. The specimen was held at 650 K for 1 h and cooled inside the furnace to room temperature. Prior to the nanoindentation testing, the side surface (1 mm × 10 mm) of the specimen was carefully polished to a mirror finish. The structures of both as-cast and annealed Zr-Cu-Ag-Al specimens were detected by X-ray diffraction (XRD) with Cu $K\alpha$ radiation. The differential scanning calorimetry (DSC) tests with heating rate of 20 K/min were performed to study the annealing effect on structure relaxation. By means of X-ray energy dispersive spectrometer (EDS) attached to the SEM, the chemical composition was detected, which is equal to the alloy ingot.

Applied stress was introduced by four-point bending through a home-made apparatus, as exhibited in Figure 1. The bending curvature r of specimen was precisely computed as 20 mm from the optical microscope image. The applied strain could be roughly estimated as z/r, in which z is the distance from the selected location to the middle line of specimen. The nanoindentation tests were performed at five regions with applied strains as -1.5%, -0.75% (compressive side is referred as $z < 0$), 0% (the neutral plane), 0.75% and 1.5% (tensile side is referred as $z > 0$) for both specimens, by selecting the locations at a distance of 0.3 mm and 0.15 mm away from the neutral line, as listed in

Table 1. The corresponding applied stress could be estimated as -1.39, -0.69, 0, 0.69 and 1.39 GPa, for the elastic modulus was reported as 92.4 GPa for the as-cast $Zr_{46}Cu_{37.6}Ag_{8.4}Al_8$ [29]. The yield stress of the as-cast specimen is about 1.8 GPa, which is apparently higher than the applied stress. It should also be noted that the 24 h-bent specimens can fully recover after unloading. In the following, we use "applied strain" to denote the five measured regions for simplicity.

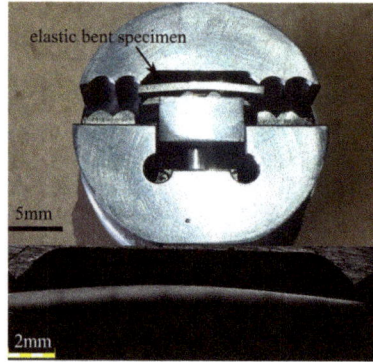

Figure 1. A home-made apparatus is used to elastically hold the specimen and the bending curvature is calculated by optical microscope.

Table 1. Hardness, contact pressure at first pop-in and strain rate sensitivity, STZ volume from steady-state creep at various pre-strained regions in nanoindentation.

Z Value, mm	Applied Strain, %	Hardness, GPa		Contact Pressure, GPa		Strain Rate Sensitivity		STZ Volume, nm	
		As-Cast	Annealed	As-Cast	Annealed	As-Cast	Annealed	As-Cast	Annealed
0.3	1.5	8	9	9.8	11.2	0.023	0.0026	2.58	20.3
0.15	0.75	8.02	8.87	9.5	10.5	0.022	0.0044	2.63	12.2
0	0	7.9	8.7	9.1	9.9	0.026	0.0093	2.31	5.87
−0.15	−0.75	7.6	8.1	8.9	9.5	0.025	0.0126	2.50	4.65
−0.3	−1.5	7.26	7.5	8.5	8.8	0.028	0.022	2.33	2.88

The nanoindentation experiments were conducted at a constant temperature of 20 °C on Agilent Nano Indenter G200. Elastic modulus, hardness and information of first pop-in were studied in load-displacement (P vs. h) curves upon a special indenter, with a nominal radius of 5 μm. The maximum load was 20 mN for as-cast specimen and 25 mN for annealed one, respectively. The loading rate was constant as 0.5 mN/s. At least 25 independent measurements were conducted at each position for both specimens. The creep tests were performed by a constant load holding method upon a standard Berkovich indenter, in which the displacement of the indenter pressed into the surface at a prescribed load was continuously recorded. The duration was 500 s at a maximum load of 100 mN and the loading rate was fixed as 1 mN/s. The reliability of the creep results was confirmed by conducting 12 independent measurements. The nanoindentation tests were carried out until thermal drift reduced to below 0.03 nm/s. Furthermore, drift correction which was calibrated at 10% of the maximum load during the unloading process would be strictly performed.

3. Results and Discussion

3.1. Structure Characterization

Figure 2a shows the typical X-ray diffraction patterns of as-cast and 650 K annealed $Zr_{46}Cu_{37.6}Ag_{8.4}Al_8$ specimens. It is clear that only a broad diffraction peak can be detected in each specimen, which represents a crystal-free structure. Figure 1 shows the DSC curves in which the glass transition temperature T_g can be observed at about 710 K, as the arrow indicates [29]. As the adopted composition has been systematically studied and confirmed to have strong glass forming ability, the XRD pattern and DSC curve may be enough to confirm the amorphous nature for both as-cast and annealed specimens without further detection by transmission electron microscopy (TEM). The inset of Figure 2b shows an enlargement of the sub-T_g region for the DSC curves, which corresponds to the enthalpy released during structure relaxation and can be strongly linked with the initial free volume content. It clearly shows that the structure relaxation process was more pronounced in the as-cast specimen compared to the 650 K annealed one, confirming the effect of annealing on eliminating free volume.

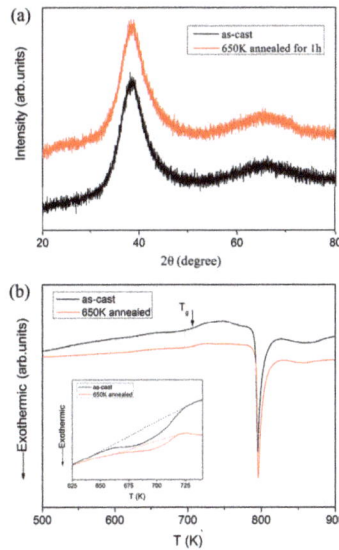

Figure 2. (a) Typical XRD patterns and (b) differential scanning calorimetry (DSC) curves for the as-cast and 650 K annealed Zr-Cu-Ag-Al bulk metallic glass, as well as the details of the sub-T_g regions of the DSC traces.

3.2. Elastic Modulus and Hardness

Spherical *P-h* curve in standard fused silica with loading rate of 0.5 mN/s was shown in Figure 3; the loading and unloading curves were fully overlapped, indicating an elastic deformation process. According to the Herztian elastic contact theory [32], the loading sequence could be perfectly fitted by:

$$P = \frac{4}{3}E_r\sqrt{R}h^{1.5},\tag{1}$$

where E_r is the reduced elastic modulus which accounts for that the elastic displacement occurs in both the tip and sample.

$$E_r = \left(\frac{1 - v_s^2}{E_s} - \frac{1 - v_i^2}{E_i}\right)^{-1},\tag{2}$$

E and *v* are the elastic modulus and Poisson's ratio, with the subscripts s and i represent the sample and the indenter, respectively. For commonly used diamond tip, E_i = 1141 GPa and v_i = 0.07 [33], these values combined with the E_s = 72 GPa and v_s = 0.18 in fused silica can be substituted in Equation (2) to calculated the E_r. Finally the effective tip radius at the front end of spherical indenter was calculated to be 2.95 μm.

Figure 4 shows the typical spherical *P-h* curves for as-cast and annealed specimens, measured at various pre-strained regions. Clearly, the *P-h* curves at the initial loading stage, namely elastic region or elastic-plastic region, were almost overlapped and can be completely fitted by the Hertzian elastic contact theory. For the as-cast specimen in Figure 4a, the power-law fitting expression is $P = 0.008\ h^{1.5}$. The Poisson's ratio is 0.36, therefore the elastic constant can be calculated as 110.6 GPa. The elastic modulus by nanoindentation is a little higher than that reported by uniaxial compression [29]. By the same analysis, power-law fitting expression is $P = 0.0081\ h^{1.5}$ for 650 K annealed specimen in Figure 4b, and elastic modulus was deduced as 111.9 GPa. The elastic modulus was independent on applied stress and slightly increased by annealing. It may be reasonable to assume that elastic modulus is directly related to the atomic structure and belongs to the intrinsic properties of a material [25].

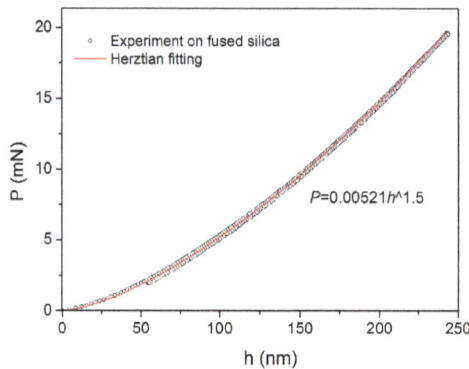

Figure 3. Elastic *P-h* curve in the standard fused silica upon a spherical indenter. The loading sequence can be perfectly fitted by Herztian elastic theory.

In a spherical-tip indentation process, hardness is defined as

$$H = P/2\pi Rh_c, \tag{3}$$

where *P* is the maximum load. The contact displacement h_c could be deduced as

$$h_c = h - \varepsilon \times P/S, \tag{4}$$

where *h* is the recorded indenter displacement, $\varepsilon = 0.75$ for a spherical tip, *S* is the stiffness which could be obtained from the unloading curve. The average hardness was listed in Table 1 for both specimens. At the compressive side, hardness was insensitive to the applied stress in the as-cast specimen and it slightly increased as increasing applied stress was applied in the annealed one. While at the tensile side, hardness showed a strong correlation with applied stress that dropped from 7.9 to 7.26 GPa in as-cast specimen and 8.7 to 7.6 GPa in annealed one, as the applied strain increased from 0 to 1.5%. Figure 5a clearly shows the variation trend of hardness as a function of applied strain. Moreover, hardness *H* obtained at the pre-strained region was compared to H_0 at the neutral plane by $(H - H_0)/H_0$, as shown in Figure 5b. With applying 1.5% tensile applied strain, hardness reduced by ~8% and ~14% in the as-cast and annealed specimens, respectively. While at the compressive counterpart, hardness increased no more than 1.5% and 3.5% for as-cast and annealed specimens. It should be noted that the asymmetric effect of applied stress on hardness has been reported previously in metallic glasses and is well explained upon the approach of excess free volume [23]. In the present study, hardness was detected at relative shallow depth and the stress field beneath the indenter was elastoplastic, i.e., severe plastic deformation did not occur. Therefore, the influence of pile-up on the true hardness may be insignificant and the applied stress-dependent height of pile-up would not be the key factor on the herein results [26]. In addition, it is the first report that an annealing treatment may enhance the effect of applied stress on hardness at both the tensile and compressive sides.

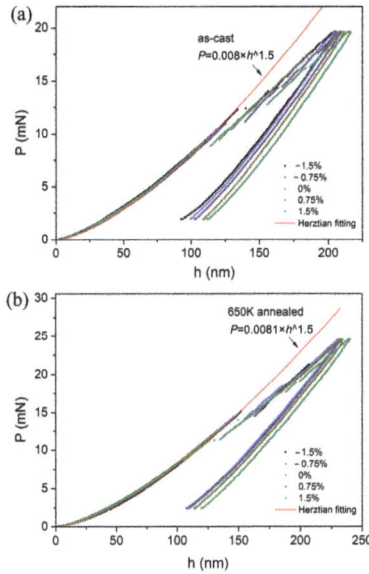

Figure 4. Representative spherical *P-h* curves at various pre-strained regions for (**a**) as-cast specimen and (**b**) 650 K annealed one. Pop-ins were clearly observed in each curve.Representative spherical *P-h* curves at various pre-strained regions for (**a**) as-cast specimen and (**b**) 650 K annealed one. Pop-ins were clearly observed in each curve.

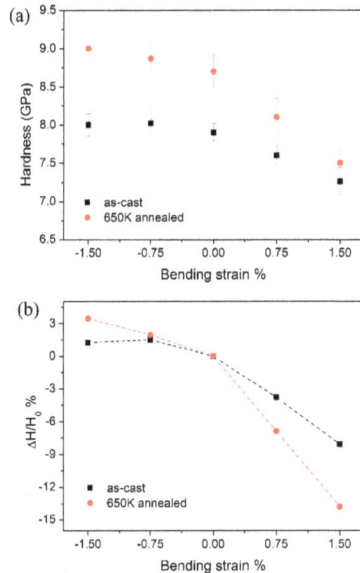

Figure 5. (**a**) Nanoindentation hardness *H* and (**b**) the variation trends compared to H_0 in the neutral plane as a function of applied bending strans for as-cast and 650 K annealed specimens.

3.3. Pop-in Shear Stress

Here, the pop-in events with the scale of ~10 nm can be observed in all the loading sequences. It is noted that the Herztian fitting line deviated from the exact *P-h* curve at the position of first pop-in in all the nanoindentations in Figure 4. This fact indicated the transition from elastic to elastic-plastic deformation once the first pop-in emerges, which also could be regarded as the onset of yielding under indentation [34]. According to Hertzian contact theory, the elastic contact radius is expressed as:

$$a = \sqrt{Rh}, \tag{5}$$

the maximum contact pressure beneath the indenter is defined as:

$$P_m = P/\pi Rh, \tag{6}$$

the maximum shear stress τ at the first pop-in could represent yield stress at the onset of plasticity in nanoindentation. For a spherical indenter, the maximum shear stress of metallic glass happens at about half the elastic contact radius according to Bei's simulation and equal to τ ~0.445 P_m [34]. The conversion coefficient is not a fixed value; while 0.31 is commonly adopted [35]. However, no matter what yield criterion is adopted, a linear function $\tau = CP_m$ could be expected, C is a constant related to the yield criterion. In the present work, we directly study the contact pressure of first pop-in to reveal the effect of applied stress on yield stress of as-cast and annealed metallic glasses.

Figure 6a shows the representative spherical *P-h* curves with various applied strains for as-cast and 650 K annealed specimens, of which the displacements have been offset for clearly viewing the positions of first pop-in. As marked by arrows, the required critical load for the first pop-in event was monotonously reduced as the decrease of the applied compressive strain and/or increase of applied tensile strain. The calculated contact pressures at first pop-in were listed in Table 1 for both specimens. In accordance with the annealing effect on hardness, the obtained contact pressure (as well as the yield stress) was effectively enhanced in the 650 K annealed specimen. As exhibited in Figure 6b, the contact pressure almost linearly decreased with an increase in the bending strain from −1.5 to 1.5% for both as-cast and annealed specimens. The strengthening effect of applied compressive stress and the softening effect of applied tensile stress on yield stress herein were found to be symmetrical. The inset in Figure 6b depicts the percentage change of P_m at the pre-strained regions compared to the neutron plane. The contact pressure was more sensitive to the applied stress in 650 K annealed specimen than the as-cast one. P_m was enhanced by ~13% by applying 1.5% compressive strain and dropped by ~12% by applying 1.5% tensile strain in annealed specimen. The percentage change of P_m in as-cast specimen was ~7% at 1.5% pre-strained regions with both stress states. Clearly, the effect of applied stress on the contact pressure (yield stress) is also structure state-dependent. Being different from the previous report [24], the obtained contact pressure was sensitive to both applied tensile and compressive stress, rather than merely at the tensile side.

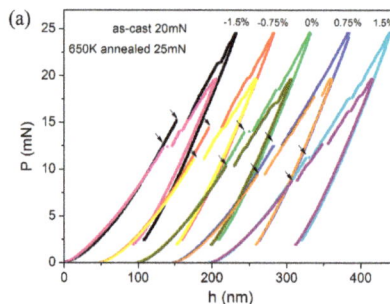

Figure 6. *Cont.*

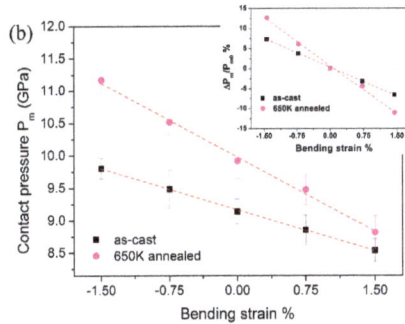

Figure 6. Representative spherical *P-h* curves for as-cast and 650 K annealed specimens. (a) Displacements of the *P-h* curves at different pre-strained regions have been offset for clearly viewing the position of first pop-in; (b) Contact pressure P_m were calculated from the first pop-in position and plotted as a function of bending strains.

3.4. Creep Behavior and Strain Rate Sensitivity

The representative creep curves were shown in Figure 7, in which creep displacements were plotted with holding time. In order to study the creep behaviors at various pre-strained regions directly, the onset of creep deformation was set to be zero in the coordinate axis. For the as-cast specimen in Figure 7a, creep deformations recorded at various pre-strained regions exhibited little difference. And obviously, such tiny differences between the creep curves were within the range of experimental errors. For the annealed specimen in Figure 7b, creep deformation was enhanced at the tensile regions and suppressed at the compressive regions, in comparison to the creep flow at the neutron plane. The total creep displacements after 500 s holding were also summarized in the insets of Figure 5, which were plotted as a function of bending strains. The mean values of total creep displacement were in a narrow range of 12–14.5 nm for the as-cast specimen. Meanwhile, it was gradually increased from 9.5 to 13.5 nm for the annealed specimen, as the bending strain increased from −1.5 to 1.5%. In Chen's work, it was claimed that creep displacement was roughly stable at the compressive side and increased by increasing applied tensile strain in a Zr-based metallic glass [25]. It needs to be pointed out that the total creep displacement was in a small range of 8 to 12 nm even under 400 mN holding with loading rate of 1 mN/s in Chen's work [25]. According to their report, creep displacement was further reduced significantly under lower holding loads and/or slower loading rates. It is possible that the variation trend of creep displacement was within the range of error bars and it would be questionable to reach a "universal law" in their study. For the creep flows of annealed specimen in of Figure 7b, they completely overlapped at the transient stage and the distinction of total creep displacements resulted from the steady-state part. It is indicated that creep behaviors were intrinsically changed by applying stress in the annealed specimen.

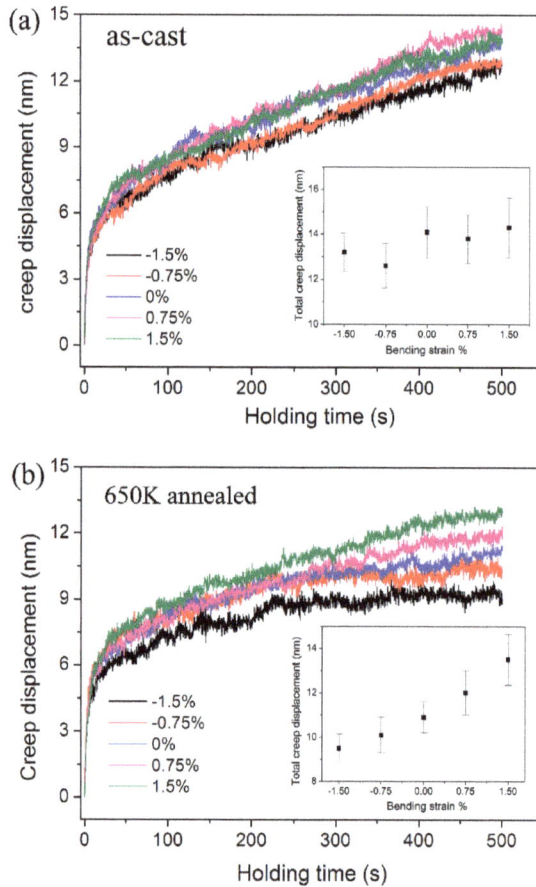

Figure 7. The representative creep displacements vs. holding time for (**a**) as-cast specimen and (**b**) 650 K annealed one at various pre-strained regions. The total creep displacements as a function of bending strains were exhibited in the insets.

Indentation creep has also been the most extended method to study strain rate sensitivity (SRS) in metals [36]. From indentation-creep tests, SRS can be directly obtained by applying time-displacement data. The relationship between hardness and indentation strain rate for a power-law creeping materials is

$$H = A\dot{\varepsilon}^m,\tag{7}$$

the value of SRS exponent m can be evaluated via:

$$m = \frac{\partial lnH}{\partial ln\dot{\varepsilon}},\tag{8}$$

for a standard Berkovich indentation process, the strain rate $\dot{\varepsilon}$ during the holding stage can be calculated as:

$$\dot{\varepsilon} = \frac{dh_c}{dt}\frac{1}{h_c},\tag{9}$$

and hardness is defined as:

$$H = \frac{P}{24.5h_c^2},$$ (10)

the plastic displacement hc beneath a Berkovich tip could be obtained as $h_c = h\text{-}0.72 \times P/S$. In the current study, it is unrealistic to detect S at each recorded creep displacement. For simplicity, the S obtained from the creep unloading curve was adopted to calculate hardness.

The experimental data could be perfectly fitted ($R^2 > 0.99$) by an empirical law:

$$h_{(t)} = h_0 + a(t - t_0)^b + kt,$$ (11)

where h_0, t_0 are the displacement and time at the beginning of holding stage. a, b, k are the fitting constants. Figure 8a shows the typical creep curve detected at the neutral plane and the fitting line in as-cast specimen. Figure 8b shows the variation of strain rate as a function of creep time deduced from the fitting line. The creep strain rate dropped precipitously from the magnitude of 10^{-2} to 10^{-4} s^{-1} within the initial 20 s. Then it was decreased gently and fell into the range of 2×10^{-5} to 1×10^{-5} s^{-1} on the last 200 s duration. The variation of creep hardness is exhibited in the inset of Figure 8b. After 500 s holding, hardness reduced from about 8.3 to 7.9 GPa. Figure 8c shows the logar-logar correlation between indentation hardness and strain rate during the holding stage. SRS can be obtained as 0.0275 by linearly fitting the part of steady-state creep. For reliability, 6~8 effective creep curves were employed to reach an average value of SRS, which were listed in Table 1. They are 0.023, 0.022, 0.026, 0.025, 0.028 for as-cast specimen and 0.0026, 0.0044, 0.0093, 0.0126, 0.022 for annealed specimen, corresponding to the regions suffered -1.5%, -0.75%, 0%, 0.75%, 1.5% applied strains, respectively. Figure 8d clearly shows the correlation between strain rate sensitivity and applied bending strain for as-cast and 650 K annealed specimens. Following the rule of creep deformation, SRS was independent on the applied stress in as-cast specimen. On the other hand, the overall SRSs declined considerably after 650 K annealing. Moreover, SRS increased with increasing applied tensile strain and decreased with increasing applied compressive strain for the annealed specimen. For nanoindentation creep, the estimated value of SRS could be varied on different test conditions for a certain material. Loading rate, holding depth and indenter type are confirmed to influence the value of SRS, which could be attributed to structure change beneath the indenter [37–39]. Besides, the value of SRS is also relying on holding time; this smaller value would be computed at the end of a shorter duration. In the present study, we emphasized the effects of applied stress and annealing on the variation of SRS, rather than revealing the SRS characteristic or creep mechanism in metallic glasses [40]. By using the self-similar Berkovich indenter, the observed applied stress effect on creep deformation would be universal, even if it is under different holding depths and/or loading rates.

Free volume evolution as applying stress was used previously to explain the variation trends of hardness in metallic glasses that the initial free volume could be largely increased in tension and insensitive to applied compressive stress [23]. Hardness is defined as the resistance to plastic deformation, which could be closely tied to the free volume content in a metallic glass. Currently, investigation on the correlation between yield stress and applied stress is relatively scarce. To the author's best knowledge, only Wang et al. systematically studied the effect of applied stress on the onset of yielding in metallic glass [24]. They suggested that the effective maximum shear stress at the first pop-in was essentially constant based on the distribution of shear stress beneath a spherical indenter by finite-element analysis. However, it could be subjective to merely ascribe the onset of yielding to the critical excess free volume [3]. Owing to the original work of Argon [18], the deformation unit with a local rearrangement of atoms, also referred as to shear transformation zone (STZ), has been widely applied to analyze the occurrence of plastic deformation in metallic glasses. Being different from structure defect, STZ is defined by its transience, i.e., it can only be identified from the atomic structures before and after deformation. According to Johnson and Samwer [12], there needs to be a critical fraction of activated STZ beyond which yielding will occur. The energy barrier for activating an STZ is in proportion to STZ volume, i.e., smaller STZs would be more easily agitated and readily accommodated to sustain the shear strain. In recent years, the measured STZ sizes displayed a strong correlation with strength and ductility in metallic glasses [41,42]. Moreover, the STZ size of metallic glass could be associated with Poisson's ratio and explains the critical size of deformation mode transition at nanoscale [42,43].

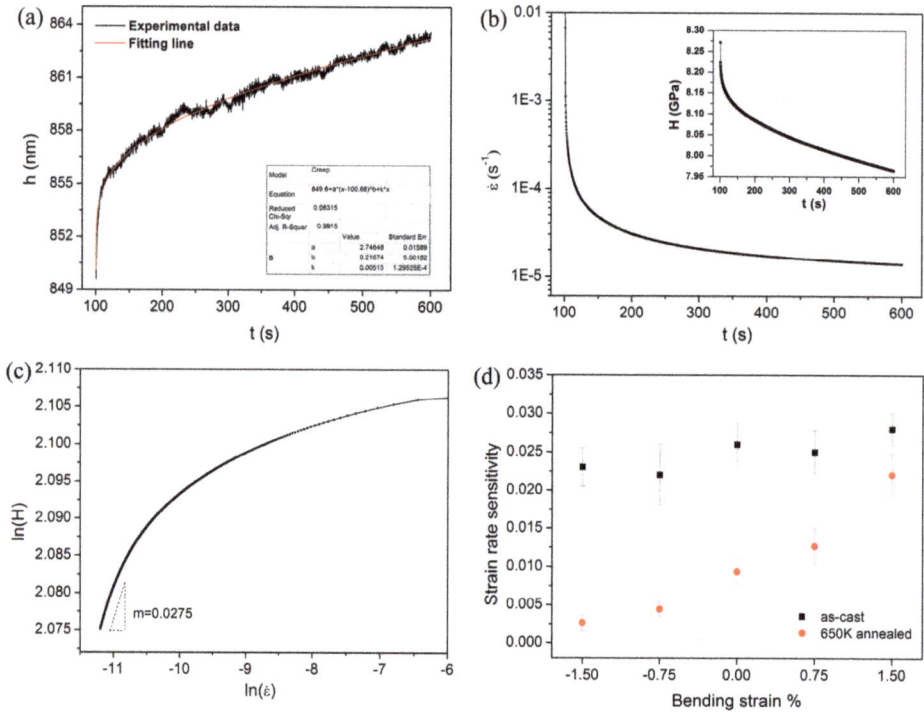

Figure 8. (**a**) The typical creep curve detected at the neutral plane and the fitting line for as-cast specimen; (**b**) The creep strain rate and hardness as a function of holding time; (**c**) The logar-logar correlation between hardness and strain rate for the creep deformation, strain rate sensitivities can be thus computed from the steady-state part; (**d**) The calculated strain rate sensitivities as a function of bending strain for both as-cast and annealed specimens.

Undoubtedly, STZ size plays an important role on mechanical properties and plastic deformation in metallic glasses. Following the cooperative shear model (CSM) by Johnson and Samwer [12], Pan et al. successfully developed an experimental method upon nanoindentation to calculate STZ volumes and atoms involved [42]. According to Pan's work, the STZ volume Ω can be expressed as:

$$\Omega = kT/C'mH, \tag{12}$$

where k is the Boltzman constant, T is the testing temperature, $C' = \frac{2R_0\zeta}{\sqrt{3}} \frac{G_0\gamma_C^2}{\tau_C} \left(1 - \frac{\tau_{CT}}{\tau_C}\right)^{1/2}$ which can be computed according to Johnson and Samwer' theory, the constants $R_0 \approx 1/4$ and $\zeta \approx 3$, the average elastic limit $\gamma_C \approx 0.027$, G_0 and τ_C are the shear modulus and the threshold shear resistance of an alloy at 0K, $\frac{\tau_C}{G_0} \approx 0.036$. The value of $\frac{\tau_{CT}}{\tau_C}$ at certain T can be estimated on the equation $\frac{\tau_{CT}}{\tau_C} = 1 - \frac{0.016}{0.036}\left(\frac{T}{T_g}\right)^{2/3}$, T_g can be obtained from the DSC curve. The STZ volume for each strained region was summarized in Table 1 for both specimens. We can reach a conclusion that STZ size is roughly stable in as-cast specimen and evidently reduced with increasing applied tensile strain and/or decreasing applied compressive strain in the annealed one. It also confirms that annealing effect could enlarge the STZ size in metallic glass [35]. As mentioned earlier, the value of SRS would be changed with different testing condition by nanoindentation creep method. Therefore, it is meaningless to discuss the specific value of STZ size in the present situation. In light of the variation trend of STZ size, the more sensitive response of mechanical properties to applied stress in 650 K annealed metallic glass could be explained qualitatively. At the regions that suffered applied tensile stress, the reduced STZ size would facilitate both instantaneous and time-dependent plastic deformations. Combined with the "softening effect" of more excess free volume, hardness and contact pressure (yield stress) consequently drop more precipitously than in the as-cast specimen. At the regions that suffered applied compressive stress, larger STZs and less excess free volume induce a faster enhancement of hardness and contact pressure (yield stress) than in the as-cast specimen. For the creep flow, the experimental result suggests that the STZ evolution might be the main creep mechanism rather than the creation and annihilation of free volume [40].

4. Conclusions

In summary, the effects of applied stress on mechanical properties in as-cast and 650 K annealed Zr-Cu-Ag-Al metallic glasses were systematically studied upon nanoindentation. Elastic modulus, hardness, contact pressure at the onset of yielding and creep resistance were measured at various pre-strained regions. Based on the experiment results, the following conclusions can be summarized:

(1) Hardness and the contact pressure for yielding evidently decreased with the application of tensile stress. At the compressive side, contact pressure enhancement was much more significant than hardness increase. Elastic modulus was unaffected by applied stress.

(2) Creep deformation was independent of applied stress in the as-cast specimen. However, it was facilitated by applied tensile stress and suppressed by applied compressive stress in the 650 K annealed specimen. Strain rate sensitivities of the annealed specimen were also applied stress-dependent.

(3) The effect of applied stress on mechanical properties was more pronounced in metallic glass with structure relaxation. This could be intrinsically due to the change of STZ size under applied stress.

Acknowledgments: The support from the National Natural Science Foundation of China (Grant Nos. 11502235, 11672356) and Zhejiang Provincial Natural Science Foundation of China (Y13E050085) are gratefully acknowledged.

Author Contributions: Yi Ma and Taihua Zhang conceived and designed the experiments; Heng Chen performed the experiments and analyzed the data; Yi Ma wrote the paper.

Conflicts of Interest: The authors declare no conflict of interest.

References

1. Klement, W.; Willens, R.; Duwez, P.O.L. Non-crystalline structure in solidified gold-silicon alloys. *Nature* **1960**, *187*, 869–870. [CrossRef]
2. Loffler, J.F. Bulk metallic glass. *Intermetallics* **2003**, *11*, 529–540. [CrossRef]
3. Schuh, C.A.; Hufnagel, T.C.; Ramamurty, U. Mechanical behavior of amorphous alloys. *Acta Mater.* **2007**, *55*, 4067–4109. [CrossRef]
4. Inoue, A.; Shen, B.; Koshiba, H.; Kato, H.; Yavari, A.R. Cobalt-based bulk glassy alloy with ultrahigh strength and soft magnetic properties. *Nat. Mater.* **2003**, *2*, 661–663. [CrossRef] [PubMed]
5. Chen, M.W. Mechanical behavior of metallic glasses: Microscopic understanding of strength and ductility. *Annu. Rev. Mater. Res.* **2008**, *38*, 445–469. [CrossRef]
6. Sun, B.A.; Wang, W.H. The fracture of bulk metallic glasses. *Prog. Mater. Sci.* **2012**, *57*, 487–656. [CrossRef]
7. Liu, Y.H.; Wang, G.; Wang, R.J.; Zhao, D.Q.; Wang, W.H. Super plastic bulk metallic glasses at room temperature. *Science* **2007**, *315*, 1385–1388. [CrossRef] [PubMed]
8. Das, J.; Tang, M.N.; Kim, K.B.; Theissmann, R.; Wang, W.H. "Work-Hardenable" ductile bulk metallic glass. *Phys. Rev. Lett.* **2005**, *94*, 205501. [CrossRef] [PubMed]

9. Schroers, J.; Johnson, W.L. Ductile bulk metallic glass. *Phys. Rev. Lett.* **2004**, *93*, 255506. [CrossRef] [PubMed]
10. Leamy, H.; Wang, T.; Chen, H. Plastic flow and fracture of metallic glass. *Metal. Trans.* **1972**, *3*, 699–708. [CrossRef]
11. Lewandowski, J.J.; Wang, W.H.; Greer, A.L. Intrinsic plasticity or brittleness of metallic glasses. *Philo. Mag. Lett.* **2005**, *85*, 77–87. [CrossRef]
12. Johnson, W.; Samwer, K. A universal criterion for plastic yielding of metallic glasses with a (T/Tg) 2/3 temperature dependence. *Phys. Rev. Lett.* **2005**, *95*, 195501. [CrossRef] [PubMed]
13. Tian, L.; Cheng, Y.Q.; Shan, Z.W.; Li, J. Approaching the ideal elastic limit of metallic glasses. *Nat. Commun.* **2012**, *3*, 609. [CrossRef] [PubMed]
14. Ghidelli, M.; Gravier, S.; Blandin, J.J.; Djemia, P. Extrinsic mechanical size effects in thin ZrNi metallic glass films. *Acta Mater.* **2015**, *90*, 232–241. [CrossRef]
15. Ghidelli, M.; Gravier, S.; Blandin, J.J.; Raskin, J.P. Size-dependent failure mechanisms in ZrNi thin metallic glass films. *Scr. Mater.* **2014**, *89*, 9–12. [CrossRef]
16. Ghidelli, M.; Idrissi, H.; Gravier, S.; Blandin, J.J. Homogeneous flow and size dependent mechanical behavior in highly ductile Zr65Ni35 metallic glass films. *Acta Mater.* **2017**, *131*, 246–259. [CrossRef]
17. Spaepen, F. A microscopic mechanism for steady state inhomogeneous flow in metallic glasses. *Acta Metal.* **1977**, *25*, 407–415. [CrossRef]
18. Argon, A.S. Plastic deformation in metallic glasses. *Acta Metal. Mater.* **1979**, *27*, 47–58. [CrossRef]
19. Hays, C.C.; Kim, C.P.; Johnson, W.L. Microstructure controlled shear band pattern formation and enhanced plasticity of bulk metallic glasses containing in situ formed ductile phase dendrite dispersions. *Phys. Rev. Lett.* **2000**, *84*, 2901. [CrossRef] [PubMed]
20. Chen, L.Y.; Fu, Z.D.; Zhang, G.Q.; Hao, X.P. New class of plastic bulk metallic glass. *Phys. Rev. Lett.* **2008**, *100*, 075501. [CrossRef] [PubMed]
21. Cao, Q.P.; Li, J.F.; Zhou, Y.H.; Horsewell, A.; Jiang, J.Z. Free-volume evolution and its temperature dependence during rolling of Cu60Zr20Ti20 bulk metallic glass. *Appl. Phys. Lett.* **2005**, *87*, 101901. [CrossRef]
22. Zhang, Y.; Wang, W.H.; Greer, A.L. Making metallic glasses plastic by control of residual stress. *Nat. Mater.* **2006**, *5*, 857–860. [CrossRef] [PubMed]
23. Wang, L.; Bei, H.; Gao, Y.F.; Lu, Z.P.; Nieh, T.G. Effect of residual stresses on the hardness of bulk metallic glasses. *Acta Mater.* **2011**, *59*, 2858–2864. [CrossRef]
24. Wang, L.; Bei, H.; Gao, Y.F.; Lu, Z.P.; Nieh, T.G. Effect of residual stresses on the onset of yielding in a Zr-based metallic glass. *Acta Mater.* **2011**, *59*, 7627–7633. [CrossRef]
25. Chen, Y.H.; Huang, J.C.; Wang, L.; Nieh, T.G. Effect of residual stresses on nanoindentation creep behavior of Zr-based bulk metallic glasses. *Intermetallics* **2013**, *41*, 58–62. [CrossRef]
26. Haag, F.; Beitelschmidt, D.; Eckert, J.; Durst, K. Influences of residual stresses on the serrated flow in bulk metallic glass under elastostatic four-point bending—A nanoindentation and atomic force microscopy study. *Acta Mater.* **2014**, *70*, 188–197. [CrossRef]
27. Chen, L.Y.; Ge, Q.; Qu, S.; Jiang, J.Z. Stress-induced softening and hardening in a bulk metallic glass. *Scr. Mater.* **2008**, *59*, 1210–1213. [CrossRef]
28. Park, K.W.; Lee, C.M.; Wakeda, M.; Shibutani, Y.; Falk, M.L. Elastostatically induced structural disordering in amorphous alloys. *Acta Mater.* **2008**, *56*, 5440–5450. [CrossRef]
29. Jiang, Q.K.; Wang, X.D.; Nie, X.P.; Zhang, G.Q.; Ma, H.; Jiang, J.Z. Zr-(Cu, Ag)-Al bulk metallic glasses. *Acta Mater.* **2008**, *56*, 1785–1796. [CrossRef]
30. Suresh, S.; Giannakopoulos, A.E. A new method for estimating residual stresses by instrumented sharp indentation. *Acta Mater.* **1998**, *46*, 5755–5767. [CrossRef]
31. Ghidelli, M.; Sebastiani, M.; Collet, C.; Raphael, G. Determination of the elastic moduli and residual stresses of freestanding Au-TiW bilayer thin films by nanoindentation. *Mater. Des.* **2016**, *106*, 436–445. [CrossRef]
32. Johnson, K.L. *Contact Mechanics*; Cambridge University Press: Cambridge, UK, 1987.
33. Oliver, W.C.; Pharr, G.M. An improved technique for determining hardness and elastic modulus using load and displacement sensing indentation experiments. *J. Mater. Res.* **1992**, *7*, 1564–1583. [CrossRef]
34. Bei, H.; Lu, Z.P.; George, E.P. Theoretical strength and the onset of plasticity in bulk metallic glasses investigated by nanoindentation with a spherical indenter. *Phys. Rev. Lett.* **2004**, *93*, 125504. [CrossRef] [PubMed]

35. Choi, I.C.; Zhao, Y.; Kim, Y.L.; Yoo, B.G.; Suh, J.H. Indentation size effect and shear transformation zone size in a bulk metallic glass in two different structural states. *Acta Mater.* **2012**, *60*, 6862–6868. [CrossRef]
36. Choi, I.C.; Yoo, B.G.; Kim, Y.J.; Jang, J.I. Indentation creep revisited. *J. Mater. Res.* **2012**, *27*, 3–11. [CrossRef]
37. Ma, Y.; Ye, J.H.; Peng, G.J.; Zhang, T.H. Loading rate effect on the creep behavior of metallic glassy films and its correlation with the shear transformation zone. *Mater. Sci. Eng. A* **2015**, *622*, 76–81. [CrossRef]
38. Yoo, B.G.; Kim, K.S.; Oh, J.H.; Ramamurty, R.; Jang, J.I. Room temperature creep in amorphous alloys: Influence of initial strain and free volume. *Scr. Mater.* **2010**, *63*, 1205–1208. [CrossRef]
39. Ma, Y.; Ye, J.H.; Peng, G.J.; Zhang, T.H. Nanoindentation study of size effect on shear transformation zone size in a Ni–Nb metallic glass. *Mater. Sci. Eng. A* **2015**, *627*, 153–160. [CrossRef]
40. Ma, Y.; Peng, G.J.; Feng, Y.H.; Zhang, T.H. Nanoindentation investigation on the creep mechanism in metallic glassy films. *Mater. Sci. Eng. A* **2016**, *651*, 548–555. [CrossRef]
41. Wu, Y.; Li, H.X.; Chen, G.L.; Hui, X.D.; Wang, B.Y. Nonlinear tensile deformation behavior of small-sized metallic glasses. *Scr. Mater.* **2009**, *61*, 564–567. [CrossRef]
42. Pan, P.; Inoue, A.; Sakurai, T.; Chen, M.W. Experimental characterization of shear transformation zones for plastic flow of bulk metallic glasses. *Proc. Natl. Acad. Sci. USA* **2008**, *105*, 14769–14772. [CrossRef] [PubMed]
43. Ma, Y.; Peng, G.J.; Debela, T.T.; Zhang, T.H. Nanoindentation study on the characteristic of shear transformation zone volume in metallic glassy films. *Scr. Mater.* **2015**, *108*, 52–55. [CrossRef]

materials

MDPI

Article

Material Flow Analysis in Indentation by Two-Dimensional Digital Image Correlation and Finite Elements Method

Carolina Bermudo *, Lorenzo Sevilla and Germán Castillo López

Department of Civil, Material and Manufacturing Engineering. EII, University of Malaga, 29071 Malaga, Spain; lsevilla@uma.es (L.S.); gcastillo@uma.es (G.C.L.)
* Correspondence: bgamboa@uma.es; Tel.: +34-951-952-427

Received: 15 May 2017; Accepted: 16 June 2017; Published: 21 June 2017

Abstract: The present work shows the material flow analysis in indentation by the numerical two dimensional Finite Elements (FEM) method and the experimental two-dimensional Digital Image Correlation (DIC) method. To achieve deep indentation without cracking, a ductile material, 99% tin, is used. The results obtained from the DIC technique depend predominantly on the pattern conferred to the samples. Due to the absence of a natural pattern, black and white spray painting is used for greater contrast. The stress-strain curve of the material has been obtained and introduced in the Finite Element simulation code used, DEFORM™, allowing for accurate simulations. Two different 2D models have been used: a plain strain model to obtain the load curve and a plain stress model to evaluate the strain maps on the workpiece surface. The indentation displacement load curve has been compared between the FEM and the experimental results, showing a good correlation. Additionally, the strain maps obtained from the material surface with FEM and DIC are compared in order to validate the numerical model. The Von Mises strain results between both of them present a 10–20% difference. The results show that FEM is a good tool for simulating indentation processes, allowing for the evaluation of the maximum forces and deformations involved in the forming process. Additionally, the non-contact DIC technique shows its potential by measuring the superficial strain maps, validating the FEM results.

Keywords: incremental forming; indentation; Digital Image Correlation; Finite Elements Method; experimental methodology

---

1. Introduction

The indentation process is considered a secondary process due to produced deformations. These deformations are localized, small, and superficial [1,2]. Indentation is generally known as a hardening test.

Nowadays, new complex functional components are demanded with reduced weight and local strength, for example, for gear elements manufacturing. Localized forming operations are an interesting alternative to conventional machining processes [3]. As a manufacturing process, indentation is increasingly adapting to the metalworking industry. New flexible processes are arising and the indentation process implementation is being analysed under different innovative approaches, such as the Incremental Forming Processes (IFP) [4]. The IFP are considered an alternative to traditional plastic forming processes. The final shapes are gradually obtained using dies smaller than the workpieces. These processes modify the material thickness in specific areas, causing permanent plastic deformation and change the material properties as well with repetitive impressions.

Additionally, the incremental techniques can be found in micro forming. Micro-bulk forming produces high quality components with no material waste and is faster than traditional techniques [5]. The aim is to export these bases to general manufacturing.

IFP present important advantages over conventional processes, highlighting the flexibility and lower forces needed [6] that improve the processes. The main challenge of the Sheet-Bulk Metal Forming (SBMF) is the material flow prediction and control. For complex components, "trial and error" is still the most feasible technique [7–11].

Currently, the manufacturing industry requires more reliable and efficient analysis tools. To optimize manufacturing and design these components, the materials' behaviour laws and consistent simulation tools are needed. This is important for every industry, but for the manufacturing industry it is urgent due to the necessity of knowing the material behaviour within high deformation ranges and, occasionally, near to failure.

Tensile strength trials using strain gauges are the most usual tests. However, this method provides the average deformation values along the strain gauge length, leading to strong discrepancies between the maximum values obtained. Usually, strains and ultimate tensile strengths are larger than the ones obtained with this system [12].

Among the different techniques for the strain measurements, Digital Image Correlation (DIC) has several outstanding advantages [13]:

- It is a non-contact method that eliminates or reduces the interactions with the specimen. This technique can be applied with the samples at high or low temperatures or other extreme conditions without changes in the measurement technique.
- There is no need for complex tools or specimen preparation. Only a charge-coupled device (CCD) camera is needed and, in case the specimen does not count with an appropriate pattern, paint for the specimen surface to create the pattern.
- Natural or white light is used. There is no need of a laser source.
- It provides a wide measurement range and great sensitivity and resolution.

The Finite Elements Method (FEM) is a numerical method widely used for the simulation of materials, components, or structures under different forces. Detecting material deformation, plastic yielding, and damage is always of great importance. On the one hand, FEM and general numerical simulations are replacing more expensive and complex experiments. Its applications enable simulations of deformation processes with an extensive range of materials. On the other hand, this methodology is as powerful as the mathematical models behind them. Therefore, it is necessary to determine the particular material behaviour law [14]. The FEM software usually includes material databases that consider the material behaviour, deformation, hardening laws, and the strain velocity dependence. Nevertheless, sometimes those models are not accurate enough under extreme conditions, which leads to experimental validations to generate new material models. Additionally, extreme simulation conditions require remeshing techniques to avoid element distortion and numerical errors, and therefore need more computational time [13,15]. FEM is still an expensive approach due to the need of an accurate simulation model.

The goal of the present research is to perform a material flow analysis in an indentation process by FEM and DIC. A single indentation is analysed as the first stage of a SBMF process. In order to be more confident with the approaches implemented, the two methods have been compared. A 99% tin material is used to achieve higher penetrations and avoid crack formation. Regarding indentation, obtaining the full-field displacement during the process improves the understanding of the mechanisms involved in the deformation process. DIC application offers the opportunity to measure displacements and strains on the sample surface in real time, obtaining the whole field with a wide measurement range [16,17].

2. Materials and Methods

The study considers two different approaches: Experimental tests and numerical analysis.

Two different experimental tests have been performed: A compression test to obtain the stress-strain law that represents the material behaviour, and indentation tests.

The Digital Image Correlation (DIC) method has been used to measure the sample deformation during indentation tests.

The compression stress-strain curves have been implemented in the FEM model. Once the material behaviour has been experimental and numerically correlated, the indentation test has been modelled. The resulting indentation forces-displacement curves are compared to validate the numerical model with the experimental procedure. After the validation, the von Mises strain distribution obtained from the FEM and DIC methods can be compared.

2.1. Digital Image Correlation

The 2D DIC technique is based on the identification and comparison of a zone on the surface of the workpiece or specimen before and after deformation (Figure 1). The image is divided by virtual subsets. This area (subset) has a unique light intensity (grey level) that stays the same during the whole deformation process.

Successive comparisons are made in order to evaluate the subset displacement. Correlation algorithms, like the Sum of Squared Differences (SSD) (Equation (1)), locate the subset in the new image. Every single subset pixel is associated with a number according to its grey level (100 for white and 0 for black), as Figure 1 shows.

$$C(x,y,u,v) = \sum_{i,j=\frac{-n}{2}}^{\frac{n}{2}} (I(x+i,y+j) - \Gamma(x+u+i,y+v+j))^2 \tag{1}$$

where, $C(x, y, u, v)$: The value of the correlation function for a given pixel in the position (x, y) that undergoes a horizontal (u) and vertical (v) displacement (reference image); n: the subset size; $I(x + i, y + j)$: The value associated to the pixel in the position $(x + i, y + j)$ (reference image); $\Gamma(x + u + i, y + v + j)$: The value associated to the pixel in the position $(x + u + i, y + v + j)$ (deformed image).

The lowest C value offers the best correlation possible, giving the new position (x, y) of the pixel in the image after deformation, as well as the horizontal and vertical displacements (u, v) [18].

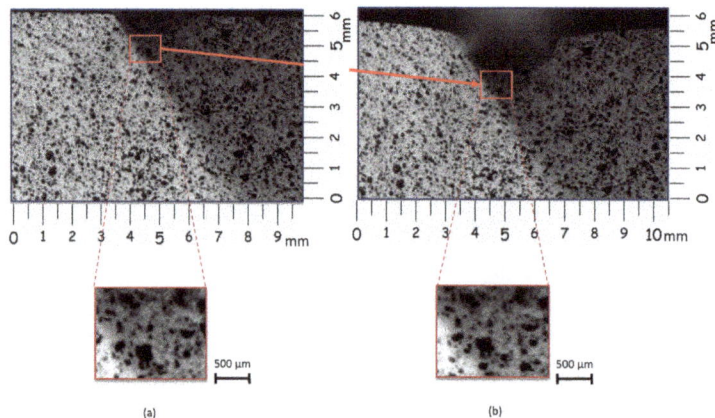

Figure 1. Subset before (**a**) and after (**b**) deformation.

However, it is necessary that the analysed specimen shows an adequate pattern. Commonly, this pattern consists of a random mottling or speckle that allows the recognition of the position of every single point before and after deformation. Frequently, the specimen under study presents a natural pattern but it is not uncommon to use other techniques, like spray paint or electrospray, to create these random patterns. Figures 1 and 2 show a pattern made with spray paint.

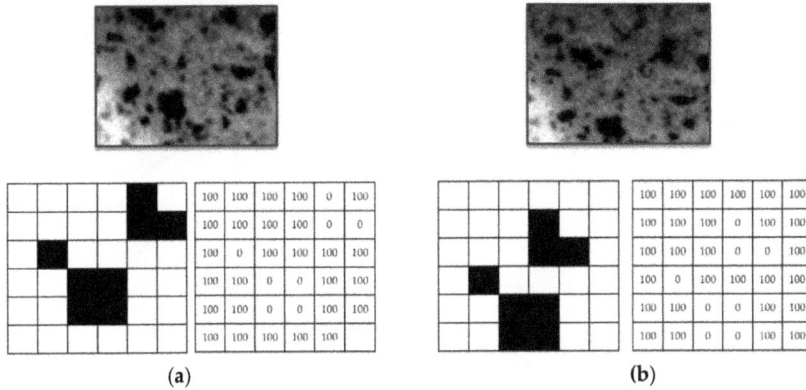

(a)   (b)

Figure 2. Subset and grey levels before (**a**) and after (**b**) deformation.

Once the displacement vectors of each pixel is obtained, it is possible to interpolate any point with the interpolation equations (Equations (2) and (3)):

$$u(x,\ y) = u_0 + u_x x + u_y y \tag{2}$$

$$v(x,\ y) = v_0 + v_x x + v_y y \tag{3}$$

where (x, y) are the desired point coordinates and (u, v) is the displacement.

Considering that between two consecutives frames the hypothesis of small strains is applicable, the Cauchy-Almansi tensor can be applied to obtain the strain field (Equation (4)):

$$\varepsilon = \begin{pmatrix} \varepsilon_{xx} & \varepsilon_{yx} \\ \varepsilon_{xy} & \varepsilon_{yy} \end{pmatrix} = \begin{pmatrix} u_{,x} & \frac{1}{2}\left(u_{,y}+v_{,x}\right) \\ \frac{1}{2}\left(u_{,y}+v_{,x}\right) & v_{,y} \end{pmatrix} \tag{4}$$

where, ε_{xx} and ε_{yy} are the longitudinal strains in the x and y directions, respectively; ε_{xy} is the angular strain; $u_{,x}, u_{,y}, v_{,x},$ and $v_{,y}$ are partial derivatives of the displacements (u, v).

2.2. Materials and Specimens

The material selected to accomplish the deep indentations is 99% tin (NB1101003), obtained from $6 \times 10 \times 70$ mm^3 bars with a 232–247 °C melting temperature (Table 1). A green sand casting process has been used to produce a $60 \times 60 \times 70$ mm^3 ingot. All the tests specimens have been machined from that tin ingot to avoid different behaviours due to cooling alterations inside the material.

Table 1. Tin composition in %.

Sn	Sb	Cu	Other
99.95	0.02	0.002	0.028

Two kinds of specimens are used for the compression and the indentation tests. The ASTM E9 standard has been followed for the compression tests. The dimensions of the specimens are $15 \times 15 \times 30$ mm^3. The length is established to allow for enough axial deformation without buckling. For the indentation tests, five 99% tin specimens of $40 \times 30 \times 30$ mm^3 have been prepared to study the material flow under a single indentation (Figure 3). In order to work in plane strain conditions, the specimen depth must be 6 to 10 times the deformation surface width [19].

In this case study, the punch is 3 mm wide, so the specimen depth is set to be 10 times bigger, 30 mm.

All workpieces need to be precisely painted to obtain an adequate pattern (Figure 4). It is important to spray the parts at a large distance to prevent the first thicker drops from falling on the surface. With finer patterns, more accuracy can be obtained. First, a white coat is sprayed at a 40–45 cm distance, generating a thin layer. Several layers can be sprayed if necessary. After the white coat dries, a black mottling is sprayed at 100 cm so that the large droplets fall before reaching the specimen (Figures 3b and 4c). Two hours later, the specimens can be tested. It is also important not to let the painting dry completely to prevent paint cracking during indentation.

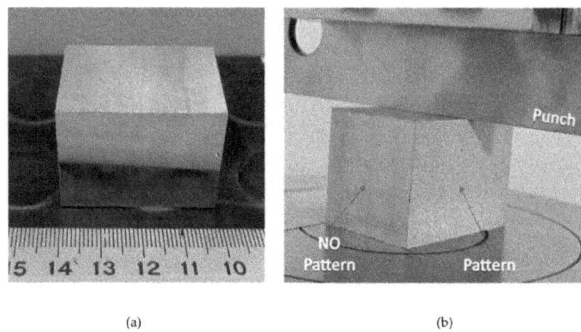

(a)   (b)

Figure 3. Specimen without pattern (**a**) and with pattern applied (**b**).

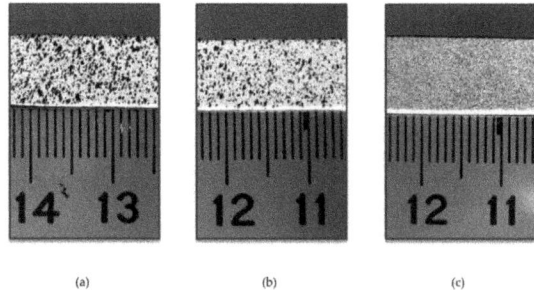

(a)   (b)   (c)

Figure 4. Evolution of the specimen pattern. First attempts (**a**,**b**) and final pattern (**c**).

2.3. Experimental Tests

The compression tests were carried out in a universal tension-compression machine Servosis ME 405, equipped with a 20 kN load cell. The test speed was set to 5 mm/min, equal to the indentation test's speed, to minimize the strain-rate influence. Due to the elasto-plastic material behaviour, it is not necessary to use the DIC technique to evaluate the samples' deformations.

For the indentation tests, a restraining tool was designed to complete the indentation process, avoiding punch inclination (Figure 5). Due to the narrow surface of the punch (3 mm) in contact with the specimen, an inclination of the punch was observed in the first samples tested. Therefore, the use of a restraining tool was necessary to prevent a non-symmetrical deformation. This tool was designed with a lateral compression force which stabilizes the punch, applying the compression with two fixed points on an in-between element. This in-between element homogeneously sets the punch and prevents its lateral displacement. Additionally, the restraining system does not affect the indentation results. A punch of steel AISI 304 is used.

(a)   (b)

(c)

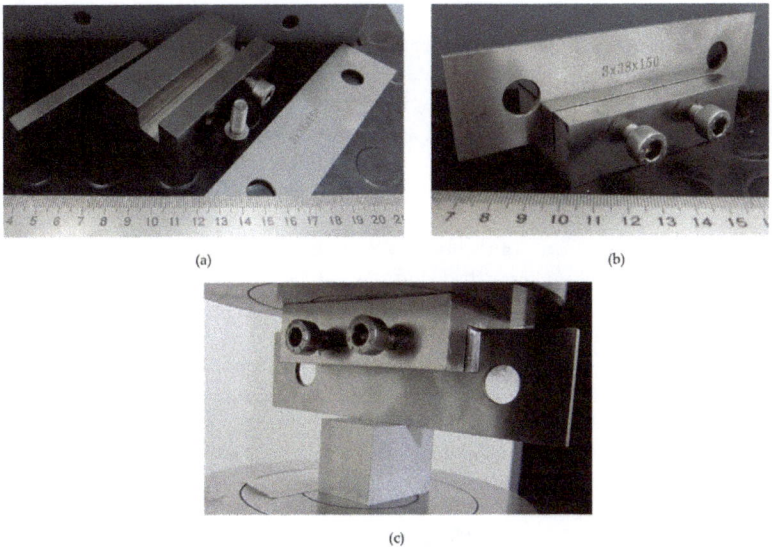

Figure 5. Restraining tool disassembled (**a**), assembled (**b**), and pre-load before test (**c**).

To obtain the image correlation, an Allied digital camera Stingray EEE 1394b of 5 megapixels (Sony, UK) was used, with a cell size of 3.45 μm × 3.45 μm. The camera was equipped with a Pentax C7528-M lens. This lens is specially designed for image processing applications. It is purposely designed to maximise the picture performance at short distances with a 75 mm focal length. With these characteristics, the pixel size is 30 μm.

For the illumination of the set, a Hedler spotlight DX 15 (metal 150 W Halide lamp, Hedler Systemlicht, Runkel, Germany) was used. The frame acquisition frequency is 2 Hz and the image acquisition is made with the software VIC SNAP [20] and VIC 2D [21] for treatment after the test is conducted. Figure 6 shows the disposition of the image acquisition system; seen in the foreground is the Data Acquisition system (DAQ) and the computer used to manage the DIC system. Placed in the background is the illumination and the camera through which the images are captured.

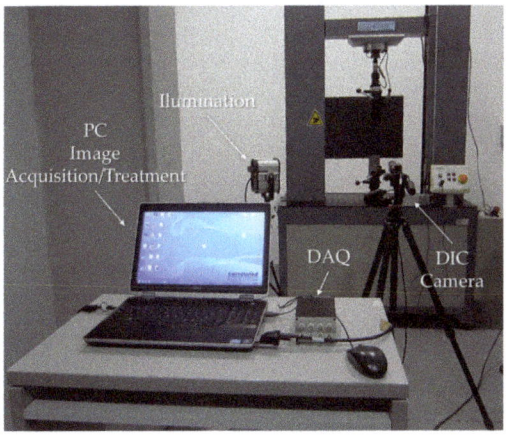

Figure 6. Image acquisition system.

Indentation tests were also carried out with the universal tension-compression machine Servosis ME 405 (Servosis Teaching Machines, Madrid, Spain), equipped with a 20 kN load cell. The speed of the indentation process (5 mm/min) is intended to be slow to capture more images and improve the precision of the displacements. With a high capture speed and low indentation speed, more images can be obtained. According to the ISO 6892-1:2010 standard [22], a minimum of five samples have been tested in order to achieve a minimum of a 95% confidence interval.

Materials like tin are strain rate sensitive [23]. For this reason, the behaviour law of the material selected has been obtained at the same speed (5 mm/min) at which the indentation tests are performed. The load forces and displacements of the tool are measured synchronously with the digital image acquisition. Thus, it is possible to know the load-time and displacement-time evolution.

For the DIC analysis, the steps are defined as the number of pixels between correlations. For the image analysis, the subsets and steps are stablished in 45 and 2 pixels, respectively, in order to achieve a confidence below 0.001 pixels. A step size of 2 means that a correlation will be carried out at every other pixel in both the horizontal and vertical directions. Note that the analysis time varies inversely with the square of the step size; i.e., a step size of one takes 25 times longer to analyse than a step size of 5. A low step number leads to a more accurate analysis, but increases the time of analysis. Steps can be on the order from 1 to 50. Figure 7 shows different captures using the DIC method.

| (a) | (b) | (c) |

Figure 7. DIC (Digital Image Correlation) analysis. Initial stage, the punch rest on top of the sample (**a**). Middle stage: The punch has penetrated 2–3 mm and the material nose is forming (**b**). Final stage: The indentation process is over and the material nose under the punch is fully created (**c**).

2.4. Numerical Simulation

The software DEFORM^{TM} 2D (version 8.1, Scientific Forming Technologies Corporation, Columbus, OH, USA) [24] was used for the FEM analysis. This software is specialized in forming process analysis. In order to ensure good results, defining a good mesh distribution is necessary. Near the punch the stress concentration is very high, so a finer mesh is necessary. In this analysis, two different zones have been defined (mesh density windows): One in the contact zone between the punch and the sample, and the other for the rest of the sample. These two zones can be seen in Figure 8.

Based on previous studies, where an indentation process was also simulated with a wider variety of materials [25,26], the optimal mesh is shown in Table 2. A 1/10 relation means that the elements of the second mesh window are 10 times the size of the elements of the first window. This relation guarantees an accurate resolution around the punch, where the main strains are taking place (first window). The rest of the workpiece (second window) is filled with greater elements, providing a shorter computational time resolution.

Table 2. Mesh applied.

Number of Elements	10,000
Number of mesh windows	2 (relation 1/10)
Remesh maximum step increment	2

To avoid the distortions of the big elements, a remesh has been established every 2 steps. Thus, the coarse size element zone becomes larger.

Two different 2D models have been developed; one in plain strain and the second one in plain stress. The two-dimensional plain strain model has been used to obtain the load-penetration curves, since most of the workpiece is near the plain strain conditions. Nevertheless, this model is not appropriate for evaluating strain maps on the surface. The material at the surface is closer to a plane-stress state. Therefore, to achieve this objective, a new 2D plain stress model has been implemented. In both cases the mesh, the type of elements (four node elements), and the boundary conditions are the same.

As boundary conditions, vertical displacements are fixed at the bottom of the sample (pink line, Figure 8). No friction has been considered for the restriction.

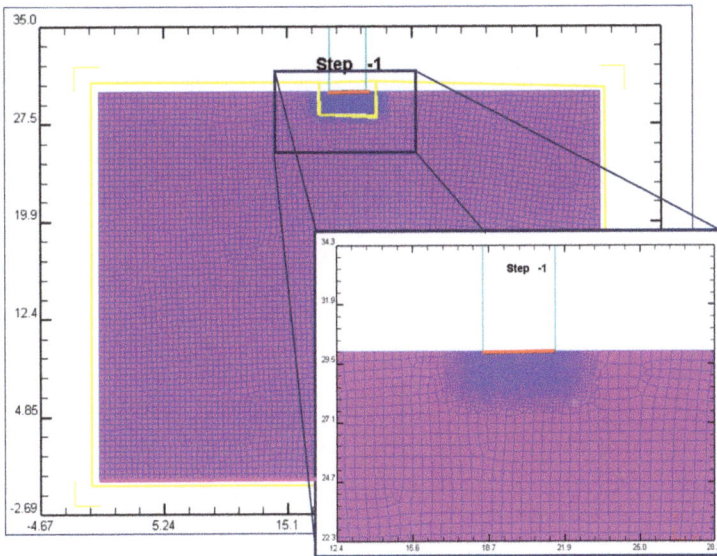

Figure 8. FEM indentation analysis.

On the other hand, a 0.12 shear type friction for cold forming has been defined during the indentation process (red line, Figure 8), between the workpiece and the punch surface. The FEM software used offers different friction values depending on the process simulated, with 0.12 being the appropriate value for cold forming [24]. Notwithstanding, in previous studies [25,26] is it established that the friction force can be neglected due to the total force that the punch needs to achieve to obtain the desired deformation.

An elasto-plastic isotropic hardening model, based on the von Misses criterion, has been used for the constitutive material model. The material data was previously obtained by the characterization tests and introduced manually in the software database, creating a new material entrance. To manufacture the samples, the tin was melt, cast, and machined. Therefore, the material that is characterized needs to be the same as the indented specimens. Five samples were obtained from these specimens and overcame compression tests in order to obtain the tin stress-strain curves to be implemented in FEM (Figure 9). The true stress and strain are obtained with Equations (5) and (6) [27]:

$$\varepsilon = \ln(1 + e) \tag{5}$$

$$\sigma = \sigma_e(1 + e) \tag{6}$$

where, e is the engineering strain; σ_e is the engineering yield tension; The Newton-Raphson method with 100 steps has been used as the iteration method. For the simulation, the general settings are shown in Table 3.

Table 3. General simulation settings.

Number of Simulation Steps	100
Step Increment to save	2
With equal die displacement	0.05

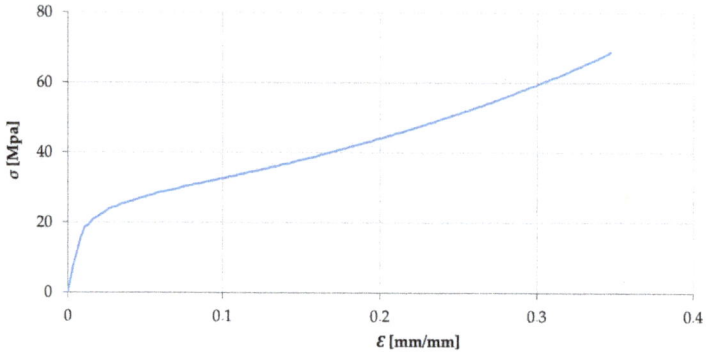

Figure 9. Tin true stress-strain curve obtained from a compression test.

3. Results and Discussion

The aim of the numerical and experimental test correlation is to validate the FEM model. Then, two approaches are used:

- The force-penetration curve along the indentation process.
- Strain maps around the indentation.

In the first approach, load-indentation curves are obtained, one measured directly by the testing machine and the other calculated by FEM. The simulation has been made in 2D plain strain. Thus, the numerical results must be modified to consider the real depth of the workpiece (30 mm).

This load-indentation curve represents the force needed by the punch to achieve the desirable penetration.

The results obtained by the numerical-experimental correlation (Figure 10) shows that the FEM results agree well with the results obtained from the experimental analysis for the cases studied, which confirms that the FEM analysis is carried out correctly and validates the implemented model. The small discontinuity presented by the curve that corresponds to the simulation is mainly due to the accumulation of errors along the different remesh cycles.

To determine whether the strain distributions around the indentation zone are also well determined by the numerical model, the DIC and FEM results are compared.

The software used for the DIC analysis, VIC-2D, offers the possibility of analyze the von Mises strains similarly to the FEM analysis code (DEFORM™). In this study, the von Mises strain has been used to represent a single value equivalent to the deformation state at each point. Figure 11 shows the image analysis with DIC for specimen 1, where the von Mises strains are displayedoverlapped. The von Mises strain is the variable through which the DIC-FEM comparison is carried out.

The area of interest (AOI) established by the 2D DIC software is close to the borders of the workpiece during the indentation process, but the information near the borders is not considered, as can be seen in Figure 11.

The gap between the punch and the workpiece that can be seen in Figure 11 is due to the elastic recovery of the material once the indentation process takes place. It can also be seen in the FEM model in Figure 10.

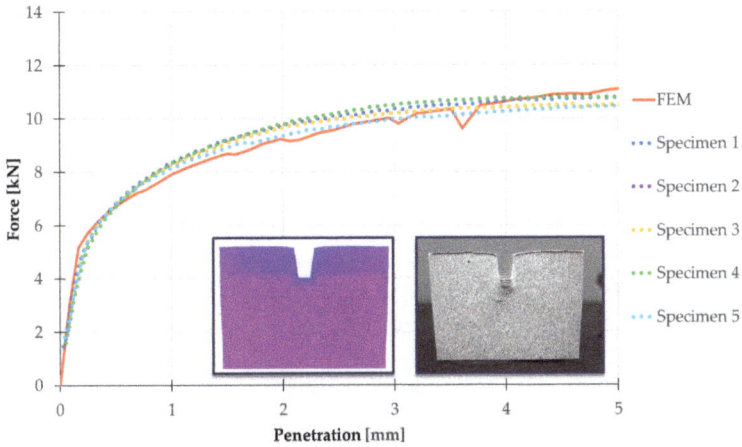

Figure 10. FEM and experimental analysis comparison.

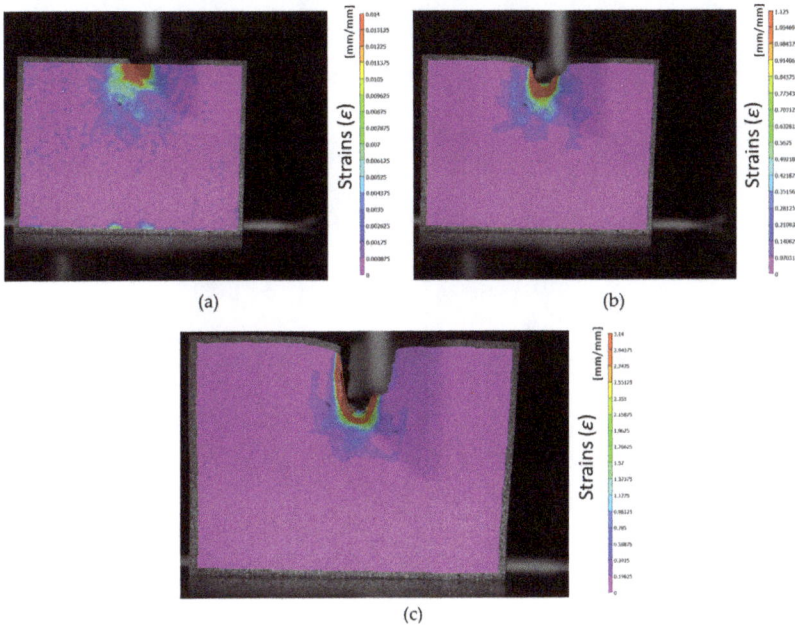

Figure 11. DIC von Mises strain (ε) analysis during the indentation process (sample 1). First stage (**a**), middle stage (**b**), and last stage (**c**).

Figure 12 shows the image correlation analysis for all of the specimens. It can be appreciated how the program filters/discards from the analysis a zone near the boundaries of the workpiece. This filter is meant to avoid the influence of the non-homogenous boundary values over the mean total values. However, the new boundary zones that are generated while the indentation is taking place (near the punch) are considered in the image analysis. It can be appreciated how the analysis can vary according to the quality of the pattern. The results from specimen 1 (Figure 12a), 4 (Figure 12d), and 5 (Figure 12e) are in good concordance, being results from specimen 4 the ones that show a better correlation . The more appropriate pattern is the one that offers a lower confidence value (c), although all of the VIC analyses are below 0.01. VIC 2D calculates the statistical confidence regions using the covariance matrix of the correlation equation. Figure 12c is rejected due to its confidence interval, with Figure 12d being the more representative in this case ($c = 0.005$). The results from specimen 2 (Figure 12b) and 3 (Figure 12c) show defects for deep indentation caused by an overly distorted pattern. This could be due to a coarse pattern. These correlation defects are only present in the latest frames, being the analysis before in good condition. For a single specimen, the total number of frames is 190. Therefore, only part of the analysis is rejected.

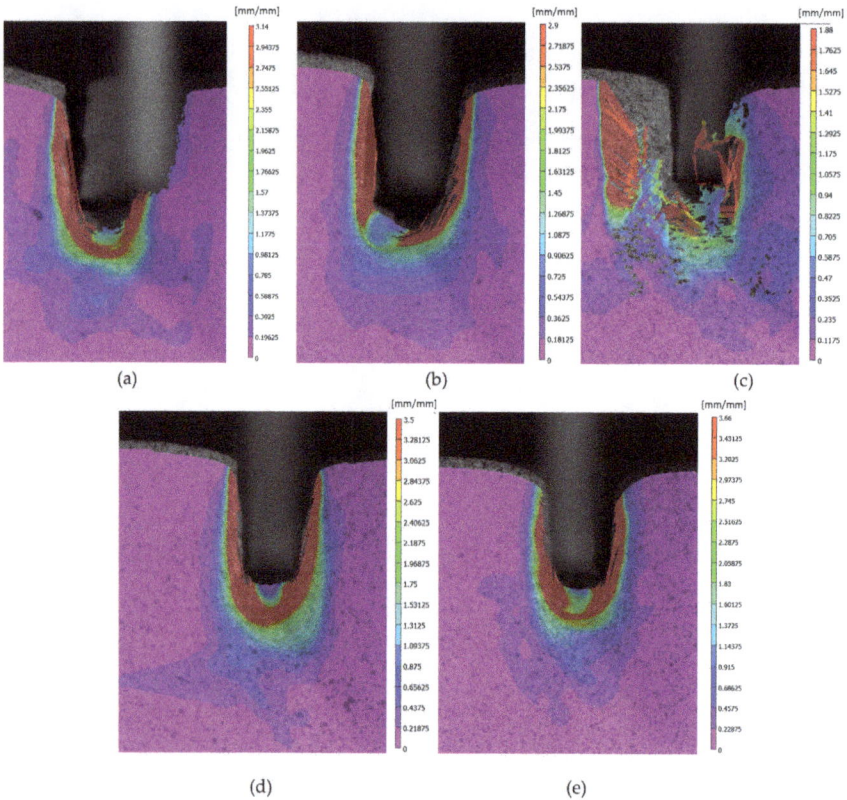

Figure 12. Von Mises strains DIC analysis for specimens 1 (**a**), 2 (**b**), 3 (**c**), 4 (**d**) and 5 (**e**).

Figure 13 shows a comparison between the FEM two dimensional plain stress model and the DIC model. It can be observed that the DIC resolution is smaller than the numerical results. This can be explained by the pattern used. For a better approximation, the necessary pattern needs to be finer (Figure 4) and focus only on the area of interest, near the punch, taking advantage of the camera resolution. Although the indentation process presents a protuberance that grows on the front surface while the punch penetrates, that area has not been taken into account for the analysis because it corresponds to the dead material area generated under the punch (Figure 14).

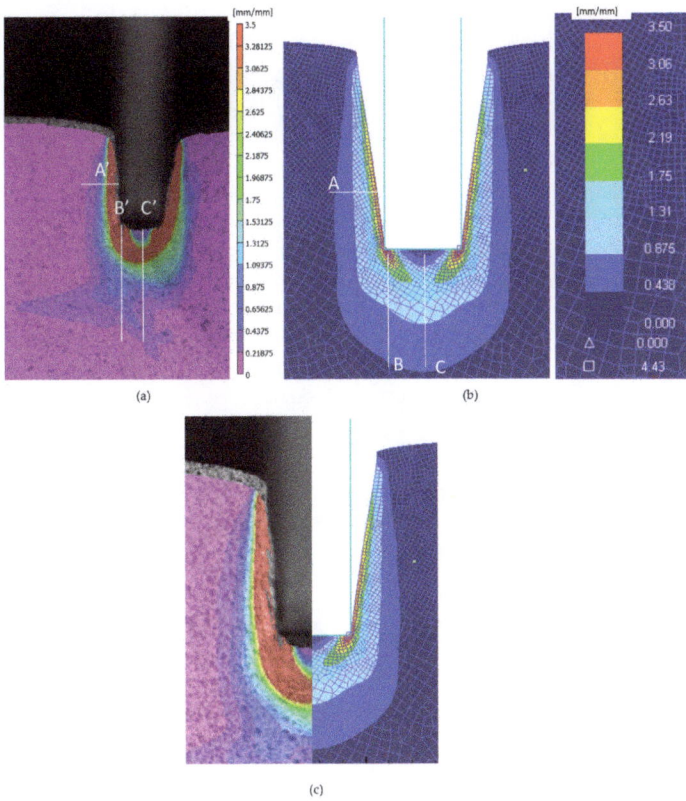

Figure 13. DIC-FEM (2D-Plain Stress) von Mises strain comparison. (**a**) Dic results, (**b**) FEM results and (**c**) comparison.

Figure 14. Detail on the sample after the indentation process.

The FEM study shows a strain concentration at the lower corners. This concentration should not be considered since it is influenced by the meshing. At the point of contact, the vertex would be infinite. In addition, a direct comparison with DIC cannot be made due to a low resolution. The low resolution is caused by the pattern and the resolution of the camera itself [28,29]. Though a direct comparison is not possible, another comparative analysis can be carried out like the one presented in this work.

Regarding the vertical displacements, the FEM model is stiffer than the workpiece. There is a vertical difference of the top of the samples when the DIC and FEM final images are compared. This difference could be explained by the fact that the 2D model is not accurate enough to reproduce the 3D part behaviour.

The comparison between both methods, in order to validate the FEM model developed, must be centred in three different regions, represented with white lines on Figure 13a,b (regions A, B, and C for FEM and A', B', and C' for DIC). Following these lines, three points are selected: Both extremes of the line (zone 1 and 3) and the middle point (zone 2). Figure 15 shows the comparison for the average strain values in those cases, presenting a good correlation between the results. For the area of the vertex of the punch, the difference is 7–11%. For the flank, there is a difference between 10% and 20%. Finally, for the middle line of the base, there is a higher disparity of 45%. This disparity on the base of the punch can be due to the dead material zone interference.

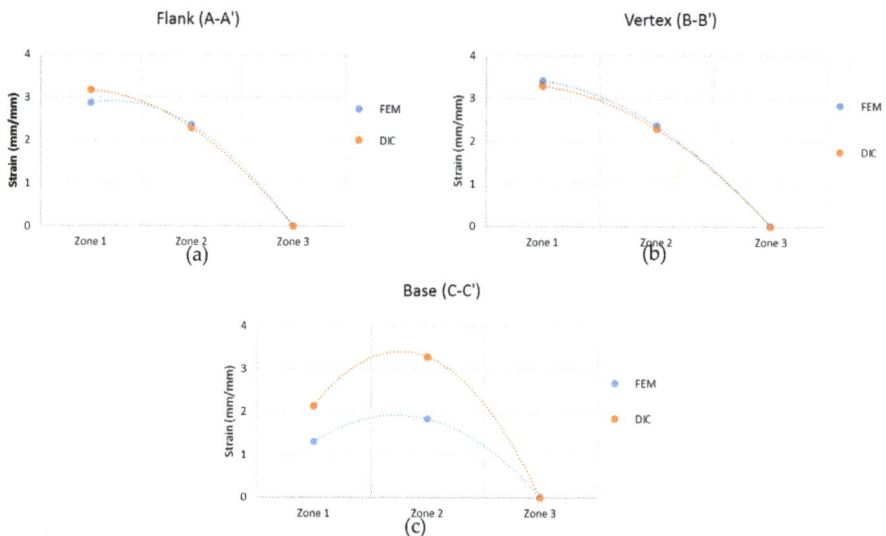

Figure 15. Von Mises strain values comparison between FEM and DIC focusing on the flank (**a**), the vertex (**b**), and the centre of the base (**c**).

Both methods present the dead nose zone (Figures 13 and 14). This zone is material that moves solidary with the punch and do not flow. This dead nose zone is produced in front of the punch in deep indentations, whether there is lubrication or not, and can affect the results obtained in that specific zone with the experimental method.

4. Conclusions

The main purpose of this work is to show that the developed FEM model allows for a correct simulation of the indentation process and can be used in future designs to evaluate the maximum forces needed in these kinds of processes and the final shape achieved before the procedure takes place.

To validate the FEM model, the DIC method was used as an experimental technique.

The material flow is analysed using FEM and DIC, showing the adequacy of each method. To guarantee a deep indentation avoiding cracking, tin (99% tin) workpieces were manufactured. Tin has characteristics of near to rigid-perfectly plastic behaviour, but it is also very sensitive to the strain rate. All the experimental tests were carried out at 5 mm/min, maintaining the same displacement rate and allowing for a wide frame capture for DIC.

A two-dimensional indentation setup was designed in FEM and DIC to study the deformation process. FEM and DIC analyses show an adequate correlation with the experimental results. The main findings that can be highlighted are:

- The stress-strain curve of the material tested, 99% tin, is experimentally obtained and satisfactorily introduced in the FEM code.
- The FEM punch force-displacement curve fits well with the experimental results.
- The non-contact technique, DIC, is an efficient method for the identification of the flow field and von Mises strains.
- The DIC and FEM Von Mises strain results show an adequate correlation taking into account the simplicity of the 2D models, presenting values with a difference between 10% to 20% (and 50% in the dead zone). It can be observed that the DIC results are always contained between the FEM results.

Notwithstanding, if precise FEM strain maps are to be obtained, a 3D model is needed. The real behaviour of the workpiece under the punch, and near the surface, cannot be studied as a two-dimensional problem. Indeed, in this zone it is necessary to take into account the three dimensional effects.

The disadvantage of FEM lies in the need to obtain the stress/strain curves of the materials being analysed. In study cases of materials like tin, that are highly influenced by strain rates, the stress/strain curve for each speed considered is needed. This is not necessary when working with materials like aluminium, as presented in previous works. This analysis confirms that DIC can provide full strain fields and can be used to examine manufacturing processes like indentation. Also, after the necessary image treatment, DIC offers the possibility to obtain different types of information such as damage, stress distribution, or the stress-strain curve.

Acknowledgments: The authors want to thank the University of Malaga—Andalucía Tech, International Campus of Excellence, for its financial support of this paper.

Author Contributions: Carolina Bermudo, Lorenzo Sevilla, and Germán Castillo López conceived and designed the numerical model and the experiments; Carolina Bermudo performed the experiments; Carolina Bermudo, Lorenzo Sevilla, and Germán Castillo López analyzed the data; Carolina Bermudo wrote the paper.

Conflicts of Interest: The authors declare no conflict of interest.

Nomenclature

IFP	Incremental Forming Processes.
SBMF	Sheet-Bulk Metal Forming.
DIC	Digital Image Correlation.
FEM	Finite Element Method.
DAQ	Data Acquisition system.
$C(x, y, u, v)$	Value of the correlation function for a given pixel in the position (x, y) that undergoes a horizontal (u) and vertical (v) displacement (reference image).
n	Subset size.
$I(x + i, y + j)$	Value associated to the pixel in the position $(x + i, y + j)$ (reference image).
$\Gamma(x + u + i, y + v + j)$	Value associated to the pixel in the position $(x + u + i, y + v + j)$ (deformed image).
(x, y)	Desired point coordinates.
(u, v)	Displacement of the desired point.
ε_{xx}	Longitudinal strains in x directions.
ε_{yy}	Longitudinal strains in y directions.
ε_{xy}	Angular strain.
u_x, u_y, v_x and v_y	Partial derivatives of the displacements (u, v).

References

1. Bernal, C.; Camacho, A.M.; Arenas, J.M.; Rubio, E.M. Analytical procedure for geometrical evaluation of flat surfaces formed by Multiple Indentation Processes. *Appl. Mech. Mater.* **2012**, *217*, 2351–2356. [CrossRef]
2. Jian, S.-R.; Chen, G.-J.; Hsu, W.-M. Mechanical properties of Cu_2O thin films by nanoindentation. *Materials* **2013**, *6*, 4505–4513. [CrossRef]
3. Kretzschmar, J.; Stockmann, M.; Ihlemann, J.; Schiller, S.; Hellfritzsch, U. Experimental-numerical investigation of the rolling process of high gears. *Exp. Tech.* **2015**, *39*, 28–36. [CrossRef]
4. Khalatbari, H.; Iqbal, A.; Shi, X.; Gao, L.; Hussain, G.; Hashemipour, M. High-speed incremental forming process: A trade-off between formability and time efficiency. *Mater. Manuf. Process.* **2015**, *30*, 1354–1363. [CrossRef]
5. Qin, Y.; Arentoft, M.; Eriksen, R.S.; Hansen, H.N. Chapter 12–Micro-bulk Forming. In *Micromanufacturing Engineering and Technology*; Elsevier: Amsterdam, The Netherlands, 2015; pp. 277–297.
6. Groche, P.; Fritsche, D.; Tekkaya, E.A.; Allwood, J.M.; Hirt, G.; Neugebauer, R. Incremental bulk metal forming. *CIRP Ann. Manuf. Technol.* **2007**, *56*, 635–656. [CrossRef]
7. Sieczkarek, P.; Wernicke, S.; Gies, S.; Tekkaya, A.E.; Krebs, E.; Wiederkehr, P.; Biermann, D.; Tillmann, W.; Stangier, D. Wear behavior of tribologically optimized tool surfaces for incremental forming processes. *Tribol. Int.* **2016**, *104*, 64–72. [CrossRef]
8. Xu, R.; Shi, X.; Xu, D.; Malhotra, R.; Cao, J. A preliminary study on the fatigue behavior of sheet metal parts formed with accumulative-double-sided incremental forming. *Manuf. Lett.* **2014**, *2*, 8–11. [CrossRef]
9. Wang, J.; Nair, M.; Zhang, Y. An efficient force prediction strategy in single point incremental sheet forming. *Procedia Manuf.* **2016**, *5*, 761–771. [CrossRef]
10. Uheida, E.H.; Oosthuizen, G.A.; Dimitrov, D. Investigating the impact of tool velocity on the process conditions in incremental forming of titanium sheets. *Procedia Manuf.* **2016**, *7*, 345–350. [CrossRef]
11. McAnulty, T.; Jeswiet, J.; Doolan, M. Formability in single point incremental forming: A comparative analysis of the state of the art. *CIRP J. Manuf. Sci. Technol.* **2017**, *16*, 43–54. [CrossRef]
12. Yang, L.; Smith, L.; Gotherkar, A.; Chen, X. *Measure Strain Distribution Using Digital Image Correlation (DIC) for Tensile Tests*; Oakland University: Rochester, MI, USA, 2010.
13. Pan, B.; Qian, K.; Xie, H.; Asundi, A. Two-dimensional digital image correlation for in-plane displacement and strain measurement: A review. *Meas. Sci. Technol.* **2009**, *20*, 62001. [CrossRef]
14. Skozrit, I.; Frančeski, J.; Tonković, Z.; Surjak, M.; Krstulović-Opara, L.; Vesenjak, M.; Kodvanj, J.; Gunjević, B.; Lončarić, D. Validation of numerical model by means of digital image correlation and thermography. *Procedia Eng.* **2015**, *101*, 450–458. [CrossRef]
15. Kim, H.S.; Kim, S.H.; Ryu, W.-S. Finite element analysis of the onset of necking and the post-necking behaviour during uniaxial tensile testing. *Mater. Trans.* **2005**, *46*, 2159–2163. [CrossRef]
16. Zhang, H.; Huang, G.; Song, H.; Kang, Y. Experimental investigation of deformation and failure mechanisms in rock under indentation by digital image correlation. *Eng. Fract. Mech.* **2012**, *96*, 667–675. [CrossRef]
17. Ho, C.-C.; Chang, Y.-J.; Hsu, J.-C.; Kuo, C.-L.; Kuo, S.-K.; Lee, G.-H. Residual strain measurement using wire EDM and DIC in aluminum. *Inventions* **2016**, *1*, 4. [CrossRef]
18. Schreier, H.; Orteu, J.-J.; Sutton, M.A. *Image Correlation for Shape, Motion and Deformation Measurements*; Springer: Boston, MA, USA, 2009.
19. Rowe, G.W. *An Introduction to the Principles of Metalworking*; Edward Arnold Limited: London, UK, 1971.
20. Correlated Solutions VIC SNAP 2009. Available online: www.correlatedsolutions.com/installs/Vic-2D-2009-Manual.pdf (accessed on 20 June 2017).
21. Correlated Solutions Vic-2D 2009, 59. Available online: correlatedsolutions.com/vic-2d/ (accessed on 20 June 2017).
22. *Metallic Materials—Tensile Testing. Part 1 Method Test Room Temp*; International Organization for Standardization: Geneva, Switzerland, 2010; ISO 6892-12009.
23. Bermudo, C.; Martín, F.; Sevilla, L. Validación experimental del Modelo Modular en la Aplicación del Teorema del Límite Superior a Procesos de indentación. *An. Ing. Mec.* **2014**, *19*, 103.
24. Fluhrer, J. *Deform. Design Environment for Forging. User's Manual*; Scientific Forminf Technologies Corporation: Columbus, OH, USA, 2010.

25. Bermudo, C.; Sevilla, L.; Martín, F.; Trujillo, F.J. Study of the tool geometry influence in indentation for the analysis and validation of the new modular upper bound technique. *Appl. Sci.* **2016**, *6*, 203. [CrossRef]
26. Bermudo, G.C. *Análisis, Desarrollo y Validación del Método del Límite Superior en Procesos de Conformado por Indentación*; Servicio de Publicaciones y Divulgación Científica: Malaga, Spain, 2015.
27. Roylance, D. Stress-Strain Curves. *Mass. Inst. Technol. Study* **2001**. Available online: https://www.saylor.org/site/wp-content/uploads/2012/09/ME1022.2.4.pdf (accessed on 20 June 2017).
28. Van Mieghem, B.; Ivens, J.; Van Bael, A. Benchmarking of depth of field for large out-of-plane deformations with single camera digital image correlation. *Opt. Lasers Eng.* **2017**, *91*, 134–143. [CrossRef]
29. Park, J.; Yoon, S.; Kwon, T.-H.; Park, K. Assessment of speckle-pattern quality in digital image correlation based on gray intensity and speckle morphology. *Opt. Lasers Eng.* **2017**, *91*, 62–72. [CrossRef]

materials

MDPI

Article

Hardening Effect Analysis by Modular Upper Bound and Finite Element Methods in Indentation of Aluminum, Steel, Titanium and Superalloys

Carolina Bermudo *, Lorenzo Sevilla, Francisco Martín and Francisco Javier Trujillo

Civil, Material and Manufacturing Engineering Department, ETSII-EPS, University of Malaga, Málaga 29071, Spain; lsevilla@uma.es (L.S.); fdmartin@uma.es (F.M.); trujillov@uma.es (F.J.T.)
* Correspondence: bgamboa@uma.es; Tel.: +34-951-952-427

Academic Editor: Ting Tsui
Received: 29 March 2017; Accepted: 17 May 2017; Published: 19 May 2017

Abstract: The application of incremental processes in the manufacturing industry is having a great development in recent years. The first stage of an Incremental Forming Process can be defined as an indentation. Because of this, the indentation process is starting to be widely studied, not only as a hardening test but also as a forming process. Thus, in this work, an analysis of the indentation process under the new Modular Upper Bound perspective has been performed. The modular implementation has several advantages, including the possibility of the introduction of different parameters to extend the study, such as the friction effect, the temperature or the hardening effect studied in this paper. The main objective of the present work is to analyze the three hardening models developed depending on the material characteristics. In order to support the validation of the hardening models, finite element analyses of diverse materials under an indentation are carried out. Results obtained from the Modular Upper Bound are in concordance with the results obtained from the numerical analyses. In addition, the numerical and analytical methods are in concordance with the results previously obtained in the experimental indentation of annealed aluminum A92030. Due to the introduction of the hardening factor, the new modular distribution is a suitable option for the analysis of indentation process.

Keywords: incremental forming; indentation; FEM; MUBT; plastic deformation; hardening effect

---

1. Introduction

Indentation is generally applied in hardening tests to characterize materials [1,2]. In this study, the indentation process is studied as an incremental process like the Single Point Incremental Forming [3], the Multiple Indentation Processes [4] or the Localized-Incremental Forming Process [5], which are now being introduced in the current industry. Previous work presented the application of the new Modular consideration for the Upper Bound Theorem (MUBT) to indentation [6–9] and validated the new model with experimental tests. In this paper, the abbreviations MUBT, as opposed to Upper Bound Theorem (UBT), will be used to refer to the modular application of the method. One of the main advantages of MUBT is that the modular configuration makes possible the introduction of the parameters that are present in forming processes without overcomplicating the analysis. This fact allows an enrichment of the study, offering a closer approximation to reality. Therefore, the introduction and study of the hardening effect is considered necessary in order to get a more accurate model and approach to the current industrial processes.

Leaning on the modularity described, several Hardening Models (HMs) are established depending on the material behavior, expanding the application of MUBT. This work aims to offer a complete understanding of the HMs developed, showing their implementation in the analysis of the indentation processes for different materials. Furthermore, the approach tries to improve previous analysis, enlarging the study and adjusting the method. The modular model is presented as an analytical tool that reduces the time and cost that generally is consumed in the workpiece analysis, knowing that these workpiece analyses are usually necessary prior the final implementation of the procedures.

The UBT, under its different variants, is an analytical approach satisfactorily suitable for obtaining the necessary force to achieve plastic deformation. Some examples can be found in the studies of Moncada et al. [10], where a special case of ring compression test with non-symmetrical neutral plane of material flow is analyzed under the UBT perspective. In addition, Yunjian et al. [11,12] present an upper bound solution of axial metal flow for rods and later, an upper bound model for strain inhomogeneity analysis in radial forging processes. Alforzan and Gunasekera [13] use the UBT as an elemental technique to design axisymmetric forging by forward and backward simulation. On the study of the indentation process, the Triangular Rigid Zone (TRZ) alternative is the kinematic-geometrical option that allows a more accurate solution, with a greater capacity of analysis, as shown in the work of Kudo [14] and, recently, proved again by Topcu [15].

Focusing on its modular application, the optimal modular model contemplates the material flow that exists under the punch and near it (Figure 1). This area of the material suffers more from the stresses and strains that occur during the forming process. The model implemented consists of 3 modules with 2 Triangular Rigid Zone (TRZ) each. The modular concept gives a better approximation to the process, allowing the inclusion of more modules if necessary. In this case, after the study of the optimal number of modules, it was demonstrated that a 3-module model offers lower forces. The study was performed taking into account different forming parameters as friction or hardening [16]. Figure 1 shows the module distribution, where L is half the width of the punch, m is the friction coefficient, b is the base of module A and B, b' is the base of module C, H_T is the height of the quarter of the workpiece analyzed, h is the height of the modules and V is the punch speed. A double symmetry is imposed to simplify the analysis. The double symmetry makes possible to focus only on a quarter of the workpiece. In addition, the proportion established between the punch and the workpiece is considered as an infinite analysis. Thus, the symmetrical punch under the workpiece can be ignored. The forces obtained for the punch above the workpiece are the same as the forces for the symmetrical punch. The analysis becomes equivalent to forming process with only one indenter like the Single Point Incremental Process or the Multiple Indentation Forming Process.

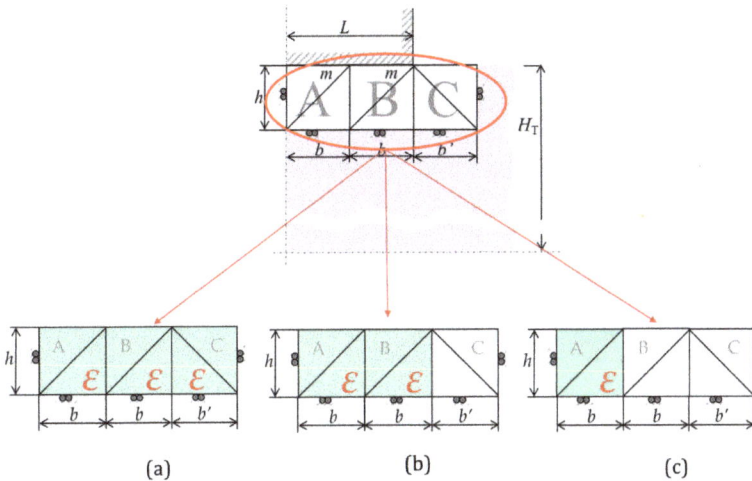

Figure 1. Optimal MUBT model (three modules with two triangular rigid zones each) and hardening effect (ε) distribution for each Hardening Model (HM): (**a**) HM1; (**b**) HM2; and (**c**) HM3.

Thanks to the modular configuration, each module can be individually analyzed, obtaining an easier $p/2k$ relations [17], being p the effort required to deform the work-piece and k the shear yield stress. After the hodograph for module A is completed (Figure 2), the $(p/2k)_A$ relation is shown in Equation (1), obtained from the UBT general expression [10].

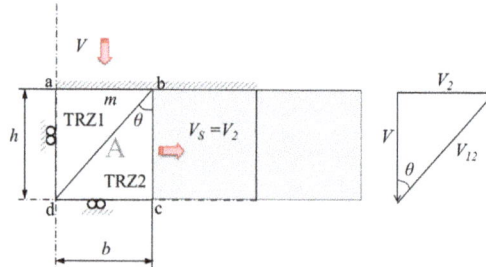

Figure 2. Module A analysis and velocity hodograph.

$$p \times b \times V \times w = k \times w \times [V_{12} \times bd + V_2 \times dc],$$ (1)

Following the diagram in Figure 2, Equation (2) is obtained:

$$\left(\frac{p}{2k}\right)_A = \frac{1}{2 \times b} \times \left[\frac{h^2 + 2b^2}{h}\right],$$ (2)

Accordingly, the same analysis is made for the other two modules. Finally, a weighted average is applied to obtain the final $p/2k$ values:

$$\left(\frac{p}{2k}\right)_T = \frac{\left(\frac{p}{2k}\right)_A \times b + \left(\frac{p}{2k}\right)_B \times b + \left(\frac{p}{2k}\right)_C \times b'}{2 \times b + b'},$$ (3)

The simplified Ludwik equation (Equation (4)) is used for the introduction of the hardening effect in the specific indentation case study, considering always plane strain conditions and cold-worked metals.

$$\sigma = Y + K \times \epsilon^n,$$ (4)

where Y is the yield strength, σ is the stress analyzed, K is the strength coefficient that depends on the material structure, and n is the strain hardening exponent, also specific for each material.

However, Equation (1) is expressed in terms of the yield stress (Y) and the true strain (ϵ). Working with the modular model, a transformation must be performed in order to adapt its application. MUBT studies deformation instants. Therefore, an approximation to the engineering strain (e) is necessary to be able to work with the original length and area. True and engineering strains are calculated as follows:

$$\epsilon = ln(1+e),$$ (5)

where, assuming the Von Mises yield criteria:

$$2k = 1.155Y,$$ (6)

$$k = 0.577Y,$$ (7)

Thus, the equation applied in the study of the hardening effect implementation in MUBT, regarding an indentation process, is:

$$k = k_0 + 0.577 \times K \times \epsilon^n,$$ (8)

with k being the yield stress in shear at that analyzed instant and k_0 the previous yield stress in shear. To validate the MUBT application considering different materials, a numerical study is performed using the Finite Element Method (FEM). This method allows the introduction of a considerable number of parameters for the proper study of plastic deformation, being the hardening effect among them. Thus, an indentation model is implemented to simulate the process over different materials and alloys. Previous analysis with aluminum A92030 [8] showed that MUBT was able to offer results close to reality. Present work implements a numerical simulation to consider a more varied range of materials and validate the HMs developed.

Thus, this study focus on the analysis and application of the HMs developed for indentation, verifying its application through a numerical analysis. With the numerical analysis, a wide range of materials is analyzed, supporting the validation of the model developed and expanding its application.

2. Materials and Methods

2.1. Finite Element Analysis

To validate the MUBT application for an extensive number of materials, a series of simulations is performed with DEFORM 2D (version 8.1, Scientific Forming Technologies Corporation, Columbus, OH, USA) [18]. This FEM software allows the implementation of different materials from its own material database. In this case study, the materials and alloys considered are aluminum, steel, titanium, and superalloys.

Steel, aluminum and their alloys have been widely introduced in the industry due to their applications. Within the aluminum and steel group, the materials chosen from the database for the FEM study are presented in Table 1. Materials are named under the Unified Numbering System (UNS) codification.

Table 1. Aluminum and steel alloys characteristics. Unified Numbering System codification (UNS).

Aluminum				Steel			
Code	Y (MPa)	K (N/mm^2)	n	Code	Y (MPa)	K (N/mm^2)	n
A92024	270	366.44	0.2	G10450	640	881.83	0.10
A95052	140	192.48	0.09	G10080	280	577.67	0.17
A96062	138	198.54	0.10	S30400	510	1073.84	0.19
A93003	120	199.54	0.12	S30200	250	1055.74	0.42
A96082	200	355.29	0.11	S32100	501,80	1084.64	0.28
A91070	68	130.45	0.21	-	-	-	-

Table 1 only shows the materials selection obtained directly from the DEFORM 2D data base. Materials that are not offered within this data base can be incorporated with the manual introduction of their properties values, like the simulation carried out with the annealed aluminum A92030 used in the experimental tests.

Likewise, working with titanium and superalloys is interesting. These materials usually have excellent mechanical strength and good resistance to creep at high temperatures. They also exhibit good resistance to corrosion and oxidation [19,20]. Within the titanium and superalloy group, the materials chosen are presented in Table 2. Even though the indentation processing of superalloys is unusual, the analysis is conducted as an extension of the study, showing the versatility of the developed model and opening the field to their consideration in further incremental processes.

Table 2. Titanium and superalloys characteristics. UNS codification.

	Superalloys				Titanium		
Material	Y (MPa)	K (N/mm^2)	n	Material	Y (MPa)	K (N/mm^2)	n
N06600	434.37	1101.03	0.20	R58010	1050	1758.12	0.17
G52986	762.62	1124.39	0.12	R50250	510	951.46	0.23
N02211	337.84	896.05	0.21	R50400	850	1398.28	0.20
G33106	660	888.91	0.008	R53400	1192,79	1315.54	0.02
-	-	-	-	R50250	510	951.46	0.23

Due to its importance and presence in the industry, the study of the indentation process for the materials in these four groups was established. All the materials were simulated with FEM and the results obtained were compared with the modular model developed. The correlation of both methods is shown in the next section.

For the boundary conditions set in the FEM simulation, an infinite case study has been considered. Although the finite and infinite cases have been covered in previous studies [21], the finite consideration is far from the indentation analyzed in this work, resembling processes such as stamping or shearing. Therefore, in this paper, only the infinite consideration will be taken into account. Figure 3 shows the 3D model implemented with FEM.

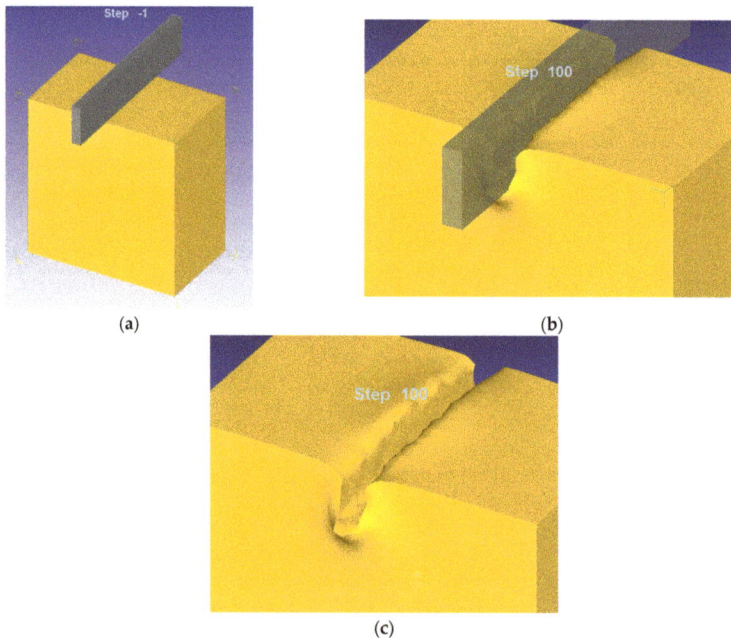

(a)

(b)

(c)

Figure 3. A 3D FEM model of the 50 × 50 × 30 mm sample and indenter: before indenting (a); during indentation process (b); and workpiece after indentation (c).

In previous studies [6], the geometry of the punch was studied under the same FEM software (version 8.1, Scientific Forming Technologies Corporation, Columbus, OH, USA). A mesh analysis was carried out in order to found the optimal mesh that will offer accurate solutions. Working with numerical methods, it is important to implement a mesh that gives solutions close to reality without increasing too much the resolution time. Accurate solutions in reasonable time can be achieved implementing mesh windows. The mesh widows allow an analysis with a high density mesh in the zone near the punch, were the deformation is taking place, leaving the rest of the workpiece with a coarse mesh. The mesh window (6 × 4 mm approximately) is located in the area where the punch performs and establishing and adequate relative element size of 1/20. That is to say that the elements placed inside the coarse mesh are 20 times larger (Figure 4). In addition, to avoid the distortions of big elements, a remesh has been established every three steps.

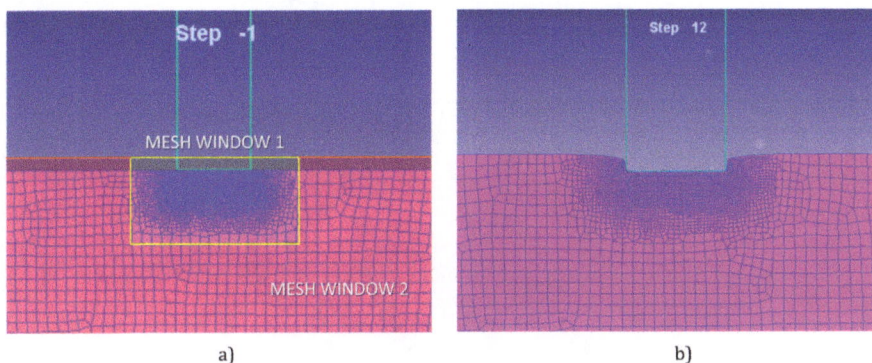

Figure 4. Mesh windows distribution: before indentation (**a**); and during indentation (**b**).

The elements used for the FEM analysis are two-dimensional plain strain elements of four nodes. Vertical displacements are fixed at the bottom of the sample as boundary conditions, without considering friction for the restriction. An elasto-plastic isotropic hardening model has been used as constitutive material model.

For the contact between the workpiece and the punch surface a shear type friction for could forming has been defined with a value of 0.12.

For each material simulated, the flow stress law attends as the form $\bar{\sigma} = \bar{\sigma} \times \left(\bar{\epsilon}, \dot{\bar{\epsilon}}, T\right)$, where $\bar{\sigma}$ is the flow stress, $\bar{\epsilon}$ is the effective plastic strain, $\dot{\bar{\epsilon}}$ is the effective strain rate and T is the temperature.

Finally, the simulations are set for 100 simulations steps, saving every five steps, with a maximum remesh increment of 3 steps and with an equal punch displacement of 0.06 mm.

Knowing that the present work is intended as a continuation of the implementation and validation of the hardening models developed for MUBT, the experimental indentation carried out with annealed aluminum A9230 is additionally introduced.

2.2. Experimental Test

Test are carried out with a universal tension-compression machine (Servosis, Madrid, Spain), Servosis ME 405 [22], which applies continuous compression force with a maximum load capacity of 100 kN. Working with materials that deform within that deformation range is essential. The force range of the tension-compression machine is limited and deep indentations were not possible. With the annealing process on the aluminum A92030, the material allows deformations with lower forces, permitting deeper penetrations. Hence, working with the materials shown in Table 1 is not possible due to the limitations present in the equipment available. Notwithstanding, the material tested, aluminum A92030 after the annealing process, was manually introduced in the software database in order to be able to work with it during FEM simulations.

For the indentation, a 3 mm width punch made of steel AISI 304 is used. The aluminum A92030 workpiece is obtained from a 50 × 50 × 2000 mm square section bar (Table 3).

To achieve plane strain conditions, the workpiece depth is 10 times the wide of the punch [23]. Thus, the workpieces final size is 50 × 50 × 30 mm.

Table 3. A92030 (UNS classification) composition.

Al (%)	Cu (%)	Pb (%)	Mg (%)	Mn (%)	Others
90.5	3.9	1.2	0.9	0.8	Rest

A tool is designed to avoid the inclination of the punch (Figure 5). The system provides a wider base for the indenter.

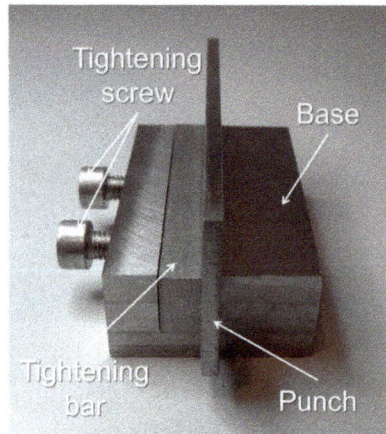

Figure 5. Assembled punch fastening tool.

The aluminum A92030 is subjected to a controlled annealing treatment, to attain an adequate depth during the indentation tests. After the annealing process, tensile test (UNE-EN ISO 6892-1 [24]) (Figure 6) were conducted to obtain the strength coefficient K and the strain-hardening exponent n of the annealed aluminum. The values obtained were $n = 0.26$ and $K = 404.66$ MPa.

Figure 6. Tested A92030 tensile samples.

An indentation of 6 mm was conducted. The overcoming of the tensile strength limit is not intended, so tests were stopped when cracks start to appear in the workpiece.

Twelve tests were performed, three for every speed range (0.6 mm/min, 4 mm/min, 60 mm/min and 400 mm/min), to analyze the influence of the speed in the hardening effect. Due to the correlation between the tests performed in the same range of speed, there was no need to increment the number of experiments. The DOE tool for experimental validation was also considered but, being non-complex tests and not having a large number of input or output variables, its application was not necessary. Each sample was given a code for identification, as follows:

EX_1-X_2-X_3-X_4-X_5

where

X_1: Specimen number;

X_2: Speed (mm/min);

X_3: Indentation depth (mm);

X_4: Specimen material (A: Annealed); and

X_5: Specimen dimension (mm).

3. Results and Discussion

Three hardening models were developed. Due to the experimental tests performed with annealed aluminum A92030, the FEM model was validated, allowing the simulation of a wider range of materials.

Finding a behavior pattern within each metal group was essential. This was possible due to the comparisons between FEM and MUBT. Being the strain hardening exponent (n) a measure of the material hardening during the forming process, it is established as a categorization parameter. The hardening models are classified according to n. For the different materials analyzed with FEM, the ASTM 646 standard is applied for the n calculation, knowing that for ductile materials at room temperature, typical values are between 0.02 and 0.5 [25,26].

The numerical application used provides only the flow stress data, i.e., the data for a material in the plastic region. Furthermore, these data represent the true stress-strain curve, making it possible to obtain the necessary information to deduce n and K for each simulated material. These results are also shown in Tables 1 and 2. Thus, a classification was established according to n (Table 4). Superalloys, due to their special characteristics, needed to be grouped separately.

Table 4. Classification according to n values.

Material	n	HM
Aluminum, steel, Titanium	$0 \leq n \leq 0.10$	HM3
and its alloys	$n > 0.10$	HM2
Superalloys	$0 \leq n \leq 0.12$	HM3
	$n > 0.12$	HM1

The modular model is compound of three different modules. Therefore, according to Table 2, three HM can be considered. The HM1 (Figure 1a) contemplates that all the material under the punch suffers the hardening effect. Therefore, the hardening equation (Equation (8)) is applied to the three modules, two of which are located below the punch (A and B) and the third outside the punch area (C). In this case, the third module that is not under the punch, experience deformation due to the push of the material under the punch and, therefore, will also experience strain hardening.

The HM2 (Figure 1b) only applies the hardening effect to the modules located under the punch. In this case, the assumption that module C does not suffer the same deformation as module A and B was made. Thus, hardening for module C may be negligible. The material considered in the modules directly under the punch receives all the compressive force. Module C is proposed with the ability to absorb this material displacement. That is to say, the hardening effect of module C could be neglected in relation to the hardening behavior that the preceding modules suffer.

For the HM3 (Figure 1c), only module A suffers from hardening. This model is suitable for materials which briefly harden due to deformation, that is, materials with small n (between 0.05 and 0.10). Therefore, the hardening effect will be concentrated only in module A, leaving the remaining modules without it. Figure 7 illustrates the difference between the three hardening models applied on the analysis of A92030, for different shape factors (H_T/L, being L constant along the graphic evolution). It can be seen that when the HM1 is implemented, the forces obtained are much higher than in the case study with the HM3. This is due to the considerations of the hardening effect module by module. The HM1 considers ε in all of the modules versus the HM3 that only implements ε in the first module, with the intention of simulate the materials with less hardening under a plastic deformation.

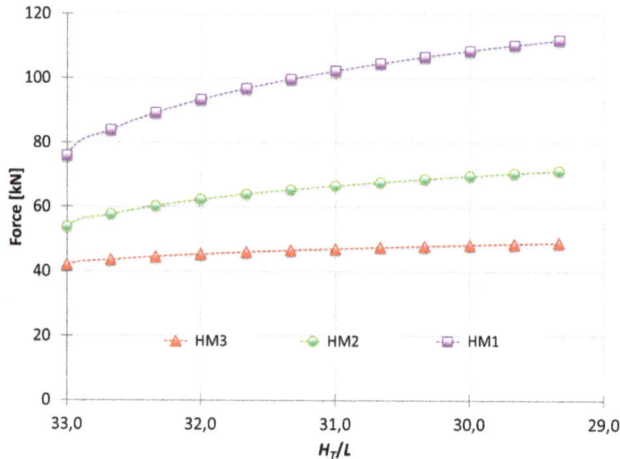

Figure 7. Results for Aluminum A92030 when the different Hardening Models apply.

As can be seen in Figure 1, the shape factor is the quotient between the total height of the quarter of the workpiece studied and half of the indenter length.

Materials presented in Tables 1 and 2 were simulated with FEM and studied with MUBT, solving each material with the HM1, HM2 and HM3. The classification presented in Table 4 was established through the comparison between MUBT and FEM. The HM3 offered accurate results for materials with a hardening exponent under 0.1. With the HM1 and HM2, the effort obtained were over the FEM range because more hardening effect was being considered. In addition, for materials with n over 0.1, the results obtained from the HM2 were of the same range as the results from the FEM, while the results from the HM3 were lower, having a higher error percentage.

Next, some examples are illustrated to visualize the proper fitting between the numerical analysis and analytical solution with the hardening effect implemented. The hardening models follow the same evolution within their corresponding group (HM1, HM2 or HM3). All the materials presented in Tables 1, 2 and 4 were simulated but due to the similar progress of the results provided graphically for each group, only part of them are presented in this paper in order to avoid similar images. Preceding studies [9,17] present a previous analysis of the case study. After an improvement of MUBT and the FEM model, results show more accurate solutions, reducing the difference between the effort values and a better concordance with each other. In addition, the results are plot only focusing on the infinite case study, avoiding higher H_T/L values.

3.1. Results for HM3

This model applies to materials which n is between 0 and 0.10. Therefore, the hardening effect is only considered in the first module (module A). Figures 8 and 9 show a MUBT-FEM comparison for two different materials, aluminum A95052 and steel G10450, being H_T/L the shape factor. In addition, the workpiece simulated is higher than the one used for the experimental tests due to the possibility to analyze an infinite case with a bigger sample in FEM.

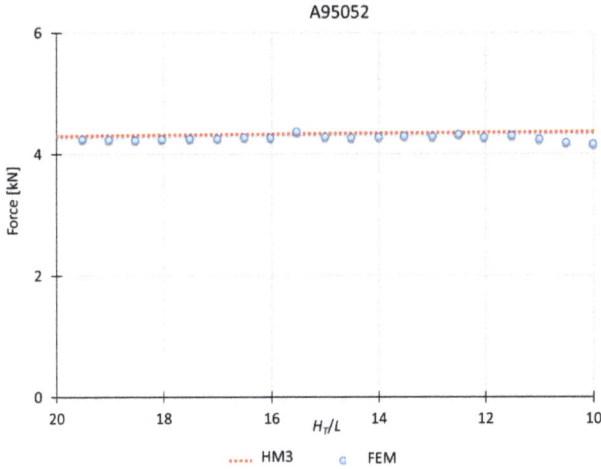

Figure 8. MUBT-FEM comparison for A95052, $n = 0.09$.

It can be seen how MUBT results present a close approximation to those given by FEM, establishing that HM3 is suitable for materials with hardening exponent between 0.05 and 0.10. With the implementation of the mesh windows and increasing remesh parameters, FEM results offer values with less alterations, like the pick that usually can be obtained due to a thick mesh or lower remesh criteria. The FEM study is carried out for plane strain conditions. The software used considers a workpiece depth of 1 mm. That will explain the low force values obtained for the different materials simulated.

Figure 9 shows the approximation between MUBT and FEM for Aluminum A96062. In this case, FEM results have not been depurated in order to present all the values that can be obtained due to the increase in the remesh criteria. After depuration, the results are offered as can be seen in the rest of the plots represented. This depuration is necessary to present clear results and be able to make an accurate comparison between methods.

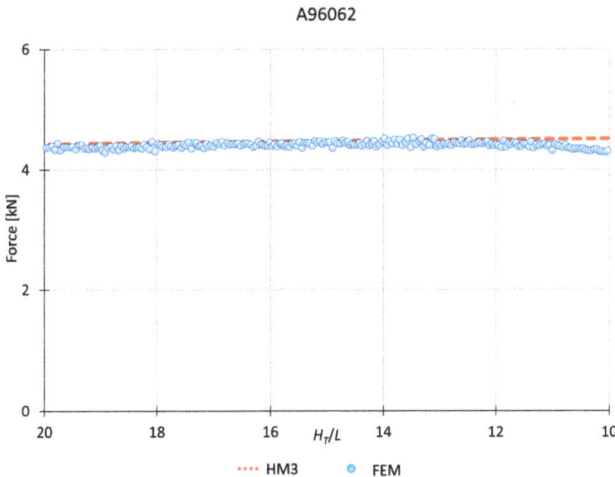

Figure 9. MUBT-FEM comparison for A96062, $n = 0.10$.

3.2. Results for HM2

This model applies to materials which n is greater than 0.10. Therefore, the hardening effect is considered in the modules under the punch (module A and B). As seen in Figure 7, to consider the hardening effect for all the modules (HM1) show results excessively far from the results obtained for this type of materials. Figures 10–12 show a MUBT-FEM comparison for aluminum A91070, steel G10080 and Titanium R50250 with n 0.21, 0.17 and 0.23 respectively.

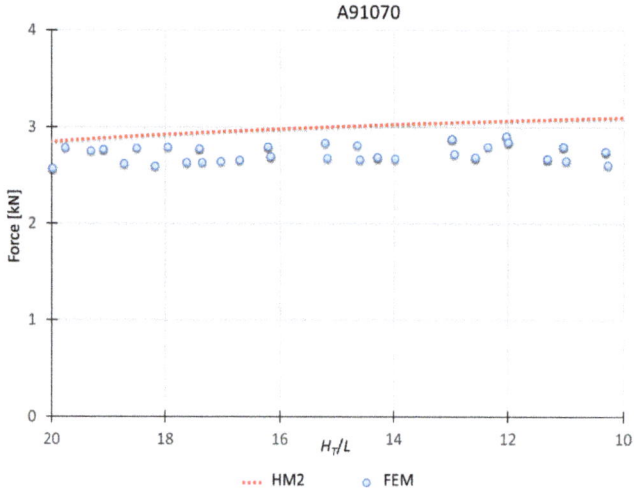

Figure 10. MUBT-FEM comparison for A91070, $n = 0.21$.

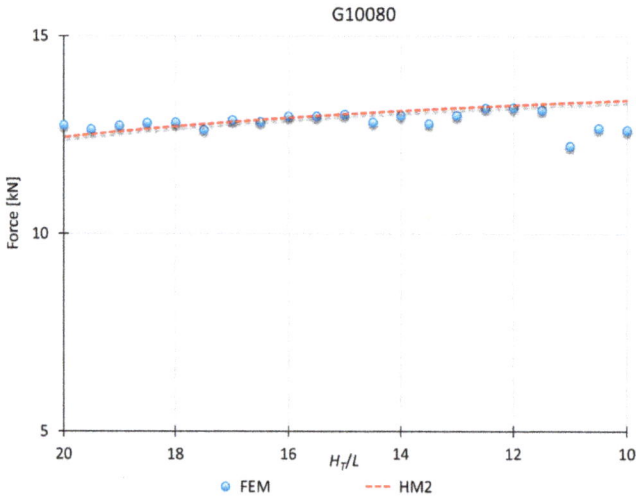

Figure 11. MUBT-FEM comparison for G10080, $n = 0.17$.

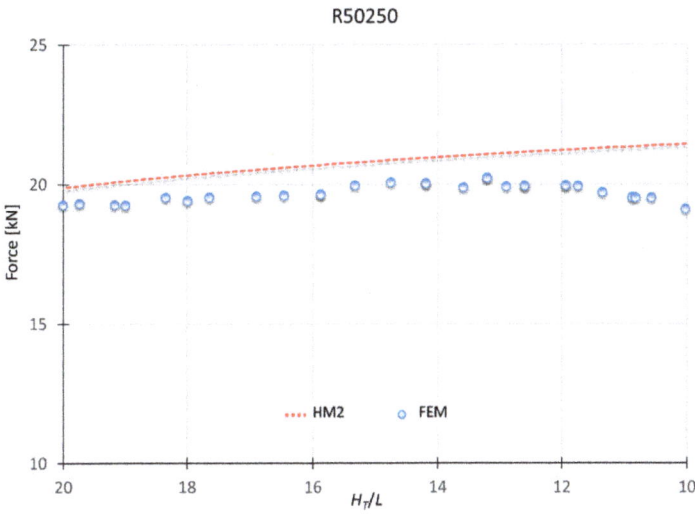

Figure 12. MUBT-FEM comparison for R50250, $n = 0.23$.

Again, a close approximation is obtained. Although a small difference is visible, this difference is not higher than 15%, which is consistent with other UBT studies [27,28].

3.3. Results for HM1

Finally, the comparison between MUBT and FEM for superalloys is shown in Figures 13 and 14 for nickel N02211 and Inconel N06600, respectively. For these materials, HM1 presents a better fit.

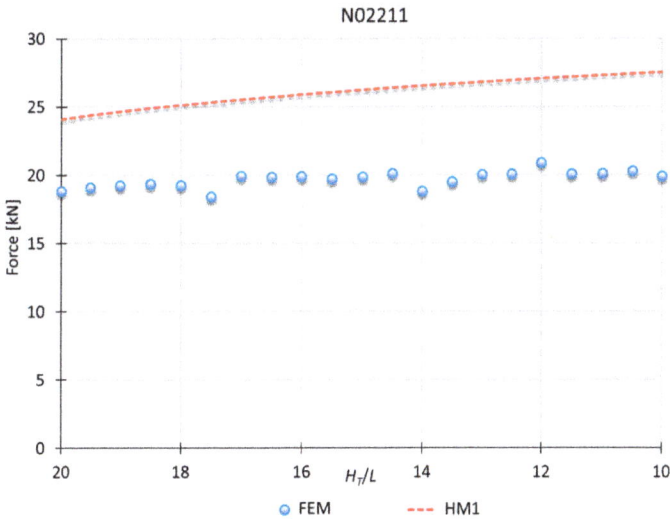

Figure 13. MUBT-FEM comparison for N02211, $n = 0.21$.

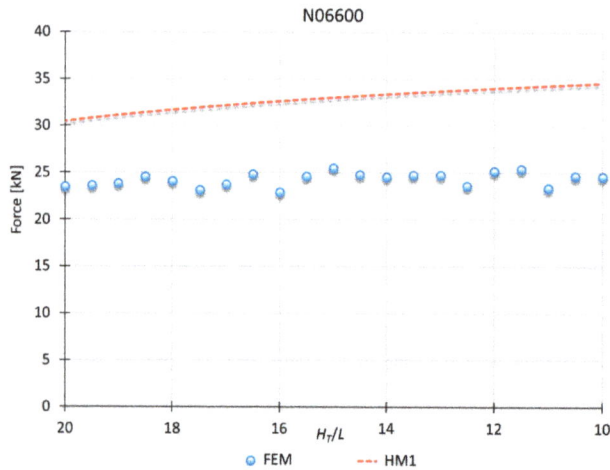

Figure 14. MUBT-FEM comparison for N06600, $n = 0.20$.

In this case, HM1 presents results more detached from those obtained with FEM. Notwithstanding, the results are consistent with the error percentage shown in other studies that UBT [29–32].

In addition, superalloys can be considered extreme cases, being out of the range of the present analysis. These types of materials are not usually used in cold indentation processes, being processed by vacuum induction melting, investment casting, powder metallurgy or spray forming/casting [33–35]. Superalloys need special treatment asides from other metals, being necessary a sub-classification and new adjustment of the model due to their behavior under great deformation. The study opens the field to the consideration of processing superalloys with this kind of process and its analysis with MUBT.

3.4. Experimental Results

Figure 15 shows the mean values of the tests E19/E20/E21-4-1.5-6-Al92030A-50 $\times$ 50 $\times$ 30 and E22/E23/E24-60-1.5-6-Al92030A-50 $\times$ 50 $\times$ 30. According to these results, speed does not produce significant changes in the final loads obtained [8].

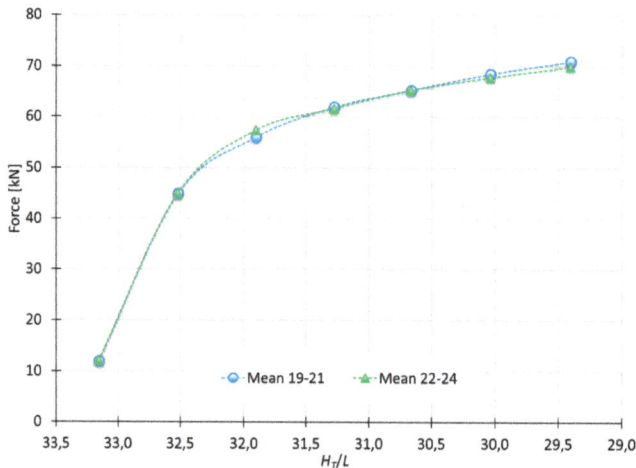

Figure 15. Indentation forces obtained from test with different speeds.

Figure 16 shows the indentation achieved for samples E17 and E23. Certain fragility in the material is appreciated near the margins of the punch. Cracks appear in the surface of the work-piece. This phenomenon justifies that deep indentations are undesirable due to possible distortions on final results.

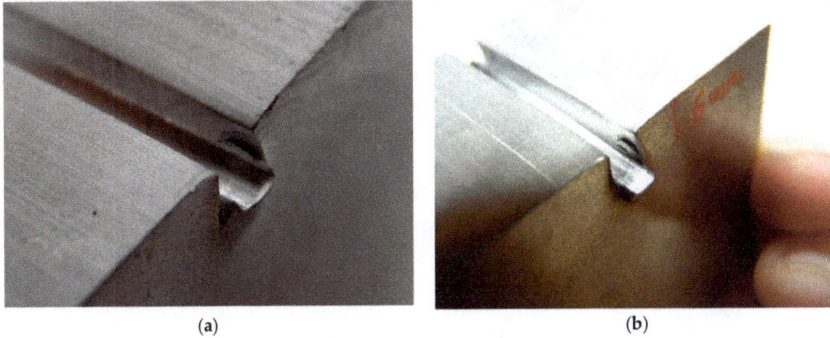

(a) (b)

Figure 16. Cracks near the indenter boundary on specimen: E17 (**a**); and E23 (**b**) after indentation.

The comparative between the performed tests and MUBT (Figure 17) shows that the model follows the same evolution as the results obtained from the indentation tests carried out.

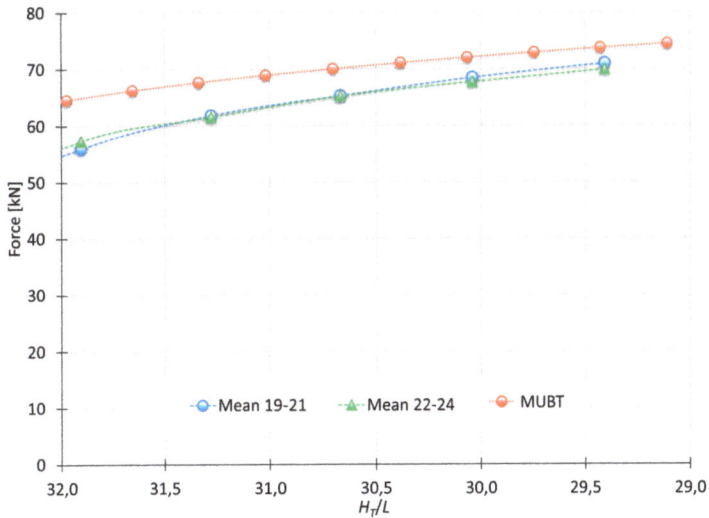

Figure 17. Comparison between MUBT and Experimental tests.

As the punch penetrates, differences between MUBT and the experimental tests decreases. Considering this study focuses on new incremental processes (Single Point Incremental Process, Localized-Incremental Forming Process or the Multiple Indentation Forming Process mentioned previously) that are being developed nowadays, shallow indentations (corresponding to the first values of the graph) can be disregarded. Hence, with a 2 mm indentation, the difference between MUBT and the test at 4 mm/min and 60 mm/min are 11.4% and 8.8%, respectively. In addition, at a 6 mm indentation, the differences are 0.6% and 2.1%, respectively, very close values that show an optimal adaptation of MUBT.

4. Conclusions

Present work shows the development of the hardening models implemented in MUBT. Three HMs were analyzed in order to achieve reasonable fitting considering different materials. The model classification is established through the hardening exponent (n). Different simulations applying numerical methods were carried out and compared with the results obtained with MUBT.

For the metals taken into account in the case studies, superalloys needed special treatment. This type of metals presents a special behavior since they do not follow the common tendencies of other regular metals. Consequently, the adjustment of the n categorization for superalloys is considered normal.

The results obtained from the hardening models developed are consistent with other analysis performed, like FEM or experimental validation with aluminum A92030. The loads obtained with the numerical method and tests are compared with those procured by MUBT, displaying a set of results in the same range.

For the n values from 0 to 0.10, where HM3 applies, MUBT results evolve similarly to the simulation results. The results only show a difference between the 3% and the 5%. For the n values above 0.10, where HM2 applies, the modular model, the numerical analysis and the experimental validation (the aluminum tested has a $n = 0.26$) also display loads with a good correlation, with differences between the 5% and the 10%, depending on the shape factor (H_T/L). In general, the model tends to stabilize, showing a progressively approach to the real load values.

Only the superalloys, for which HM1 applies, present a bigger disparity in the results. However, the superalloys case is presented as an extension of the analysis. This extension shows the capacity and versatility of MUBT. It exposes the possibility of a new subcategorization for the superalloys group to better adapt the model to their evolution. Notwithstanding, the results obtained are within the acceptable range according to other studies in the application of the Upper Bound Element Method referenced in this paper work. In addition, these types of materials are not usually used in cold indentation processes, being processed by vacuum induction melting, investment casting, powder metallurgy or spray forming/casting. Thus, the difference between the results obtained is not relevant in this case study, but opens the field to the consideration of processing superalloys with this kind of process.

Therefore, this paper is presented as a further step in the validation of the MUBT application to an indentation process considering the hardening effect, showing once more the suitability of the model and the accuracy of the results that can be obtained working with conventional materials. The investigation offers three hardening models that permit the study of a wide range of materials (aluminum, steel, titanium, etc.). The delimitation of the three hardening models allows obtaining results in concordance with preceding studies and close to reality foe each material.

Acknowledgments: The authors want to thank the University of Malaga—Andalucía Tech, International Campus of Excellence, for its financial support on this paper.

Author Contributions: Carolina Bermudo, Lorenzo Sevilla and Francisco Martín conceived and designed the experiments; Carolina Bermudo performed the experiments; Carolina Bermudo, Lorenzo Sevilla and Francisco Martín analyzed the data; Francisco Javier Trujillo contributed in the analysis tools; and Carolina Bermudo wrote the paper.

References

1. De Filippis, L.; Serio, L.; Facchini, F.; Mummolo, G.; Ludovico, A. Prediction of the vickers microhardness and ultimate tensile strength of AA5754 H111 friction stir welding butt joints using artificial neural network. *Materials* **2016**, *9*, 915. [CrossRef]
2. Akatsu, T.; Numata, S.; Shinoda, Y.; Wakai, F. Effect of the elastic deformation of a point-sharp indenter on nanoindentation behavior. *Materials* **2017**, *10*, 270. [CrossRef]

3. McAnulty, T.; Jeswiet, J.; Doolan, M. Formability in single point incremental forming: A comparative analysis of the state of the art. *CIRP J. Manuf. Sci. Technol.* **2017**, *16*, 43–54. [CrossRef]
4. Bernal, C.; Camacho, A.M.; Arenas, J.M.; Rubio, E.M. Analytical procedure for geometrical evaluation of flat surfaces formed by Multiple Indentation Processes. *Appl. Mech. Mater.* **2012**, *217–219*, 2351–2356. [CrossRef]
5. Camacho, A.M.; Marín, M.M.; Rubio, E.M.; Sebastián, M.A. Modeling strategies for efficient FE simulation of localised-incremental forging processes. *AIP Conf. Proc.* **2012**, *1431*, 725–732.
6. Bermudo, C.; Sevilla, L.; Martín, F.; Trujillo, F. Study of the tool geometry influence in indentation for the analysis and validation of the new modular upper bound technique. *Appl. Sci.* **2016**, *6*, 203. [CrossRef]
7. Bermudo, C.; Martín, F.; Sevilla, L. Analysis and selection of the modular block distribution in indentation process by the Upper Bound Theorem. *Procedia Eng.* **2013**, *63*, 388–396. [CrossRef]
8. Bermudo, C.; Martín, F.; Sevilla, L.; Martín, M.J. Experimental validation of the new modular application of the upper bound theorem in indentation. *PLoS ONE* **2015**, *10*, 1–15. [CrossRef] [PubMed]
9. Bermudo, C.; Martín, F.; Sevilla, L. Hardening study on the application of the upper bound theorem in indentation processes by means of modules of triangular rigid zones. *Procedia Eng.* **2015**, *132*, 282–289. [CrossRef]
10. Moncada, A.; Martín, F.; Sevilla, L.; Camacho, A.M.; Sebastián, M.A. Analysis of ring compression test by upper bound theorem as special case of non-symmetric part. *Procedia Eng.* **2015**, *132*, 334–341. [CrossRef]
11. Wu, Y.; Dong, X.; Yu, Q. An upper bound solution of axial metal flow in cold radial forging process of rods. *Int. J. Mech. Sci.* **2014**, *85*, 120–129. [CrossRef]
12. Wu, Y.; Dong, X. An upper bound model with continuous velocity field for strain inhomogeneity analysis in radial forging process. *Int. J. Mech. Sci.* **2016**, *115–116*, 385–391. [CrossRef]
13. Alforzan, A.; Gunasekera, J.S. An upper bound elemental technique approach to the process design of axisymmetric forging by forward and backward simulation. *J. Mater. Process. Technol.* **2003**, *142*, 619–627. [CrossRef]
14. Kudo, H. An upper-bound approach to plane-strain forging and extrusion-II. *Int. J. Mech. Sci.* **1960**, *1*, 229–252. [CrossRef]
15. Topcu, N. Numerical, Analytical and Experimental Analysis of Indentation. Ph.D. Thesis, Midlle East Technical University, Ankara, Turkey, 2005.
16. Bermudo, C.; Martín, F.; Sevilla, L. Optimización del modelo modular en procesos de indentación mediante el teorema del límite superior. *An. Ing. Mecánica* **2014**, *19*, 97.
17. Bermudo, C. Análisis, Desarrollo y Validación del Método del Límite Superior en Procesos de Conformado por Indentación. Ph.D. Thesis, University of Malaga, Servicio de Publicaciones y Divulgación Científica, Malaga, Spain, 17 September 2015.
18. Fluhrer, J. *Deform. Design Environment for Forging. User's Manual*; Scientific Forming Technologies Corporation: Columbus, OH, USA, 2010.
19. Seiong, S.; Younossi, O.; Goldsmith, B.W.; Corporation, R. *Titanium. Industrial Base, Price Trends and Technology Initiatives*; RAND Corporation: Santa Monica, CA, USA, 2009.
20. Reed, R.C. *The Superalloys: Fundamentals and Applications*; Cambridge University Press: Cambridge, UK, 2008.
21. Bermudo, C.; Martín, F.; Sevilla, L. Selection of the optimal distribution for the Upper Bound Theorem in indentation processes. *Mater. Sci. Forum* **2014**, *797*, 117–122. [CrossRef]
22. Servosis SL Serie ME-405 2012, 1–8. Available online: http://www.servosis.com/assets/me-405.pdf (accessed on 3 January 2013).
23. Rowe, G.W. *An Introduction to the Principles of Metalworking*; Edward, A., Ed.; St. Martin's Press: New York, NY, USA, 1971.
24. AENOR UNE-EN ISO 6892-1. Metallic materials—Tensile testing. In *Part 1 Method Test Room Temp. (ISO 6892-1: 2009)*; AENOR: Madrid, Spain, 2010.
25. Dieter, G.E.; Kuhn, H.A.; Semiatin, S.L. *Handbook of Workability and Process Design*; ASM International: New York, NY, USA, 2003.
26. Matusevich, A.; Mancini, R.; Massa, J. Determinación del exponente "n" de endurecimiento por deformación mediante un algoritmo de mínimos cuadrados ponderados. *Revista Iberoamericana de Ingeniería Mecánica* **2013**, *17*, 171–182.
27. Parvizi, A.; Abrinia, K. A two dimensional Upper Bound Analysis of the ring rolling process with experimental and FEM verifications. *Int. J. Mech. Sci.* **2014**, *79*, 176–181. [CrossRef]

28. Hosseinabadi, H.G.; Serajzadeh, S. Hot extrusion process modeling using a coupled upper bound-finite element method. *J. Manuf. Process.* **2014**, *16*, 233–240. [CrossRef]
29. Altinbalik, T.; Ayer, O. Effect of die inlet geometry on extrusion of clover sections through curved dies: Upper Bound analysis and experimental verification. *Trans. Nonferrous Met. Soc. China* **2013**, *23*, 1098–1107. [CrossRef]
30. Hwang, B.C.; Hong, S.J.; Bae, W.B. An UBET analysis of the non-axisymmetric extrusion/forging process. *J. Mater. Process. Technol.* **2001**, *111*, 135–141. [CrossRef]
31. Hwang, B.C.; Lee, H.I.; Bae, W.B. A UBET analysis of the non-axisymmetric combined extrusion process. *J. Mater. Process. Technol.* **2003**, *139*, 547–552. [CrossRef]
32. Khoddam, S.; Farhoumand, A.; Hodgson, P.D. Upper-Bound analysis of axi-symmetric forward spiral extrusion. *Mech. Mater.* **2011**, *43*, 684–692. [CrossRef]
33. Prasad, N.E.; Wanhill, R.J. H. Processing of aerospace metals and alloys: Part 2—Secondary Processing. In *Aerospace Materials and Material Technologies*; Springer: Singapore, 2016; pp. 25–38.
34. Prasad, V.V.S.; Rao, A.S. Electroslag melting for recycling scrap of valuable metals and alloys. In *Recycling of Metals and Engineered Materials*; John Wiley & Sons, Inc.: Hoboken, NJ, USA, 2000.
35. Gessinger, G.H.; Gessinger, G.H. Chapter 8–Joining techniques for P/M superalloys. In *Powder Metallurgy of Superalloys*; Brown Boveri & Co. Ltd.: Baden, Switzerland, 1984; pp. 295–327.

materials

MDPI

Article

A Comparison of Microscale Techniques for Determining Fracture Toughness of LiMn$_2$O$_4$ Particles

Muhammad Zeeshan Mughal [1,*], **Hugues-Yanis Amanieu** [2,3], **Riccardo Moscatelli** [1] and **Marco Sebastiani** [1]

[1]   Engineering Department, University of Rome "ROMA TRE", Via della Vasca Navale 79, 00146 Rome, Italy; riccardo.moscatelli@uniroma3.it (R.M.); marco.sebastiani@uniroma3.it (M.S.)
[2]   Institute for Materials Science and Center for Nanointegration Duisburg-Essen (CENIDE), University of Duisburg-Essen, Universitätsstr. 15, 45141 Essen, Germany; hy.amanieu@uni-due.de or HuguesYanis.Amanieu@leclanche.com
[3]   Robert Bosch GmbH, Robert-Bosch-Platz 1, 70839 Gerlingen-Schillerhoehe, Germany
*   Correspondence: zeeshan.mughal@uniroma3.it; Tel.: +39-06-57333496

Academic Editor: Matt Pharr
Received: 14 March 2017; Accepted: 7 April 2017; Published: 12 April 2017

Abstract: Accurate estimation of fracture behavior of commercial LiMn$_2$O$_4$ particles is of great importance to predict the performance and lifetime of a battery. The present study compares two different microscale techniques to quantify the fracture toughness of LiMn$_2$O$_4$ particles embedded in an epoxy matrix. The first technique uses focused ion beam (FIB) milled micro pillars that are subsequently tested using the nanoindentation technique. The pillar geometry, critical load at pillar failure, and cohesive FEM simulations are then used to compute the fracture toughness. The second technique relies on the use of atomic force microscopy (AFM) to measure the crack opening displacement (COD) and subsequent application of Irwin's near field theory to measure the mode-I crack tip toughness of the material. Results show pillar splitting method provides a fracture toughness value of ~0.24 MPa.m$^{1/2}$, while COD measurements give a crack tip toughness of ~0.81 MPa.m$^{1/2}$. The comparison of fracture toughness values with the estimated value on the reference LiMn$_2$O$_4$ wafer reveals that micro pillar technique provides measurements that are more reliable than the COD method. The difference is associated with ease of experimental setup, calculation simplicity, and little or no influence of external factors as associated with the COD measurements.

Keywords: fracture toughness; atomic force microscopy; pillar splitting; lithium-ion batteries; nanoindentation; focused ion beam

---

1. Introduction

Low cost and low toxicity of spinel LiMn$_2$O$_4$ makes them a good cathode material for the lithium-ion batteries. Unfortunately, its commercialization is limited by its short lifetime. The primary reason for its short lifetime is extensively documented and related to the dissolution of manganese atoms in the electrolyte, which is the main source for capacity fade [1,2]. Mechanical failure upon cycling produces more surface area that could lead to the loss of material through dissolution. It can also result in the loss of adhesion with the current collector [3]. Internal pressure associated with the intercalation and deintercalation results in the fracture of the crystal. It was previously reported for LiMn$_{1.95}$Al$_{0.05}$O$_4$, also a spinel that two parallel phenomena occurs: (1) brittle cracking at the first electrochemical cycle and (2) fatigue leading to fracture [4]. More recently, the same issues were observed for commercial LiMn$_2$O$_4$, where oxygen deficiency is witnessed; a phenomenon which can be reduced with stoichiometric spinel [5]. It has been shown that the favorite cracking plane is {111} [6].

It is expected for LiMn$_2$O$_4$ as {111} planes have the lowest solid-to-vapor surface energy, but {101} faceting occurs as well [6]. The main challenge associated with the fracture toughness measurement of LiMn$_2$O$_4$ spinel materials is that a bulk single crystal of this type of spinel cannot be easily grown above a few micrometers [7]. Therefore, it is difficult to measure the fracture properties of single grains without the use of microscale techniques.

Fracture toughness measurement using the indentation testing has been widely used over the past three decades for brittle materials such as glasses and ceramics [8–11]. Lawn et al. [8] (and then Anstis et al. [9]) suggested the classic relationship of fracture toughness assessment using Vickers indentation based on the half-penny crack configuration

$$K_c = \alpha \left(\frac{E}{H} \right)^{1/2} \frac{P}{c^{3/2}}$$ (1)

where P is the indentation load, c is the radial crack length from indentation center to the crack tip, E is the Young's modulus, H is the hardness, and α is the constant that depends upon indenter geometry. Anstis et al. empirically determined the value of α as 0.016 ± 0.004 for Vickers indentation [9]. With the development of nanoindentation testing in the early 1990s [12], it was revealed that Equation (1) also applies to the three sided Berkovich indenter commonly used in the nanoindentation testing. Later Jang and Pharr [13] suggested that the indenter angle has an effect on the cracking behavior and can influence the fracture toughness values. Their study using Si and Ge shows that by simply changing the indenter shape from cube corner (35.3°) to Berkovich (65.3°) indenter, the coefficient value decreases by ~50%.

In the present article, we compare two microscale techniques, namely pillar splitting method [14] and crack opening displacement (COD) [15], to characterize the fracture toughness values of micrometric particles of LiMn$_2$O$_4$. In the first technique, nanoindentation is used to split the focused ion beam (FIB) milled micro pillars. Fracture is realized by splitting at reproducible loads that are experimentally quantified from displacement bursts in the loading segment of the load-displacement curve [14,16]. The fracture toughness (K_C) can be evaluated by using the following simple equation [14]

$$K_c = \gamma \frac{P_c}{R^{3/2}}$$ (2)

where K_c is the fracture toughness (MPa.m$^{1/2}$), P_c is the critical load at failure (mN) and R the pillar radius (μm). γ is a dimensionless coefficient and has been calculated for a wide range of materials properties in a recent paper [16]. It is worth noting that the γ coefficient contains the influences of elastic and plastic properties and is consequently material specific. The usefulness of Equation (2) lies in its simplicity, as both the critical load and pillar radius are easily measured quantities. Recent papers have demonstrated the applicability of the pillar indentation splitting method for a wide range of material properties, which includes most ceramic materials [17–19].

The second method uses atomic force microscopy (AFM) to measure the crack opening displacements (COD) after nanoindentation. Irwin's near field solution was then applied to evaluate the mode I crack tip toughness as first introduced by Rödel et al. [15] by means of scanning electron microscopy. Additionally, indentations are performed on a wafer of spinel LiMn$_2$O$_4$ with its top surface parallel to the {111} plane. This highly-oriented crystal enabled reproducible crack patterns around the indents and Anstis solution for half-penny cracks [9] was used to evaluate the fracture toughness and is also used as a reference in this work. Crack evolution below the surface of the indent was also observed for the wafer using the FIB cross sections.

Lastly, both microscale techniques are compared based on their merits and demerits and recommendations are provided on the use of suitable technique for the fracture toughness assessment of this challenging material.

2. Materials and Methods

2.1. Sample Preparation

Active particles of LiMn$_2$O$_4$ based cathode material extracted from commercial cells and cycled three times at one C-rate from 2.5 to 4.2 V. These cathode materials were prepared for nanoindentation as described in the previous work [20]. A wafer of {111}-oriented lithium manganese (III, IV) oxide was prepared using a wafer of {111}-oriented manganese (II) monoxide as precursor (SurfaceNet GmbH, Rheine, Germany). The preparation method developed by Kitta et al. [21] was used.

2.2. Pillar Splitting Experiments

Fabrication of the micro-pillars were performed using the focused ion beam (FIB) procedure based on the ring-core milling approach developed by some of the authors [22,23]. The milling was performed in a single outer to inner pass using the FEI Helios NanoLab 600 at a current of 0.92 nA. At least five pillars were milled to an aspect ratio (h/d) of >1.2, where h is the pillar height and d is the top diameter. It has been shown previously [22] that this geometrical design provides complete residual stress relaxation in the upper part of the pillar. It is worth highlighting that using the correct combination of current and dwell time; a single pillar can be milled within 10 minutes. Pillars were only milled on the particles that are wide and deep enough to accommodate a 5 μm pillar. Special care was taken to avoid the porous particles. All pillars were tested using a Berkovich indenter on a Keysight G200 nanoindenter at a constant strain rate of 0.05 s^{-1} and an indentation depth set to 400 nm into the top surface. The instrument frame stiffness and indenter area function were calibrated before and after testing on a certified fused silica reference sample. The continuous stiffness measurement (CSM) mode was turned off during the tests.

2.3. Crack Opening Displacement Measurement

A thin layer (~1 nm) of Pt/Pd alloy was sputter coated on each sample for easy imaging by scanning electron microscope (SEM). The samples were indented with cube corner tips down to 400 nm without continuous stiffness measurement and with a strain rate target of 0.05 s^{-1}, leaving an indentation print about 600 nm wide. The SEM was used to determine the suitability of the cracks for COD measurements (see Figure 1a). The cracks were selected if they were long enough to be mapped and did not grow too close to a particle edge or a defect. Tapping mode atomic force microscope (Dimension 3100) using very sharp tips imaged each crack (TESP-SS, Bruker, 42 N/m, 320 kHz, 5 nm max radius) (see Figure 1b). The challenge was not to smooth the tips while scanning as the stiff particle edges can easily damage them. Approach and scan were done using a very small initial force set point (2%) in order to find the cracks. Once a crack was found, the force was increased until trace and retrace lines were similar. TGX1 test grating samples (NT-MDT) were used before and after measurements to characterize the AFM tip sharpness using the same procedure. In order to minimize sub-critical crack growth between crack formation and COD measurement, indentations were carried out in the early morning and the rest of the procedure was carried out within one day, sometimes extending to the next day. For a crack along the x-axis and its tip located at (x = 0, y = 0), Irwin describes the crack displacement as follows [24]:

$$u_x(X) = \begin{cases} \dfrac{(1-v^2)K_{IC}}{E}\sqrt{\dfrac{8X}{\pi}}, & X \geq 0 \\ 0, & X < 0 \end{cases} \tag{3}$$

where $u_x(X)$ represents the near-tip crack opening displacement at position (x = X, y = 0), E is the elastic modulus, v represents the Poisson's ratio, and K_{IC} is the mode I fracture toughness. For each image, u_x was measured on 12 cross-sections perpendicular to the crack for different distances X to the crack tip. Special care was taken to avoid reverse tip imaging at the crack walls. K_{IC} was calculated by measuring the slopes of X vs. u_x and inputting the value into Equation (3). Measurement reliability is assessed by the correlation coefficient R^2 of the linear regression.

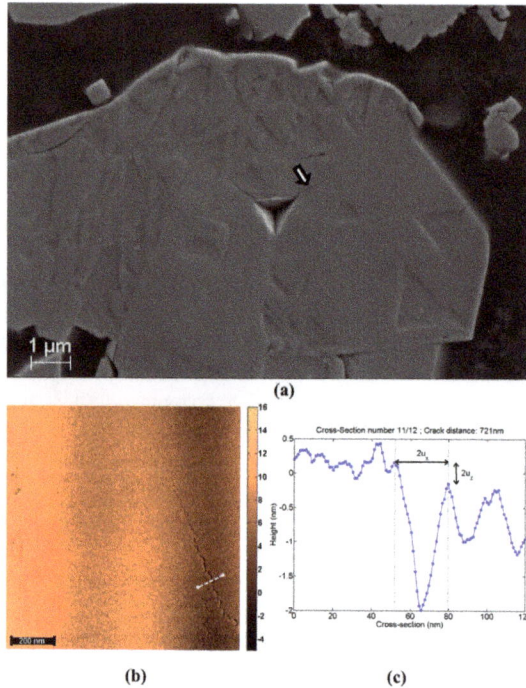

(a)

(b)

(c)

Figure 1. (a) SEM micrograph of a cube-corner indent in a $LiMn_2O_4$ particle. The cracks do not propagate straight from the corners of the indent. They are also not longer than the indent; the arrow indicates the crack mapped by means of AFM and visible in (b,c) is the cross-section (indicated by a white line in (b)) used to estimate u_x for $x = 721$ nm.

3. Results

3.1. Crack Length Measurement

The fracture toughness measurement using the classical Anstis solution for half-penny cracks is highlighted in Equation (1) and largely relies on the configuration of crack pattern. This is the most widely used method for fracture toughness calculations. For the case of $LiMn_2O_4$ particles (Figure 1a), the configuration of the crack pattern does not allow the application of the crack-length measurement method traditionally used after indentation [25]. However, the cracking pattern of the reference wafer sample (pictured in Figure 4a) seems to allow such measurements. Using Anstis solution for half-penny crack [9] and E and H values of 95.70 GPa and 6.70 GPa respectively, a fracture toughness of 0.33 ± 0.07 MPa.m$^{1/2}$ was calculated for the wafer sample. It is worth mentioning that a α value of 0.0569 was used for the calculation of fracture toughness as proposed by Jang and Pharr [13] for the cube corner indenters. One might wonder at this point that the coefficient value of 0.0569 was originally proposed for Si and Ge material. The usability of this coefficient value is justified by comparing the E/H ratio of Si and the wafer material used in this study. Both Si and wafer material have an E/H ratio of 14.60 and 14.30, respectively. This value of fracture toughness (0.33 MPa.m$^{1/2}$) will be used as a reference and is not further detailed in this work.

3.2. Pillar Compression

Figure 2a shows a SEM micrograph of a pillar after splitting along with the load displacement curve on the reference wafer sample (Figure 2b). The fracture toughness was calculated using Equation (2) and the γ value used for this LiMn$_2$O$_4$ sample is 0.25, obtained by finite element modeling (FEM) as described in a previous study [14], assuming a value of 95.73 GPa for the elastic modulus and 6.71 GPa for the hardness [26]. Regarding the tests on the real commercial cathode samples, an example of a pillar before and after splitting (discharged fresh cell) is reported in Figure 2c,d. Results from splitting experiments on a series of FIB-milled pillars are shown on Figure 2e. Using the elastic modulus of 86.67 GPa and a hardness of 6.95 as reported in a previous paper [26], a γ coefficient of 0.22 was calculated. Note that this value also includes the correction, obtained through CZ-FEM, for the effects coming from the compliant polymer substrate. A critical failure load of 3.90 ± 0.22 mN gives a toughness value of 0.24 ± 0.01 MPa.m$^{1/2}$, which is in very good agreement with the estimations obtained on the LiMn$_2$O$_4$ reference wafer. Table 1 summarizes the results for both samples.

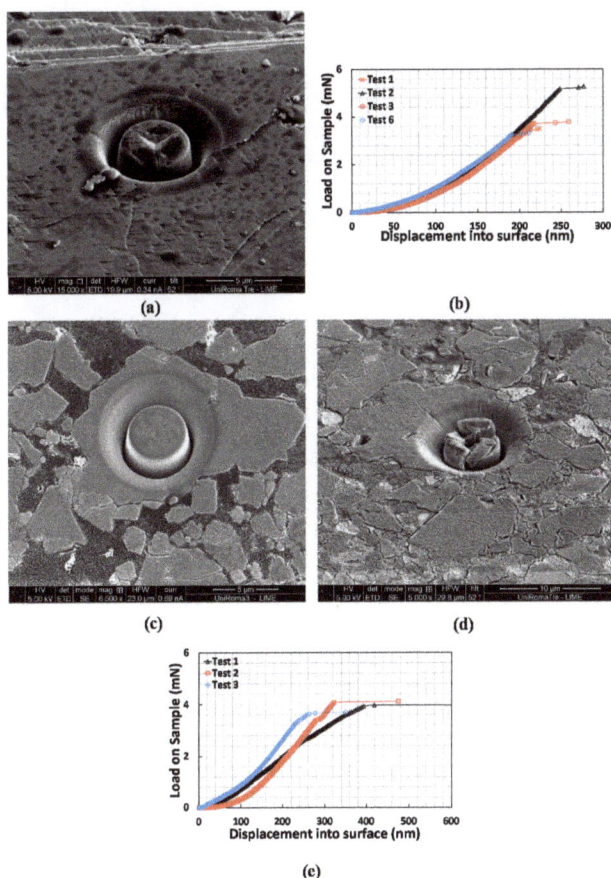

Figure 2. (**a**) SEM micrograph of a split pillars on the wafer and (**b**) a representative load-displacement curve highlighting the critical splitting load by a pop-in event; (**c**) SEM micrograph of a pillar on commercial LiMn$_2$O$_4$ particle before and (**d**) after splitting; (**e**) Representative pillar splitting data obtained on the commercial LiMn$_2$O$_4$ particles.

Table 1. Results summary for pillar compression tests on LiMn$_2$O$_4$ wafer and particles.

Parameter	Wafer	Particles
E-modulus, E (GPa), [26]	95.73 ± 3.93	86.67 ± 11.29
FE Poisson's ratio, ν	0.25	0.25
Hardness, H (GPa), [26]	6.71 ± 0.44	6.95 ± 0.76
Substrate corrected finite element γ (Equation (1))	0.25	0.22
Experimental pillar radius, R (µm)	2.36 ± 0.10	2.36 ± 0.10
Experimental instability load, P_c (mN)	3.88 ± 0.85	3.90 ± 0.22
Fracture toughness, K_c (MPa.m$^{1/2}$)	0.27 ± 0.06	0.24 ± 0.01

3.3. Crack Opening Displacement

The COD and the height difference versus the distance to the crack tips of seven cracks are plotted in Figure 3. It is evident that not all measurements present the same quality; some are relatively linear while others are very irregular. This is due to the roughness of the particles, such as scratches from polishing, leading to imprecisions. Experimental measurements were fitted with linear functions forced to zero. The crack tips were previously positioned from the AFM error images. The slope is inserted in Equation (3) to find K_{IC} using an elastic modulus of 90 GPa [20] and a Poisson's ratio of 0.3. The quality of the measurements was determined from the coefficient of determination R^2. Weighted means were applied to obtain a quantitative value, the weights being the R^2 of each fit. Table 2 summarizes all the results. For reference, the same method was used for the cracks on wafer (Figure 4a) and a crack-tip toughness of about 0.7 MPa.m$^{1/2}$ was found.

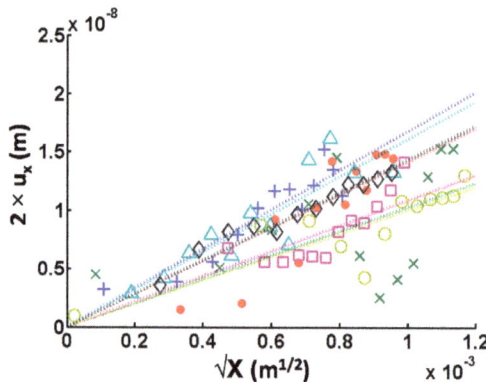

Figure 3. Crack Opening Displacement (2 × u_x) versus square root of distance to crack tip ($\sqrt{X}$) for seven different cracks. Each plot is fitted with a linear function forced to zero (dotted lines). Colors of markers from the experimental measurements correspond to colors of fitting lines.

Table 2. Mode I fracture toughness values of seven different cracks on LiMn$_2$O$_4$ particles using crack opening displacement technique.

Sample	K_{IC} (MPa.m$^{1/2}$)	R^2
+	1.00	0.84
⎩	0.89	0.95
⌈	0.62	0.69
©	0.67	0.66
⌠	0.99	0.77
|	0.88	0.70
×	0.64	0.13
Weighted means	0.81% ± 18%	N/A

3.4. Crack Orientation

Figure 4a is an SEM image of the wafer after indentation. The 60° facet edges indicate the highly-oriented crystal as previously obtained (see Reference [21]). It is not a single crystal as can be seen from EBSD measurements (Figure 4b–e) but all the grains have their top surface parallel to the {111} plane and have only a little misorientation. The edges of the triangular pattern are perpendicular to the <121> direction. All the indents formed cracks growing along the <121> direction (Figure 4a), hence forming {101} planes if perpendicular to the top surface. The cracks always grow perpendicularly to the edges of the triangular pattern regardless of the orientation of the indenting diamond tip. Figure 5 schematically depicts this phenomenon along with FIB cross-sections to see the cracks development below the surface. They indicate that the cracks grew first perpendicularly to the top surface, opening {101} plane within a depth of 100 nm. Then they deviate at an angle of 30° to 40°. There are two possible explanations for this deviation. Another material could be present below the spinel material. In fact, X-ray diffraction measurements published in a previous work [26] showed that Bixbyite Mn_2O_3 could be present and it is possible that there is a deficit of oxygen ions between the cubic MnO substrate and the spinel $LiMn_2O_4$ top surface. Otherwise, only the spinel material could be present and the propagation deviates because surface formation is easier in this new plane.

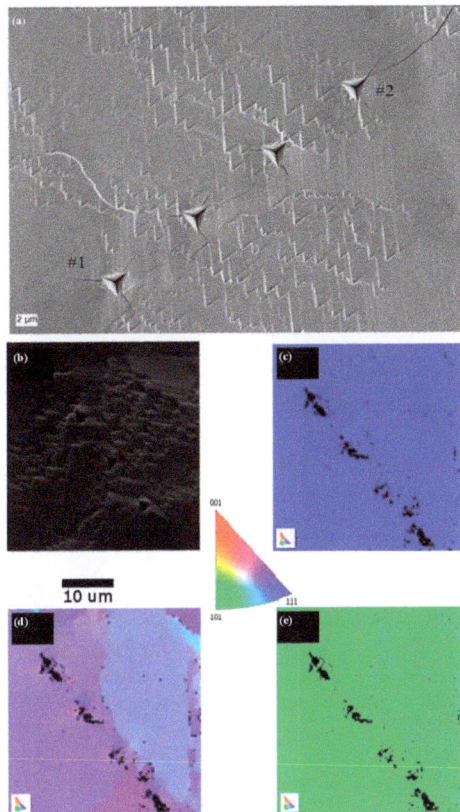

Figure 4. (a) SEM micrograph of four 400-nm deep indents performed by depth control on the wafer sample. Cracks of particle labeled #1 and #2 were observed by FIB cross-sections; (b) Secondary electron image of the mapped area; (c) Z direction; (d) Y direction; and (e) X direction. Color-coding indicated in the center.

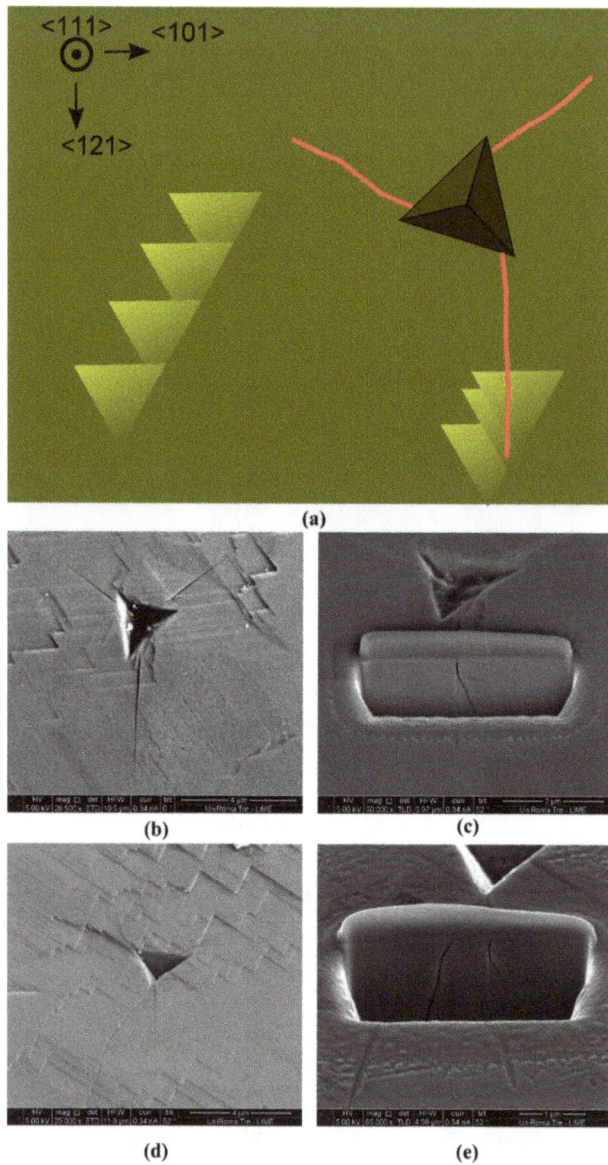

Figure 5. (a) Schematics of the wafer. Yellow areas correspond to the wafer materials where the triangles represent the patterns observed on Figure 4a, the darker triangle represents an indent print with a 'random' rotation and the red lines represent typical cracks, growing perpendicularly to the edge of the pattern triangles. (b,d) SEM images of two 400 nm deep indents before and (c,e) after FIB milling. The first indent (b,c), labeled #1 in Figure 4a, produced a reproducible crack pattern where three cracks formed perpendicularly (or 30°) to the edges of the triangular pattern during crystal growth, regardless of the indenting tip rotation. The second indent (d,e), labeled #2 in Figure 4a, produced two cracks at its corner, one similar to the previous one and another one which is much longer and certainly due to defects. Under the surface (c–e), the cracks grew perpendicularly to the top surface for about 100 nm before deviating of an angle of 30–40°.

4. Discussion

4.1. Reliability of the Methods

4.1.1. Pillar Splitting Method

A critical analysis of reliability of the pillar splitting method along with its applicability for a series of reference coating and bulk materials is presented in previous papers [14,16]. The main advantage of this technique is that the effect of FIB damage on the measurements is almost negligible. This is because the crack nucleation and growth usually happens within the core volume of the pillars, whilst the FIB damage is only present in the very 10–100 nm at the pillars edges. In comparison with the COD method, the pillar splitting method seems more reliable as it is easier to set-up experimentally and provides highly reproducible pillar splitting loads. Another advantage of this technique over the conventional methods which used the nanoindentation technique for fracture toughness assessment is that there is no need to measure the crack length, hence not only significantly enhancing the test time but also favoring less experimental hazard. One particular challenge regarding the lithium-based composite electrodes is the influence of the surrounding compliant substrate. This issue is addressed in a recent paper [16], which shows that the effect of complaint matrix on the critical splitting load is ~11% in the worst case scenario and can be corrected by evaluating specific values of γ coefficient. The calculated value of fracture toughness (0.27 ± 0.06 MPa.m$^{1/2}$ for wafer and 0.24 ± 0.01 MPa.m$^{1/2}$ for particles) using pillar splitting method already include the substrate corrected coefficient values. These values of fracture toughness are in very good agreement with the Anstis half-penny crack method, which gives a value of 0.33 ± 0.08 MPa.m$^{1/2}$. There is no quantitative report of fracture toughness on these challenging materials in literature with the exception of one paper recently published by the authors [17]. In that article, authors studied the change in fracture toughness vs. the state of charge and observed a decrease in fracture toughness (0.49 to 0.26 MPa.m$^{1/2}$) as the state of charge increases. Finally, these values of fracture toughness are in very good agreement with the reported values of similar cathode materials, e.g., Wolfenstein et al. reported the fracture toughness of Li-olivine cathodes ($LiCoPO_4$) between 0.4–0.5 MPa.m$^{1/2}$ [27]. This further confirms that the pillar splitting technique is a more reliable method for measuring the fracture toughness of $LiMn_2O_4$ particles.

4.1.2. Crack Opening Displacement

For COD measurements, it is difficult to apply higher loads without destroying the particles, rendering measurements impossible. Inevitably, the crack lengths come close to the typical size of an indent as shown in Figure 1a. It is important to highlight that the strong hypothesis of Irwin's near-field solution is that the crack walls are traction free [24]. This could be the main reason for the systematic error in COD measurements, because some residual stress may be present in the particles due to their processing. On the contrary, residual stresses are fully relieved in the case of pillar geometry as demonstrated in previous papers [22,23]. A second issue is the measurement uncertainty coupled with user's interpretation of AFM measurements. It is difficult to decipher inverse AFM tip imaging from crack wall imaging as the measured vertical displacements u_z have the same order of magnitude as the surface roughness. This can be observed from real measurements in Figure 1c. This behavior was simulated using Equation (3), tip radius effects and artificial images were generated which are then interpreted by the group of scientists. Figure 6a highlights the mode I fracture toughness values obtained from the group experiment. It can be noticed how close the computed values are to the input values in the case of the simulated cracks. Yet a user's interpretation can drastically change the measurement from simple to double, as is the case for sample 1. Overall, the averaged values from the group approach the author's measurements, which shows that most users interpreted the COD the same way. These results show that the COD measurements using Irwin's near field theory are not the reliable method to quantify fracture properties of micrometric particles particularly because of the systematic error associated with the plastic zone.

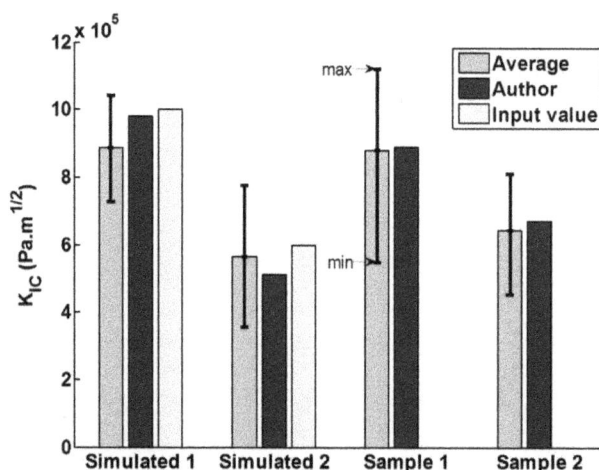

Figure 6. Mode I fracture toughness values obtained from the group experiment. The first set of bars show the average toughness measured by the group. The error bars associated with it indicate the highest and lowest values obtained from the group. The darker set of bars show the values obtained by the first author. The lighter set of bars show the values inputted to simulate the surfaces. Sample **1** and sample **2** correspond to Ⓛ and Ⓒ in Table 2 and Figure 3.

5. Conclusions

Fracture toughness of commercial $LiMn_2O_4$ particles embedded in a polymer matrix are evaluated using two micro-scale techniques. Pillar splitting method gives a fracture toughness value of about 0.24 MPa.m$^{1/2}$, while crack-opening displacement gives a value of ~0.85 MPa.m$^{1/2}$. The first method appeared to be more reliable for determining the fracture toughness properties of these materials because of the ease of experimental setup, well-defined pillar geometry, simplicity, and reproducibility of results. In the case of the second method, the size of the particles does not allow crack growth to a size where stress-free crack walls and elastic–brittle theory can be considered for COD measurement. Concerning materials properties, this ceramic seems very brittle as its toughness lies in the lower range of ceramics toughness (below 1 MPa.m$^{1/2}$). Additional tests on an oriented $LiMn_2O_4$ wafer showed that fracturing is also very dependent on the crystal orientation of the indented grains as the cracks always grow in the <121> direction on {111}-oriented surfaces, with a possible favorite cleavage plane being {101}. This further point to the possibility of producing engineered structure and/or the particle shapes to reduce brittleness.

Acknowledgments: This work and the PhD research of H.Y. Amanieu is supported by the European Commission within the FP7 Marie Curie Initial Training Network "Nanomotion" (Grant # 290158, www.nanomotion.eu). M.Z. Mughal and M. Sebastiani acknowledge the financial support through the European FP7 Project, iSTRESS (Grant # 604646, www.istress.eu). Authors are also grateful for the support of the German public funded project ReLiOn (contract No. 03X4619C), funded by the German Federal Ministry of Education and Research (BMBF) and managed by the Project Management Agency Forschungszentrum Jülich (PTJ). Authors also acknowledge the technical assistance from Kitta for the production of wafers. Special thanks to Fabrizia Vallerani for her support during FIB pillar milling at the interdepartmental laboratory of electron microscopy (LIME) of University of "Roma TRE", Rome, Italy.

Author Contributions: M.Z. Mughal performed pillar-splitting experiments, and wrote and edited most sections of the manuscript. H.Y. Amanieu performed COD and EBSD measurements. R. Moscatelli did nanoindentation tests and M. Sebastiani critically reviewed the pillar-splitting data and collaborated in the editing process.

Conflicts of Interest: The authors declare no conflict of interest.

References

1. Lei, J.; Li, L.; Kostecki, R.; Muller, R.; McLarnon, F. Characterization of sei layers on $LiMn_2O_4$ cathodes with in situ spectroscopic ellipsometry. *J. Electrochem. Soc.* **2005**, *152*, A774–A777. [CrossRef]
2. Vetter, J.; Novák, P.; Wagner, M.R.; Veit, C.; Möller, K.C.; Besenhard, J.O.; Winter, M.; Wohlfahrt-Mehrens, M.; Vogler, C.; Hammouche, A. Ageing mechanisms in lithium-ion batteries. *J. Power Sources* **2005**, *147*, 269–281. [CrossRef]
3. Moon, H.-S.; Lee, W.; Reucroft, P.J.; Park, J.-W. Effect of film stress on electrochemical properties of lithium manganese oxide thin films. *J. Power Sources* **2003**, *119–121*, 710–712. [CrossRef]
4. Chen, D.; Indris, S.; Schulz, M.; Gamer, B.; Mönig, R. In situ scanning electron microscopy on lithium-ion battery electrodes using an ionic liquid. *J. Power Sources* **2011**, *196*, 6382–6387. [CrossRef]
5. Hao, X.; Lin, X.; Lu, W.; Bartlett, B.M. Oxygen vacancies lead to loss of domain order, particle fracture, and rapid capacity fade in lithium manganospinel ($LiMn_2O_4$) batteries. *ACS Appl. Mater. Interfaces* **2014**, *6*, 10849–10857. [CrossRef] [PubMed]
6. Huang, M.-R.; Lin, C.-W.; Lu, H.-Y. Crystallographic facetting in solid-state reacted $LiMn_2O_4$ spinel powder. *Appl. Surf. Sci.* **2001**, *177*, 103–113. [CrossRef]
7. Akimoto, J.; Takahashi, Y.; Gotoh, Y.; Mizuta, S. Single crystal x-ray diffraction study of the spinel-type $LiMn_2O_4$. *Chem. Mater.* **2000**, *12*, 3246–3248. [CrossRef]
8. Lawn, B.R.; Evans, A.; Marshall, D. Elastic/plastic indentation damage in ceramics: The median/radial crack system. *J. Am. Ceram. Soc.* **1980**, *63*, 574–581. [CrossRef]
9. Anstis, G.; Chantikul, P.; Lawn, B.R.; Marshall, D. A critical evaluation of indentation techniques for measuring fracture toughness. I.-direct crack measurements. *J. Am. Ceram. Soc.* **1981**, *64*, 533–538. [CrossRef]
10. Meschke, F.; Alves-Riccardo, P.; Schneider, G.A.; Claussen, N. Failure behavior of alumina and alumina/silicon carbide nanocomposites with natural and artificial flaws. *J. Mater. Res.* **1997**, *12*, 3307–3315. [CrossRef]
11. Pharr, G. Measurement of mechanical properties by ultra-low load indentation. *Mater. Sci. Eng. A* **1998**, *253*, 151–159. [CrossRef]
12. Oliver, W.C.; Pharr, G.M. Improved technique for determining hardness and elastic modulus using load and displacement sensing indentation experiments. *J. Mater. Res.* **1992**, *7*, 1564–1580. [CrossRef]
13. Jang, J.-I.; Pharr, G.M. Influence of indenter angle on cracking in si and ge during nanoindentation. *Acta Mater.* **2008**, *56*, 4458–4469. [CrossRef]
14. Sebastiani, M.; Johanns, K.; Herbert, E.; Carassiti, F.; Pharr, G. A novel pillar indentation splitting test for measuring fracture toughness of thin ceramic coatings. *Philos. Mag.* **2015**, *95*, 1928–1944. [CrossRef]
15. Rödel, J.; Kelly, J.F.; Lawn, B.R. In situ measurements of bridged crack interfaces in the scanning electron microscope. *J. Am. Ceram. Soc.* **1990**, *73*, 3313–3318. [CrossRef]
16. Sebastiani, M.; Johanns, K.E.; Herbert, E.G.; Pharr, G.M. Measurement of fracture toughness by nanoindentation methods: Recent advances and future challenges. *Curr. Opin. Sol. State Mater. Sci.* **2015**, *19*, 324–333. [CrossRef]
17. Mughal, M.Z.; Moscatelli, R.; Amanieu, H.-Y.; Sebastiani, M. Effect of lithiation on micro-scale fracture toughness of $LixMn_2O_4$ cathode. *Scr. Mater.* **2016**, *116*, 62–66. [CrossRef]
18. Mughal, M.Z.; Moscatelli, R.; Sebastiani, M. Load displacement and high speed nanoindentation data set at different state of charge (soc) for spinel $LixMn_2O_4$ cathodes. *Data Br.* **2016**, *8*, 203–206. [CrossRef] [PubMed]
19. Renzelli, M.; Mughal, M.Z.; Sebastiani, M.; Bemporad, E. Design, fabrication and characterization of multilayer cr-crn thin coatings with tailored residual stress profiles. *Mater. Des.* **2016**, *112*, 162–171. [CrossRef]
20. Amanieu, H.-Y.; Rosato, D.; Sebastiani, M.; Massimi, F.; Lupascu, D.C. Mechanical property measurements of heterogeneous materials by selective nanoindentation: Application to $LiMn_2O_4$ cathode. *Mater. Sci. Eng. A* **2014**, *593*, 92–102. [CrossRef]
21. Kitta, M.; Akita, T.; Kohyama, M. Preparation of a spinel $LiMn_2O_4$ single crystal film from a mno wafer. *J. Power Sources* **2013**, *232*, 7–11. [CrossRef]
22. Korsunsky, A.M.; Sebastiani, M.; Bemporad, E. Residual stress evaluation at the micrometer scale: Analysis of thin coatings by fib milling and digital image correlation. *Sur. Coat. Technol.* **2010**, *205*, 2393–2403. [CrossRef]
23. Sebastiani, M.; Eberl, C.; Bemporad, E.; Pharr, G.M. Depth-resolved residual stress analysis of thin coatings by a new fib–Dic method. *Mater. Sci. Eng. A* **2011**, *528*, 7901–7908. [CrossRef]
24. Lawn, B.R. *Fracture of Brittle Solids*; Cambridge University Press: Melbournce, Australia, 1993.

25. Cuadrado, N.; Casellas, D.; Anglada, M.; Jiménez-Piqué, E. Evaluation of fracture toughness of small volumes by means of cube-corner nanoindentation. *Scr. Mater.* **2012**, *66*, 670–673. [CrossRef]
26. Amanieu, H.-Y.; Aramfard, M.; Rosato, D.; Batista, L.; Rabe, U.; Lupascu, D.C. Mechanical properties of commercial LixMn$_2$O$_4$ cathode under different states of charge. *Acta Mater.* **2015**, *89*, 153–162. [CrossRef]
27. Wolfenstine, J.; Allen, J.L.; Jow, T.R.; Thompson, T.; Sakamoto, J.; Jo, H.; Choe, H. Licopo4 mechanical properties evaluated by nanoindentation. *Ceram. Int.* **2014**, *40*, 13673–13677. [CrossRef]

materials

MDPI

Article

Investigation on Indentation Cracking-Based Approaches for Residual Stress Evaluation

Felix Rickhey, Karuppasamy Pandian Marimuthu and Hyungyil Lee *

Department of Mechanical Engineering, Sogang University, Seoul 04107, Korea; felix@sogang.ac.kr (F.R.); pandian@sogang.ac.kr (K.P.M.)
* Correspondence: hylee@sogang.ac.kr; Tel.: +82-2705-8636

Academic Editor: Ting Tsui
Received: 14 March 2017; Accepted: 6 April 2017; Published: 12 April 2017

Abstract: Vickers indentation fracture can be used to estimate equibiaxial residual stresses (RS) in brittle materials. Previous, conceptually-equal, analytical models were established on the assumptions that (i) the crack be of a semi-circular shape and (ii) that the shape not be affected by RS. A generalized analytical model that accounts for the crack shape and its change is presented. To assess these analytical models and to gain detailed insight into the crack evolution, an extended finite element (XFE) model is established. XFE analysis results show that the crack shape is generally not semi-circular and affected by RS and that tensile and compressive RS have different effects on the crack evolution. Parameter studies are performed to calibrate the generalized analytical model. Comparison of the results calculated by the analytical models with XFE results reveals the inaccuracy inherent in the previous analytical models, namely the neglect of (the change of) the crack aspect-ratio, in particular for tensile RS. Previous models should therefore be treated with caution and, if at all, used only for compressive RS. The generalized model, on the other hand, gives a more accurate description of the RS, but requires the crack depth.

Keywords: residual stress; indentation fracture; fracture toughness; extended finite element analysis

---

1. Introduction

Residual stresses (RS) exist in many structures. They may have been induced intentionally (e.g., shot peening, chemical strengthening) or inevitably (e.g., cold working due to polishing, thermal treatment accompanied by phase transformation) and significantly affect fatigue life, corrosion or wear resistance, in a positive or negative way [1]. Tempering, for example, is a very effective means to improve the strength and contact damage resistance in glass ceramics. Here, compressive RS are introduced into the surface, thereby increasing the effective stress for damage initiation and propagation. RS can also enhance the mobility of charge carriers in semiconductor devices [2]. Hence, RS play a central role regarding the performance of brittle structures, and their determination has been of considerable interest [3,4].

Techniques for RS determination can be categorized into destructive and non-destructive techniques [5]. Destructive methods, such as hole-drilling, saw-cutting, curvature and layer removal [6], rely on the deformation due to the (partial) relief of RS upon removal of material. When employing non-destructive techniques (NDT), RS are usually inferred indirectly. NDT include ultra-sonic methods, micromagnetic methods, Raman spectroscopy, neutron or X-ray diffraction [7–9]. Many of these methods are however rather expensive or limited in their applicability.

A mechanical NDT, indentation, is a convenient, inexpensive and quick means for RS estimation and can be applied to ductile [10,11], as well as brittle materials [12]. Generally, RS support (tensile RS) or work against (compressive RS) the penetration of the material by the indenter, resulting in a downward (tensile RS) or upward shift (compressive RS) of the characteristic indentation force-indentation depth curve. Further, RS influence the pile-up/sink-in at the impression border. In the case of ductile materials, RS are determined directly from these RS-induced changes in resistance to indentation and pile-up/sink-in behavior. However, in the case of brittle materials, cracks may emanate from the corners of the impression or inside the material, depending on the indenter shape and material [13], and grow into a half penny-shaped crack. This method has been frequently used to measure the fracture toughness of ceramics and glasses. To evaluate RS, the sensitivity of the final crack dimensions to in-plane RS is made use of. This method is particularly advantageous for local subsurface RS determination and in cases where optical and other conventional methods such as fracturing in flexure are not employable. So far, spherical [14–17], conical [18], cube-corner [19,20], Vickers [15,21], Berkovich [22] and Knoop indenters [18] have been employed.

Tandon and Cook [23] and Koike et al. [24] investigated the differences in sharp indentation crack initiation and propagation between annealed, tempered and ion-exchange-strengthened glasses and noted that compressive RS yield a decrease in the propensity to initiation of radial and median cracks. Zeng and Rowcliffe [25] presented a Vickers indentation method to analyze the RS field around a highly-stressed region in a glass specimen (the RS field itself was generated by a Vickers indenter). Later, Kese and Rowcliffe [19] did the same with cube-corner indenters, but they assumed a semi-elliptical crack geometry, which is different from those assumed by Zeng and Rowcliffe [25] and Zeng et al. [26]; a new crack geometry factor was proposed based on a cube-corner indenter. However, the calculated RS were 2~4-times higher than those calculated by Zeng and Rowcliffe [25], Zeng et al. [26] and Peitl et al. [27]. Roberts et al. [16] and Bisrat and Roberts [28] exploited the shift in the threshold load for the propagation of a pre-existing surface crack to estimate RS. The Vickers indentation study by Peitl et al. [27] revealed that the radial-median cracks in the glass ceramic they employed are not of semi-circular, but of a semi-elliptical shape, which means that the appropriateness of assuming a semi-circular crack, as was done before, is questionable. To account for the departure from the semi-circular shape, they introduced a correction factor in the formulation of Zeng and Rowcliffe [27,29]. Rodríguez-López et al. [30] applied the method of Peitl et al. to evaluate the RS in laser-cladded glass-ceramic sealants on Crofer22APU steel. Therefore, the experimental part of this approach was limited to one material, i.e., glass ceramics (glass matrix).

Today, indentation fracture is, for example, used for determining RS in dental ceramics [31–37] or in shot-peened ceramics [38]. Thus far, indentation-based studies were limited to compressive RS and only a few materials (mainly glass). Previous analytical models were established on assumptions whose appropriateness has not been verified by experiments, which were qualitative rather than quantitative or numerical techniques. The goal here is therefore (i) to provide better insight into the influence of both compressive and tensile equibiaxial RS on subsurface crack evolution and final crack shape and size; (ii) how crack evolution is affected by relevant material properties and (iii) to scrutinize previous models and their underlying assumptions by comparison with a generalized analytical model and numerical results obtained by the extended finite element method (XFEM). To the authors' knowledge, this is the first XFEM-based study on indentation of pre-stressed specimens.

2. Analytical Models

The difference in size of indenter-induced cracks in a stressed structure as compared with the RS-free equivalent can be explained by considering the RS field to be superimposing onto the indentation stress field [14]. The RS-induced change in the stress field affects crack initiation and propagation. Tensile RS increase the tensile wedging forces and thus cause earlier damage initiation and a larger final crack. Conversely, compressive RS result in restrained (or even totally suppressed) crack formation and thus a smaller final crack (or no crack) [15,39].

2.1. Shape Factor for a Semi-Elliptical Surface Crack Subject to Remote Tension

Based on linear elastic fracture mechanics (LEFM), the mode I-stress intensity factor (SIF) for a crack subject to remote tensile stress σ^∞ (Figure 1) can be expressed as:

$$K = Y\sigma^\infty \sqrt{\pi a_{char}},\tag{1}$$

where a_{char} is a characteristic crack dimension. The shape factor Y accounts for the crack configuration and the change of the SIF along the crack front. Here, we assume the crack to be semi-elliptical. $a_{char} = c_z$ (crack depth), and the aspect ratio of the ellipse is $\rho \equiv c_z/c$, where c is the length of the crack on the surface (Figures 1 and 2). Y for a semi-elliptical surface crack with $\rho \leq 1$ (in this study, ρ was always < 1) under mode I-loading conditions is [40]:

$$Y(\rho, \omega) = \frac{M f_\omega g}{\sqrt{Q}},\tag{2}$$

$$Q = 1 + 1.464\rho^{1.65} \ ; \ M = 1.13 - 0.09\rho,$$
$$f_\omega = \left[\sin^2\omega + \rho^2\cos^2\omega\right]^{1/4} \ ; \ g = 1 + 0.1\left(1 - \sin^2\omega\right),\tag{3}$$

where Q, M, f_ω and g are geometry factors and the parametric angle ω is $0°$ at the surface and $90°$ at the apex of the crack. The variation of Y along the crack front is plotted in Figure 3 (left) for diverse ρ. For $\rho > \rho_{eq} = 0.826$, crack growth takes place at the surface (Point B), and for $\rho < \rho_{eq}$, K becomes maximum at the apex (A). Regarding the change of Y at B with ρ in the range [0.7, 1], the maximum difference in Y is 2.5% (Figure 3, right).

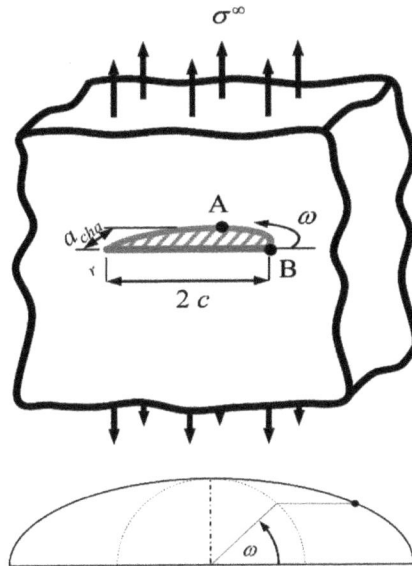

Figure 1. Semi-elliptical surface crack in a semi-infinite medium subject to remote tensile stress σ^∞ (**top**) and the definition of the parametric angle (**bottom**) (following Anderson [41]).

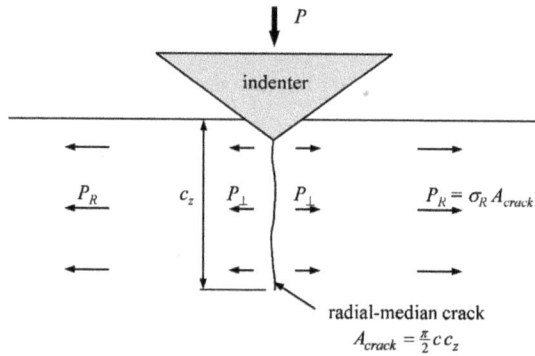

Figure 2. Residual force P_R superimposes on wedging forces $P_\perp$, both acting normal to the crack surface A_{crack}.

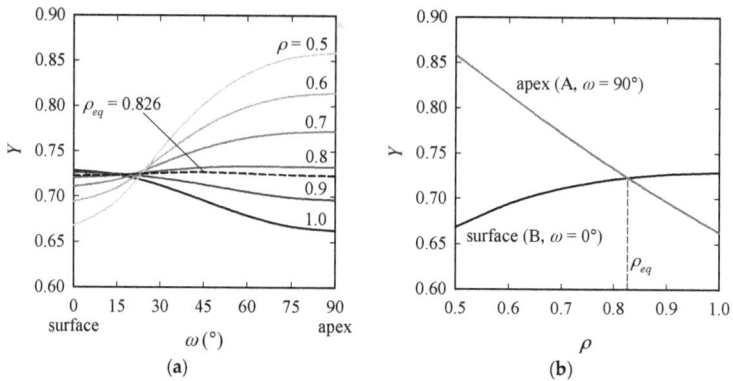

Figure 3. Change of shape factor Y with ω for (**a**) diverse ρ (Equation (2)) and (**b**) with ρ at Points A and B; for $\rho = 0.826$, Y has equal values at A and B.

2.2. Previous Approaches

Marshall and Lawn [42] and, shortly afterwards, Swain et al. [43] presented LEFM-based models, which are conceptually equal. The equibiaxial RS field was superimposed on the indentation stress field, as shown in Figure 2. Self-similarity requires that the indentation load P scale with the wedging force acting normal to the crack surfaces, $P_\perp$. Since loading modes are consistent (mode I), the contributions to the total SIF K_{tot} are additive; thus:

$$K_{tot} = K_{ind} + K_R, \tag{4}$$

where K_{ind} and K_R are the SIF due to the indentation stress field and the RS field, respectively. The crack was assumed to be of a semi-circular shape, i.e., $\rho = 1$, independent of material or RS. The SIF solution for a semi-circular surface crack in a semi-infinite medium subject to tensile stress σ_R has the form:

$$K_R = Y(1, \omega)\sigma_R \sqrt{\pi c}, \tag{5}$$

which is maximum at the surface ($\omega = 0°$). For indentation on an RS-free specimen (e.g., Lawn et al. [44]):

$$K_{ind} = \chi \frac{P_{max}}{c^{3/2}}, \tag{6}$$

where χ is a material- and indenter-dependent constant. At the crack tip during crack propagation, $K_{tot} = K_c$, where K_c is the critical SIF (=fracture toughness). Inserting Equations (5) and (6) into Equation (4), Marshall and Lawn [42] arrived at the following expression for σ_R:

$$\sigma_R = \frac{\sqrt{\pi}\chi}{2m\sqrt{c}}\left[\frac{K_c}{\chi} - \frac{P_{max}}{c^{3/2}}\right], \tag{7}$$

Correction m was set to unity, which means a neglect of free surface-effects and a homogeneous RS field. Indentations to diverse P_{max} were carried out on as-annealed soda-lime glass specimens (to obtain the material's reference K_c/χ) and tempered soda-lime glass specimens, and results gave an approximately linear relation between $P/c^{3/2}$ and $c^{1/2}$. Zeng and Rowcliffe [25] also applied Equation (7), but with $m = \pi/2$.

Swain et al. [43] suggested indenting RS-free and specimens subject to RS, so that equal final crack lengths c_n are obtained. The compressive RS, σ_R, was calculated from the load difference by:

$$\sigma_R = \frac{\chi}{Y\sqrt{\pi}c_n^2}[P_{o\ max} - P_{max}], \tag{8}$$

where subscript 'o' denotes the (reference) RS-free case. χ is obtained from the indentation of the annealed (i.e., RS-free) specimen, and P_{max} is the maximum indentation load necessary to produce a crack of nominal length c_n in the stressed material. Experiments on tempered and annealed soda-lime glass specimens verified the proportionality of c_n^2 and the load difference. While Swain et al. assumed $Y = \pi^{-1/2}$, Chaudhri and Phillips [15] later accounted for the free surface and corrected the shape factor to $Y = 1.16\pi^{-1/2}$.

When modifying the model by Swain et al. [43] so that it can be used for indentations to a nominal P_{max} (instead of c_n), we can write the above analytical models in the following standard form:

$$\sigma_R = \frac{\chi}{Y}\frac{P_{max}}{\sqrt{\pi}c^2}\left[\left(\frac{c}{c_o}\right)^{3/2} - 1\right]; \quad Y = \begin{cases} 0.56 & \text{Swain } et\ al.\ [43] \\ 0.64 & \text{Marshal and Lawn } [42] \\ 0.65 & \text{Chaudhri and Phillips } [15] \\ 1 & \text{Zeng and Rowcliffe } [25] \end{cases}, \tag{9}$$

The analytical models thus differ only by Y, which however ranges, depending on the source, between 0.56 and one. Two indentations, one on the RS-free and one on the specimen subject to RS, to equal P_{max}, are required to get c/c_o. For two indentations to equal h_{max} (as in our numerical analyses), Equation (9) becomes:

$$\sigma_R = \frac{K_c}{Y\sqrt{\pi c}}\left[1 - \frac{C}{C_o}\left(\frac{c}{c_o}\right)^{3/2}\right], \tag{10}$$

where C/C_o denotes the relative change of Kick's law coefficient. Kick's law denotes the linear relation between indentation load P and squared indentation depth h^2, $P = C\,h^2$, for indentation with symmetric sharp indenters on elastic-perfectly plastic materials [39]. Kick's law coefficient C is then a material constant. Note that Kick's law holds independent of RS or cracking. As stated before, Equation (9) assumes that $\rho = 1$ holds independent of material properties and indenter shape. However, ρ is expected to be material- and indenter-dependent (compare Rickhey et al. [45] for Knoop indentation of the RS-free specimen) and may, moreover, change under the influence of RS, so that $\rho/\rho_o \neq 1$. The consequences of these assumptions will be scrutinized in Section 4.

2.3. Generalized Analytical Model

To assess the appropriateness of these assumptions, a generalized analytical model, similar to Peitl et al. [27], is suggested. The stresses in the elastic far-field are approximated by the Boussinesq solution for the stress distribution in an elastic half-space subject to a point load acting normal to the surface. The crack-driving circumferential stresses σ_θ in a median plane (i.e., a plane normal to the specimen surface) are:

$$\sigma_\theta \propto \frac{P_\perp}{\pi R^2} g_\omega(\omega), \tag{11}$$

where $P_\perp$ is the force acting normal to the crack plane (Figure 2) and g_ω is an angular function of ω, which is defined in Figure 1. At the surface ($R = c$, $\omega = 0°$; Point B in Figure 1), where the crack propagates, the following relation exists for the SIF:

$$K_{ind} \propto Y\frac{P_\perp}{\pi c^2}\sqrt{\pi c_z} = Y\sqrt{\rho}\frac{P_\perp}{\sqrt{\pi}c^{3/2}}, \tag{12}$$

and the shape factor becomes:

$$Y = \frac{1.243 - 0.1\rho}{\sqrt{1 + 1.464\rho^{1.65}}}\sqrt{\rho}, \tag{13}$$

In accordance with self-similarity, the crack-driving force scales with the indentation load P_{max}, i.e.,

$$P_\perp = kP_{max}, \tag{14}$$

where k is an indenter shape- and material-dependent scale factor. k is expected to be closely related to the residual field intensity factor χ. Plugging Equation (14) into Equation (12), we get:

$$K_{ind} = Y\sqrt{\rho}\frac{kP_{max}}{\sqrt{\pi}c^{3/2}}, \tag{15}$$

The SIF contribution from the in-plane RS, which act as remote tensile stresses, i.e., $\sigma^\infty = \sigma_R$, is:

$$K_R = Y\sqrt{\rho}\sigma_R\sqrt{\pi c}, \tag{16}$$

During crack propagation $K = K_c$ must hold, that is:

$$\sigma_R = 0 \rightarrow K_{tot} = K_{ind} = K_c; \sigma_R \neq 0 \rightarrow K_{tot} + K_R = K_c, \tag{17}$$

Comparison of the equibiaxial RS case with the RS-free case gives:

$$\sigma_R = \frac{k}{\sqrt{\pi}\chi}\frac{K_c}{\sqrt{\pi c}}\left[\frac{Y_o}{Y}\sqrt{\frac{\rho_o}{\rho}} - \left(\frac{c}{c_o}\right)^{3/2}\right] = \frac{kP_{max}}{\pi c^2}\left[\frac{Y_o}{Y}\sqrt{\frac{\rho_o}{\rho}}\left(\frac{c}{c_o}\right)^{3/2} - 1\right], \tag{18}$$

(It was eventually found that the variation of Y_o/Y is very small, so that the term Y_o/Y might as well be removed from Equations (18) and (19) without loss of accuracy. However, we do not remove it here for consistency.) In the case of indentations, to equal h_{max}, Equation (18) changes to:

$$\sigma_R = \frac{kK_c}{\pi\chi\sqrt{c}}\left[\frac{Y_o}{Y}\sqrt{\frac{\rho_o}{\rho}} - \frac{c}{c_o}\left(\frac{c}{c_o}\right)^{3/2}\right], \tag{19}$$

Note that for $\rho = 1$ and $\rho/\rho_o = 1$, Equation (18) reduces to:

$$\sigma_R = \frac{kK_c}{\pi\chi\sqrt{c}}\left[1 - \left(\frac{c_o}{c}\right)^{3/2}\right] = \frac{kP_{max}}{\pi c^2}\left[\left(\frac{c}{c_o}\right)^{3/2} - 1\right], \tag{20}$$

which is equal to Equation (9) with $k/\chi = \pi^{1/2}/Y$.

In summary, the assumptions common to both models, henceforth termed 'simple' (Equation (9) or (10)) and 'generalized' (Equation (18) or (19)), are that SIF are additive and that the crack shape can be sufficiently described by ρ. The 'simple' model further requires that the crack be semi-circular ($\rho = 1$) and that RS do not affect ρ ($\rho/\rho_0 = 1$). The 'generalized' model, on the other hand, accounts for the material- and indenter-dependent crack aspect ratio and its possible change with RS.

The analytical models are not restricted to a particular sharp indenter. The non-interaction of radial-median cracks argues for Knoop indentation. However, since for general in-plane biaxial RS, the Vickers indenter will be advantageous (the non-equality of RS will be reflected by non-equal crack lengths in the median planes through the indenter diagonals, and provided sufficient sensitivity, only one indentation will be necessary), it is chosen here to make the study extendible to general biaxial RS.

3. FE Model and Imposition of Equibiaxial RS

Numerical analyses are performed with Abaqus/Standard. As the XFE model shown in Figure 4 is quite similar to the validated one in Rickhey et al. [45], this section is limited to essential information and to highlight differences. Owing to symmetry, modeling of one quarter specimen is sufficient. The model consists of $\approx 10^5$ eight-node brick elements [46]. All nodes on the outer surfaces of the model are fixed, so that RS can be introduced through the "initial conditions" option in Abaqus. Bottom nodes are fixed, as well. For simplicity, the indenter is restricted to movements in the z-direction and assumed rigid. The latter assumption is, however, not expected to influence the accuracy of the results to a degree that invalidates the approach, because the indenter's elastic modulus is usually much higher than that of the material to be indented.

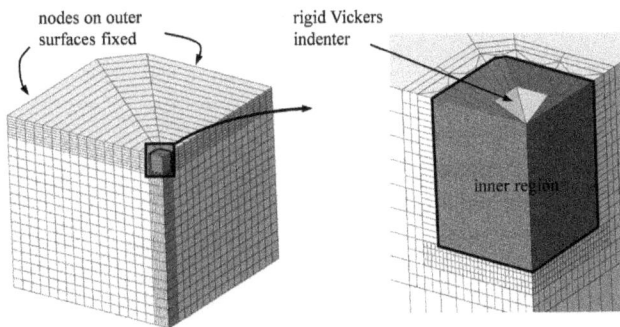

Figure 4. Quarter FE model for evaluation of residual stresses (RS) in brittle materials by Vickers indentation cracking.

The material to be indented is assumed to exhibit elastic-perfectly plastic material behavior and to yield according to the von Mises yield condition (yield strength σ_y). In fact, the compressive behavior of many brittle materials can be accurately described by this material model [47]. Damage is assumed to obey a bilinear continuum traction-separation law, governed by a damage-initiating stress threshold $\hat{\sigma}$ (=0.8 GPa) and the fracture energy Γ, which is tentatively set to 3.0 MPa μm. The dimensionless viscosity parameter, introduced to mitigate convergence problems associated with material softening, is set to $\zeta = 5 \times 10^{-5}$. The role and choice of damage model parameters was discussed in detail in [45,48]. Friction between the indenter and specimen is considered with $\mu = 0.2$. If not stated otherwise, all indentations were performed with a prescribed $h_{max} = 1.5$ μm.

4. FE Results and Observations for Equibiaxial RS

Parameter studies were carried out by varying material properties and equibiaxial RS to see whether and how equibiaxial RS affect cracking in general and ρ/ρ_0 in particular. Finally, the RS calculated by the analytical models are compared with the RS input to the numerical analyses to discuss the importance of ρ and ρ/ρ_0 regarding the accuracy of the results. First, we analyze the influence of RS on crack evolution for a reference material with properties close to polycrystalline silicon ($E = 200$ GPa, $v = 0.3$, $\sigma_y = 5$ GPa).

4.1. Observations Made for Reference Material

XFE results for the reference material reveal the following: The impression half-diagonal remains unaffected by RS, which agrees with Swain et al. [43]. Kick's law ($P \propto h^2$) holds irrespective of RS and cracking. The decrease of C caused by cracking is negligible; as expected, tensile RS support the wedging process performed by the indenter so that C decreases with increasing RS (Figure 5). Compressive RS impede both the crack propagation in the depth direction during loading and the opening-up during unloading, whereas tensile RS have the opposite effect. The relative change in c and c_z (normalized by 'RS-free' values c_o and c_{zo}, respectively) is however more pronounced for tensile RS than for compressive RS.

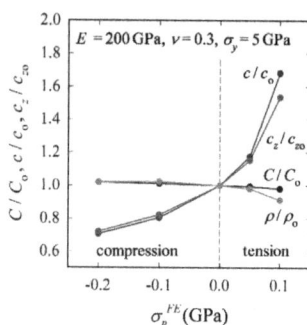

Figure 5. Relative change of Kick's law coefficient C/C_o, crack length c/c_o, crack depth c_z/c_{zo} and crack aspect-ratio ρ/ρ_o with σ_R^{FE} (FE input).

Figure 6 demonstrates how RS influence the crack configurations at load reversal ($P = P_{max}$) and after unloading ($P = 0$) for $\sigma_R = -0.2$, 0 and 0.1 GPa. We observe that while for the RS-free and compressive RS cases, c_z does not change during unloading, there is an increase in c_z in the tensile RS case. Further, compressive RS do not cause an evident change in ρ (i.e., $\rho/\rho_o \approx 1$), but ρ decreases with increasing tensile RS, indicating that the influence of tensile RS on radial-median cracking is fundamentally different from that of compressive RS. Note that a fundamental difference between tensile and compressive RS has also been observed in indentation of ductile materials (e.g., Sines and Carlson [49]; Suresh and Giannakopoulos [50]; Rickhey et al. [11]).

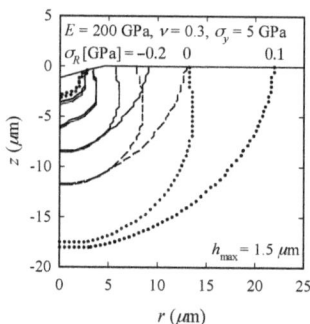

Figure 6. Influence of compressive and tensile RS on crack evolution at load reversal (semi-transparent lines) and after unloading.

From sharp indentation of the RS-free specimen, we know that upon sufficient loading, χ becomes constant; the generated crack is then called well-developed. We now investigate the influence of RS on the residual field intensity at diverse h_{max}. Let us for this purpose introduce an apparent residual field intensity coefficient χ^{app} as follows:

$$\chi^{app} \equiv \frac{K_c}{P_{max}/c^{3/2}}, \tag{21}$$

For the RS-free case, $\chi^{app} = \chi$. Hence, with Equation (18), we get:

$$\sigma_R = \frac{k}{\sqrt{\pi\chi}} \frac{K_c}{\sqrt{\pi c}} \left[\frac{Y_o}{Y} \sqrt{\frac{\rho_o}{\rho}} - \frac{\chi}{\chi^{app}} \right] = \frac{kP_{max}}{\pi c^2} \left[\frac{Y_o}{Y} \sqrt{\frac{\rho_o}{\rho}} \frac{\chi^{app}}{\chi} - 1 \right], \tag{22}$$

for load-controlled indentations and, with Equation (19),

$$\sigma_R = \frac{kK_c}{\pi\chi\sqrt{c}} \left[\frac{Y_o}{Y} \sqrt{\frac{\rho_o}{\rho}} - \frac{C}{C_o} \frac{\chi}{\chi^{app}} \right], \tag{23}$$

for depth-controlled indentations (note that K_c is a material property and as such independent of RS). In a similar fashion, we introduce:

$$\chi_z^{app} \equiv \frac{K_c}{P_{max}/c_z^{3/2}}, \tag{24}$$

When plotting χ^{app} obtained from indentations to h_{max} = [0.75, 2.25] μm, Figure 7 further demonstrates the different nature of compressive and tensile RS. (Note that some combinations of h_{max} and σ_R could not be considered because of the limited size of the inner region (high h_{max}, high σ_R) or because of suppressed radial-median crack formation (low h_{max}, low σ_R). The size of the inner region is limited by computational costs, which are determined by h_{max}, the chosen element size and $\hat{\sigma}$ [45].) χ^{app} and χ_z^{app} remain constant for compressive RS (as known from the RS-free case), but increase clearly with h_{max} for tensile RS; ρ [= $(\chi^{app}/\chi_z^{app})^{2/3}$] is independent of RS for compressive RS, but decreases with increasing RS for tensile RS.

The coefficient k in Equation (19) is found by calculating $\sigma_R^{FE}/(\sigma_R^{Eq}/k)$, where σ_R^{FE} is the RS imposed in the FE analysis and σ_R^{Eq} is the RS calculated by Equation (19). Plotting k from Equation (19) against RS (Figure 8), we see that k is approximately constant for the reference material ($\approx$0.24). This will be discussed in more detail in the next section.

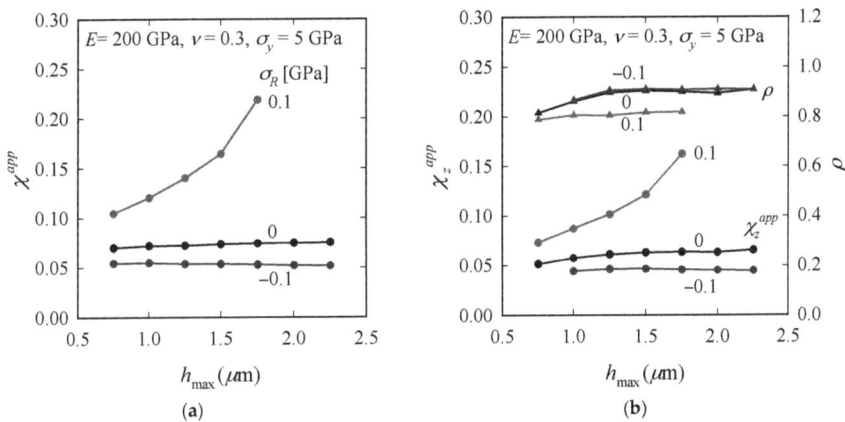

Figure 7. Change of χ^{app} (**a**); χ_z^{app} and ρ (**b**) with h_{max} for σ_R = 0.1, 0 and −0.1 GPa.

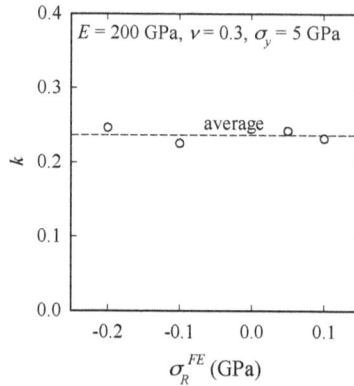

Figure 8. Coefficient k in Equation (19) vs. σ_R.

4.2. Influence of Material Properties

To determine how the relation between k and χ is influenced by material properties, we performed parameter studies varying E, ν and σ_y, so that a large range of brittle materials is covered (Table 1). It is to be noted that the pair (E, σ_y) represents the usually given pair (E, H); the conversion, which is necessary for application in numerical analysis, is described in [45]. Indentation cracking tests were simulated with $\sigma_R = \{-0.2, -0.1, 0, 0.05, 0.1, 0.15\}$ GPa. Due to the high sensitivity to RS for low modulus materials (as we will see later), $\sigma_R = 0.15$ GPa was applied only to materials with $E = 600$ GPa. RS lower than -0.2 GPa were not considered because radial-median cracks could not be generated (not even when increasing h_{max} to 3.0 μm). Higher RS were not imposed because they would have required a very large inner region (Figure 4) and thus significantly higher computational effort.

Figure 9 shows the change of ρ/ρ_0 for four materials. ρ is found to be affected by compressive RS to a relatively low degree, which means that regarding ρ, a material-dependent, but RS independent constant may be acceptable. However, in the case of tensile RS, ρ becomes quite sensitive and decreases rapidly with increasing RS. The decrease of ρ/ρ_0 is tantamount to a more pronounced opening-up of the crack at the surface (B).

Further, independent of material, tensile RS cause the crack to grow in the depth direction during unloading. With increasing RS, the crack shape approaches that of the equilibrium ellipse (ρ_{eq}), whereupon the crack grows also in the depth direction during unloading because the SIF reaches K_c at both A and B. For sufficiently high tensile RS, ρ is expected to be that of the equilibrium ellipse. For low RS, on the other hand, c_z remains constant (only influenced by elastic unloading), which is consistent with theory (because $\rho < \rho_{eq}$).

Table 1. Range of material properties applied in the parametric numerical simulations.

Material Properties	Values
E (GPa)	100, 200, 300, 400, 600
ν	0.1, 0.2, 0.3
σ_y (GPa)	3, 5, 8

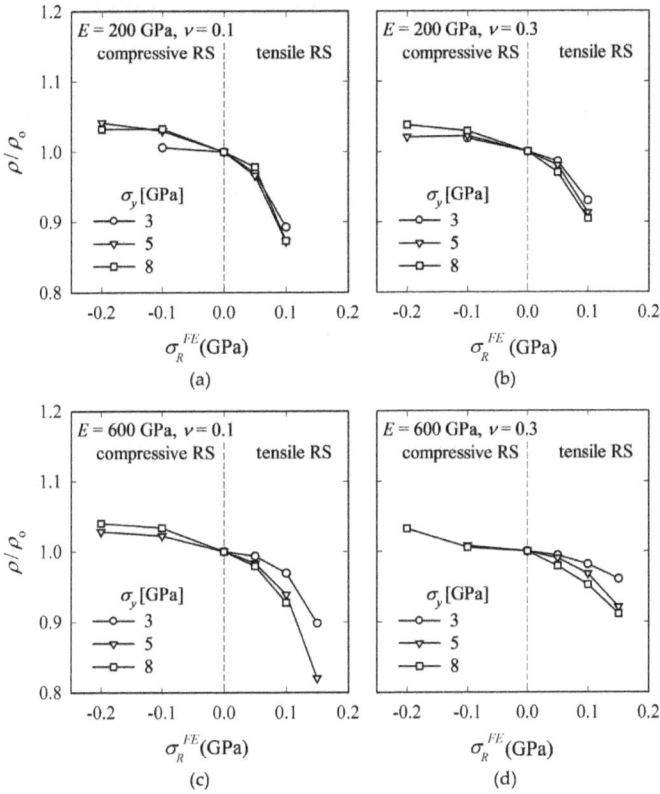

Figure 9. Normalized crack ratio ρ/ρ_0 for all combinations of $E = \{200, 600\}$ GPa, $\nu = \{0.1, 0.3\}$ and $\sigma_y = \{3, 5, 8\}$ GPa (note that not for all materials have radial-median cracks formed).

As mentioned in Section 2, the coefficient k is expected to be closely related to χ. Plotting results for k/χ, we find that, despite some scatter, k/χ shows no systematic dependence on σ_y and ν, yet a low linear dependence on E (Figure 10). To show that results are independent of h_{max}, the procedure is performed again, but now with $h_{max} = 2.0$ μm (instead of 1.5 μm), which, according to Kick's law, means a 1.8-times increase in P_{max}. The results in Figure 10 clearly show that k/χ is not affected by h_{max} (or P_{max}). Based on the FE results for both h_{max}, k/χ can be related to E through:

$$\frac{k}{\chi} = a + bE, \tag{25}$$

where $a = 3.37$ and $b = -9.07 \times 10^{-4}$ GPa^{-1}. The corresponding curve is shown by the dashed lines in Figure 10. The scatter may be explained, at least partly, by the inaccuracies associated with obtaining c and c_z through interpolation.

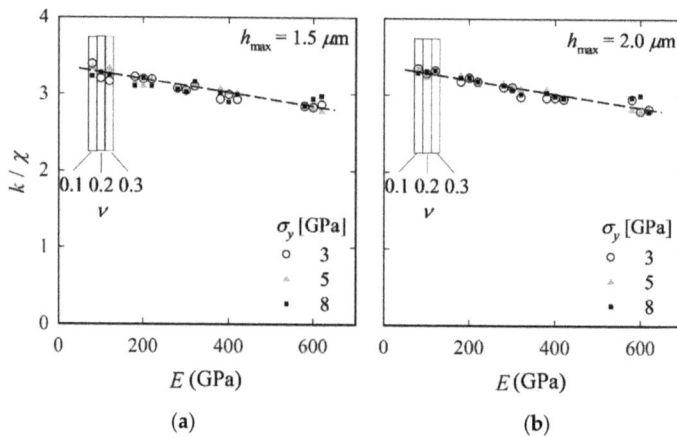

Figure 10. Mean average values of k/χ over the whole range of equibiaxial RS states vs. E; $h_{max} = 1.5$ (a) and 2.0 μm (b); data points for $\nu = 0.1$ and 0.3 are plotted slightly left and right, respectively, of their real location for better visibility.

To show that the 'generalized' model is further independent of Γ, Γ is varied in the range [1.5, 5.0] MPa μm (to reduce computational expense, not all combinations are analyzed). Two materials are considered: the reference material ($h_{max} = 1.5$ and 2.0 μm) and the material with $E = 600$ GPa, $\nu = 0.1$, $\sigma_y = 8$ GPa ($h_{max} = 1.5$ μm). Plugging k/χ from Equation (25) into Equation (19), RS ($\sigma_R{}^{Eq}$) are calculated and compared with the FE input values ($\sigma_R{}^{FE}$). The results plotted in Figure 11 reveal that deviations of $\sigma_R{}^{Eq}$ from $\sigma_R{}^{FE}$ are not systematic and are in the range obtained for $\Gamma = 3$ MPa μm. Equation (19) is thus independent of h_{max} and Γ.

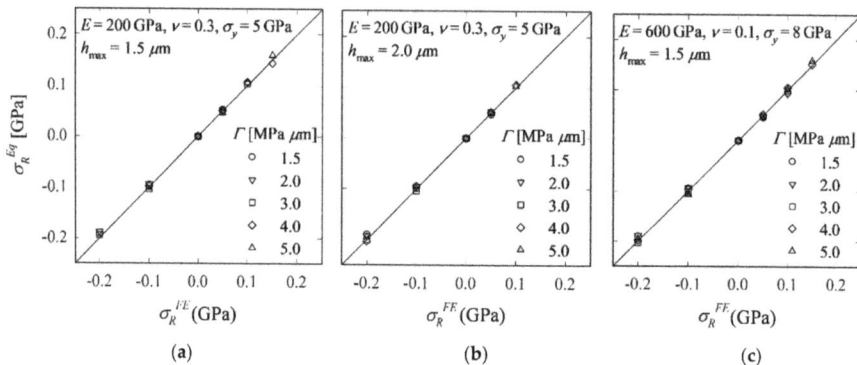

Figure 11. Comparison of RS calculated by Equation (19) ($\sigma_R{}^{Eq}$) with FE input values ($\sigma_R{}^{FE}$) for materials with different Γ; $E = 200$ GPa, $\nu = 0.3$, $\sigma_y = 5$ GPa; $h_{max} = 1.5$ (a) and 2.0 μm (b) and $E = 600$ GPa, $\nu = 0.1$, $\sigma_y = 8$ GPa, $h_{max} = 1.5$ μm (c).

4.3. Comparison of Analytical Models and Conclusion

The RS calculated by the simple and generalized analytical models are compared with FE input values. As can be seen from Figure 12, the 'simple' model with $Y = 0.65$ underestimates RS systematically with errors close to 30%. The error is particularly large for tensile RS. When reducing the average deviation to zero by applying $Y = 0.55$, the error, albeit not anymore systematic, still amounts to up to 20%. Results obtained by the 'generalized' model are in a much better agreement with FE results. Maximum and average errors are reduced from 20% and 5% to 8% and 2%, respectively.

The improved accuracy hints at the importance of the crack aspect-ratio ρ and its change with RS, in particular for tensile RS. We conclude that for the case of compressive RS, ρ and its change caused by RS may be disregarded at the expense of some loss of accuracy. The 'simple' model with a corrected Y of 0.55 can thus be used in cases where (i) a quick evaluation of compressive equibiaxial RS is needed and (ii) the crack depth is unknown or difficult to obtain. However, one should be careful when applying it to the estimation of tensile RS because the results become inaccurate owing to the significant deviation of ρ from ρ_0, as shown in Figure 9. The 'generalized' model, on the other hand, is more accurate, but only applicable when the crack depth can be measured, e.g., by focused ion beam (FIB) tomography [51].

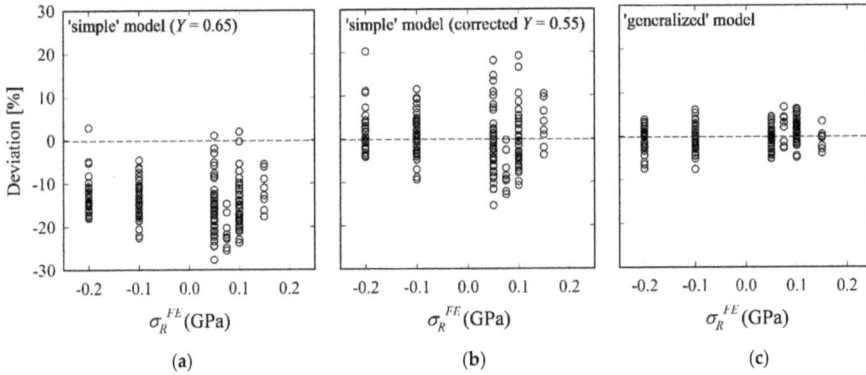

Figure 12. Deviations $(=(\sigma_R{}^{Eq} - \sigma_R{}^{FE})/\sigma_R{}^{FE})$ of results calculated by 'simple' (a); corrected 'simple' (b) and 'generalized' analytical models (c) from FE input RS.

5. Conclusions

The high sensitivity of crack formation to RS makes indentation fracture a powerful non-destructive tool for the evaluation of local subsurface RS. Previous analytical models presupposed (i) a semi-circular crack shape, i.e., $\rho = 1$; and (ii) that the ratio ρ/ρ_0 does not change in the presence of RS, i.e., $\rho/\rho_0 = 1$. However, results revealed that ρ (including ρ_0) is smaller than one and that ρ/ρ_0 clearly decreases with increasing tensile RS. Further, the influence of tensile RS on the crack evolution is notably different from that of compressive RS. While for compressive RS, the change of ρ/ρ_0 is rather small and may be neglected, it should be considered in the case of tensile RS, which demonstrates the weakness of the 'simple' model. The proposed 'generalized' model, which accounts for (the change of) the crack aspect ratio ρ predicted RS much more accurately than the previous models, indicating the importance of (the change of) ρ, in particular for the tensile RS case. It is, however, to be noted that indentation cracking is not applicable in the case of very high compressive RS because the RS inhibit the development of the median crack into a well-developed radial-median crack. Higher tensile RS were not treated here because of the high sensitivity of the final crack size to even small increases in RS. We intend to tackle this issue by experiment in the near future.

Despite the simplicity of the previous approach (especially the independence of c_z), it contains assumptions that lead to inaccuracies and should therefore be used, if at all, only for compressive RS. In connection with the previous model, a shape factor of $Y = 0.55$ was found to give an acceptable prediction of compressive RS. The model proposed here is recommended in the case of tensile RS and where the crack depth can be measured.

Materials **2017**, *10*, 404

Acknowledgments: This research was supported by the Basic Science Research Program through the National Research Foundation of Korea funded by the Ministry of Science, ICT & Future Planning (No. NRF-2015 R1A2A1A 15056163).

Author Contributions: Felix Rickhey outlined the theoretical model, established the FE model, performed FEA, analyzed the data and wrote the paper. Karuppasamy Pandian Marimuthu established the FE model, analyzed the data and co-wrote the paper Hyungyil Lee analyzed the data and co-wrote the paper.

Conflicts of Interest: The authors declare no conflict of interest.

References

1. Warren, A.W.; Guo, Y.B.; Weaver, M.L. The influence of machining induced residual stress and phase transformation on the measurement of subsurface mechanical behavior using nanoindentation. *Surf. Coat. Technol.* **2006**, *200*, 3459–3467. [CrossRef]
2. Golovin, Y.I. Nanoindentation and mechanical properties of solids in submicrovolumes, thin near-surface layers, and films: A Review. *Phys. Solid State* **2008**, *50*, 2205–2236. [CrossRef]
3. Groth, B.P.; Langan, S.M.; Haber, R.A.; Mann, A.B. Relating residual stresses to machining and finishing in silicon carbide. *Ceram. Int.* **2016**, *42*, 799–807. [CrossRef]
4. Wang, C.; Jiang, C.; Cai, F.; Zhao, Y.; Zhu, K.; Chai, Z. Effect of shot peening on the residual stresses and microstructure of tungsten cemented carbide. *Mater. Des.* **2016**, *95*, 159–164. [CrossRef]
5. Rossini, N.S.; Dassisti, M.; Benyounis, K.Y.; Olabi, A.G. Methods of measuring residual stresses in components. *Mater. Des.* **2012**, *35*, 572–588. [CrossRef]
6. Skouras, A.; Paradowska, A.; Peel, M.J.; Flewitt, P.E.J.; Pavier, M.J. Residual stress measurements in a ferritic steel/In625 superalloy dissimilar metal weldment using neutron diffraction and deep-hole drilling. *Int. J. Press. Vessel. Pip.* **2013**, *101*, 143–153. [CrossRef]
7. Hauk, V. *Structural and Residual Stress Analysis by Nondestructive Methods: Evaluation-Application-Assessment*; Elsevier Science B.V.: Amsterdam, The Netherlands, 1997.
8. Kandill, F.A.; Lord, D.J.; Fry, A.T.; Grant, P.V. *A Review of Residual Stress Measurement Methods: A Guide to Technique Selection*; NPL Report/MATC/A: NPL Report; National Physical Laboratory: Teddington, UK, 2001.
9. Jannotti, P.; Subhash, G. Measurement of residual stresses in B4C-SiC-Si ceramics using Raman spectroscopy. In Residual Stress, Thermomechanics & Infrared Imaging, Hybrid Techniques and Inverse Problems, Volume 9. In Proceedings of the 2015 Annual Conference on Experimental and Applied Mechanics, Costa Mesa, CA, USA, 8–11 June 2015; Bossuyt, S., Schajer, G., Carpinteri, A., Eds.; Springer International Publishing: Cham, Switzerland, 2016; pp. 341–345.
10. Jang, J.I. Estimation of residual stress by instrumented indentation: A review. *J. Ceram. Process. Res.* **2009**, *10*, 391–400.
11. Rickhey, F.; Lee, J.H.; Lee, H. A contact size-independent approach to the estimation of biaxial residual stresses by Knoop indentation. *Mater. Des.* **2015**, *84*, 300–312. [CrossRef]
12. Sebastiani, M.; Bemporad, E.; Carassiti, F.; Schwarzer, N. Residual stress measurement at the micrometer scale: Focused ion beam (FIB) milling and nanoindentation testing. *Philos. Mag.* **2011**, *91*, 1121–1136. [CrossRef]
13. Rickhey, F.; Kim, M.; Lee, H. XFEM simulation of radial-median crack formation in brittle medium underneath sharp indenters. In Proceedings of the KSME 2014 Spring Annual Meeting, Ansan, Korea, 22–23 May 2014; The Korean Society of Mechanical Engineers: Seoul, Korea, 2014; pp. 2143–2148.
14. Lawn, B.; Wilshaw, R. Indentation fracture: Principles and applications. *J. Mater. Sci.* **1975**, *10*, 1049–1081. [CrossRef]
15. Chaudhri, M.M.; Phillips, M.A. Quasi-static indentation cracking of thermally tempered soda lime glass with spherical and Vickers indenters. *Philos. Mag. A* **1990**, *62*, 1–27. [CrossRef]
16. Roberts, S.G.; Lawrence, C.W.; Bisrat, Y.; Warren, P.D.; Hills, D.A. Determination of surface residual stresses in brittle materials by Hertzian indentation: Theory and experiment. *J. Am. Ceram. Soc.* **1999**, *82*, 1809–1816. [CrossRef]
17. Petit, F.; Sartieaux, A.C.; Gonon, M.; Cambier, F. Fracture toughness and residual stress measurements in tempered glass by Hertzian indentation. *Acta Mater.* **2007**, *55*, 2765–2774. [CrossRef]

18. Chandrasekar, S.; Chaudhri, M.M. Indentation cracking in soda-lime glass and Ni-Zn ferrite under Knoop and conical indenters and residual stress measurements. *Philos. Mag. A* **1993**, *67*, 1187–1218. [CrossRef]

19. Kese, K.; Rowcliffe, D.J. Nanoindentation method for measuring residual stress in brittle materials. *J. Am. Ceram. Soc.* **2003**, *86*, 811–816. [CrossRef]

20. Scares, P.C.; Lepienski, C.M. Residual stress determination on lithium disilicate glass-ceramic by nanoindentation. *J. Non. Cryst. Solids* **2004**, *348*, 139–143.

21. Ahn, Y.; Chandrasekar, S.; Farris, T.N. Determination of surface residual stresses in machined ceramics using indentation fracture. *J. Manuf. Sci. Eng.* **1996**, *118*, 483–489. [CrossRef]

22. Malzbender, J.; de With, G.; den Toonder, J.M. Elastic modulus, indentation pressure and fracture toughness of hybrid coatings on glass. *Thin Solid Films* **2000**, *366*, 139–149. [CrossRef]

23. Tandon, R.; Cook, R.E. Indentation Crack Initiation and Propagation in Tempered Glass. *J. Am. Ceram. Soc.* **1993**, *76*, 885–889. [CrossRef]

24. Koike, A.; Akiba, S.; Sakagami, T.; Hayashi, K.; Ito, S. Difference of cracking behavior due to Vickers indentation between physically and chemically tempered glasses. *J. Non. Cryst. Solids* **2012**, *358*, 3438–3444. [CrossRef]

25. Zeng, K.; Rowcliffe, D. Experimental measurement of residual stress field around sharp indentation in glass. *J. Am. Ceram. Soc.* **1994**, *77*, 524–530. [CrossRef]

26. Zeng, K.; Giannakopoulos, A.E.; Rowcliffe, D.J. Vickers indentations in glass-II. Comparison of finite element analysis and experiments. *Acta Metall. Mater.* **1995**, *43*, 1945–1954. [CrossRef]

27. Peitl, O.; Serbena, F.C.; Mastelaro, V.R.; Zanotto, E.D. Internal residual stress measurements in a bioactive glass–ceramic using Vickers indentation. *J. Am. Ceram. Soc.* **2010**, *93*, 2359–2368. [CrossRef]

28. Bisrat, Y.; Roberts, S.G. Residual stress measurement by Hertzian indentation. *Mater. Sci. Eng. A* **2000**, *288*, 148–153. [CrossRef]

29. Serbena, F.C.; Zanotto, E.D. Internal residual stresses in glass-ceramics: A review. *J. Non. Cryst. Solids* **2012**, *358*, 975–984. [CrossRef]

30. Rodríguez-López, S.; Comesaña, R.; del Val, J.; Durán, A.; Justo, V.M.; Serbena, F.C.; Pascual, M.J. Laser cladding of glass-ceramic sealants for SOFC. *J. Eur. Ceram. Soc.* **2015**, *35*, 4475–4484. [CrossRef]

31. Taskonak, B.; Mecholsky, J.J.; Anusavice, K.J. Residual stresses in bilayer dental ceramics. *Biomaterials* **2005**, *26*, 3235–3241. [CrossRef] [PubMed]

32. Fischer, H.; Hemelik, M.; Telle, R.; Marx, R. Influence of annealing temperature on the strength of dental glass ceramic materials. *Dent. Mater.* **2005**, *21*, 671–677. [CrossRef] [PubMed]

33. Anunmana, C.; Anusavice, K.J.; Mecholsky, J.J. Residual stress in glass: Indentation crack and fractography approaches. *Dent. Mater.* **2009**, *25*, 1453–1458. [CrossRef] [PubMed]

34. Choi, J.E.; Waddell, J.N.; Swain, M.V. Pressed ceramics onto zirconia. Part 2: Indentation fracture and influence of cooling rate on residual stresses. *Dent. Mater.* **2011**, *27*, 1111–1118. [CrossRef] [PubMed]

35. Baldassarri, M.; Stappert, C.F.J.; Wolff, M.S.; Thompson, V.P.; Zhang, Y. Residual stresses in porcelain-veneered zirconia prostheses. *Dent. Mater.* **2012**, *28*, 873–879. [CrossRef] [PubMed]

36. Al-Amleh, B.; Neil Waddell, J.; Lyons, K.; Swain, M.V. Influence of veneering porcelain thickness and cooling rate on residual stresses in zirconia molar crowns. *Dent. Mater.* **2014**, *30*, 271–280. [CrossRef] [PubMed]

37. Wendler, M.; Belli, R.; Petschelt, A.; Lohbauer, U. Characterization of residual stresses in zirconia veneered bilayers assessed via sharp and blunt indentation. *Dent. Mater.* **2015**, *31*, 948–957. [CrossRef] [PubMed]

38. Pfeiffer, W.; Frey, T. Strengthening of ceramics by shot peening. *J. Eur. Ceram. Soc.* **2006**, *26*, 2639–2645. [CrossRef]

39. Johnson, K.L. *Contact Mechanics*; Cambridge University Press: Cambridge, UK, 1985.

40. Newman, J.C.; Raju, I.S. An empirical stress-intensity factor equation for the surface crack. *Eng. Fract. Mech.* **1981**, *15*, 185–192. [CrossRef]

41. Anderson, T.L. *Fracture Mechanics: Fundamentals and Applications*; CRC Press: Boca Raton, FL, USA, 2005.

42. Marshall, D.B.; Lawn, B.R. An indentation technique for measuring stresses in tempered glass surfaces. *J. Am. Ceram. Soc.* **1977**, *60*, 86–87. [CrossRef]

43. Swain, M.V.; Hagan, J.T.; Field, J.E. Determination of the surface residual stresses in tempered glasses by indentation fracture mechanics. *J. Mater. Sci.* **1977**, *12*, 1914–1917. [CrossRef]

44. Lawn, B.R.; Evans, A.G.; Marshall, D.B. Elastic/plastic indentation damage in ceramics: The median/radial crack system. *J. Am. Ceram. Soc.* **1980**, *63*, 574–581. [CrossRef]

45. Rickhey, F.; Lee, J.H.; Lee, H. XFEM investigation on Knoop indentation cracking: Fracture toughness and aspect-ratio of radial-median cracks. *Mater. Des.* **2016**, *107*, 393–405. [CrossRef]

46. *Abaqus User's Manual-Version 6.14*, Dassault Systems Simulia Corp: Providence, RI, USA, 2014.

47. Francois, P.; Lefebvre, A.; Vanderschaeve, G. Low temperature plasticity of brittle materials. A new device for compressive testing under confining pressure. *Phys. Status Solidi* **1988**, *109*, 187–192. [CrossRef]

48. Lee, J.H.; Gao, Y.F.; Johanns, K.E.; Pharr, G.M. Cohesive interface simulations of indentation cracking as a fracture toughness measurement method for brittle materials. *Acta Mater.* **2012**, *60*, 5448–5467. [CrossRef]

49. Sines, G.; Carlson, R. Hardness measurements for determination of residual stresses. *ASTM Bull.* **1952**, *180*, 35–37.

50. Suresh, S.; Giannakopoulos, A.E. A new method for estimating residual stresses by instrumented sharp indentation. *Acta Mater.* **1998**, *46*, 5755–5767. [CrossRef]

51. Cuadrado, N.; Seuba, J.; Casellas, D.; Anglada, M.; Jiménez-Piqué, E. Geometry of nanoindentation cube-corner cracks observed by FIB tomography: Implication for fracture resistance estimation. *J. Eur. Ceram. Soc.* **2015**, *35*, 2949–2955. [CrossRef]

materials

Article

Effect of the Elastic Deformation of a Point-Sharp Indenter on Nanoindentation Behavior

Takashi Akatsu [1,2,*], Shingo Numata [2], Yutaka Shinoda [2] and Fumihiro Wakai [2]

[1] Faculty of Art and Regional Design, Saga University, 1 Honjo-machi, Saga 840-8502, Japan
[2] Laboratory for Materials and Structures, Institute of Innovative Research, Tokyo Institute of Technology, R3-24 4259 Nagatsuta, Midori, Yokohama 226-8503, Japan; numata.s.zz@m.titech.ac.jp (S.N.); shinoda.y.ac@m.titech.ac.jp (Y.S.); wakai.f.aa@m.titech.ac.jp (F.W.)
* Correspondence: akatsu@cc.saga-u.ac.jp; Tel.: +81-952-28-8667

Academic Editors: Ting Tsui and Matt Pharr
Received: 25 January 2017; Accepted: 2 March 2017; Published: 7 March 2017

Abstract: The effect of the elastic deformation of a point-sharp indenter on the relationship between the indentation load P and penetration depth h (P-h curve) is examined through the numerical analysis of conical indentations simulated with the finite element method. The elastic deformation appears as a decrease in the inclined face angle β, which is determined as a function of the elastic modulus of the indenter, the parabolic coefficient of the P-h loading curve and relative residual depth, regardless of h. This indicates that nominal indentations made using an elastic indenter are physically equivalent to indentations made using a rigid indenter with the decreased β. The P-h curves for a rigid indenter with the decreased β can be estimated from the nominal P-h curves obtained with an elastic indenter by using a procedure proposed in this study. The elastic modulus, yield stress, and indentation hardness can be correctly derived from the estimated P-h curves.

Keywords: nanoindentaion; elastic deformation; finite element method; numerical analysis

---

1. Introduction

Nanoindentation is a form of mechanical testing characterized as a depth-sensing indentation [1] to evaluate local mechanical properties through the analysis of the indentation load P versus the penetration depth h (P-h curve, hereafter). The analysis is principally based on a geometrical definition in which the indentation is carried out on a flat surface using an indenter geometrically defined such as flat-ended, spherical, ellipsoidal, point-sharp (e.g., conical, Berkovich, Vickers, cube corner, etc.). The point-sharp indentation has an advantage in local mechanical testing owes to the analytical simplicity for the geometrical similarity [2].

The bluntness of the indenter tip is one of the inevitable problems of undesirable tip geometry, especially for the point-sharp indentations, because it is impossible to make an ideally sharp indenter. The degree of the bluntness of a point-sharp indenter has been expressed in terms of the radius of curvature at the tip [3–5], but the actual geometry of a blunt tip is not guaranteed to be spherical. An area function [6,7] which gives the projected contact area at the maximum indentation load is another approach to express the bluntness of a point-sharp indenter, but the area function is theoretically valid only for hardness evaluation. A truncated tip which represents a blunt tip in an extremely poor situation [8] is a suitable model for a strict discussion on the effect of the tip bluntness on indentation behavior. According to the appendix of this paper, where a truncated tip is considered, the undesirable effect of the bluntness of a point-sharp indenter can be removed out simply if the P-h curve is shifted with Δh_{tip} in the h direction for indentations deeper than $2\Delta h_{tip}$, where Δh_{tip} is the distance between ideally sharp and blunt tips (see Figures A1–A4). In addition, Δh_{tip} can be estimated through an extrapolation of the linear relationship between h and P^{***} observed in the large P and h region to $P = 0$ (see Figure A5).

The elastic deformation of an actual point-sharp indenter, which has been conventionally taken into account on the basis of Hertzian contact [6]; Hertzian contact was basically used for spherical indentations as a modification of the elastic modulus evaluation. It is also an inevitable problem of undesirable tip geometry, especially for indentations on a very hard material, and there is still some controversy whether the modification based on the Hertzian contact can be applied to point-sharp indentations. Moreover, there are no reports on the modifications of the indenter elastic deformation for other mechanical properties such as the indentation hardness or yield stress. The geometrical changes of a point-sharp indenter due to elastic deformation should be considered when evaluating local mechanical properties with the nanoindentation technique.

In this paper, the effect of the elastic deformation of a point-sharp indenter on a *P-h* curve is quantified in a numerical analysis of conical nanoindentation behaviors simulated with the finite element method (FEM). In addition, a procedure of deriving physically meaningful *P-h* curves, which should be utilized for mechanical property evaluation. Finally, the validity and accuracy of this method is examined.

2. FEM Simulation of Nanoindentation

A conical indentation on a cylindrical elastoplastic solid was modeled in order to avoid the difficulty of modeling a pyramidal indenter widely used for actual nanoindentations. The FEM simulation exploited the large strain elastoplastic capability of ABAQUS code (Version 5.8.1) in the same way as reported in the literature [9,10]. Indentation contact was simulated by the use of elastic cone indenters with two different inclined face angles β (19.7° and 30°). Young's modulus of the elastic indenter was in the range of 300–1140 GPa. The finite-element mesh in the elastic indenter with β of 19.7° was composed of 775 4-node quadrilateral axisymmetric elements with 2443 nodes. The elastic indenter with β of 30° had 704 elements with 2258 nodes.

The FEM simulation used elastoplastic linear strain hardening rules, i.e., $\sigma = E\varepsilon$ for $\sigma < Y$, and $\sigma = Y + E_p\varepsilon_p$ for $\sigma \geq Y$, where σ is the stress, E the Young's modulus and ε the strain. Here, Y is the yield stress and E_p ($\equiv d\sigma/d\varepsilon_p$) is the plastic strain hardening modulus, where $d\sigma$, $d\varepsilon$, $d\varepsilon_e$, and $d\varepsilon_p$ are, respectively, the incremental values of stress, total, elastic, and plastic strains. Indentations were simulated for E, Y and E_p ranges of 50–1000 GPa, 0.1–60 GPa, and 0–200 GPa, respectively. The von Mises criterion with isotropic hardening was used to determine the onset of yielding flow.

3. Results and Discussion

A quadratic relationship between P and h on loading is theoretically guaranteed for a point-sharp indentation on the flat surface of a homogeneous elastoplastic solid [11,12]. The quadratic relationship was also observed in simulated *P-h* curves made with an elastic cone indenter. This indicates that the elastic deformation of a cone indenter can be described as a decrease in β determined regardless of h. Therefore, nominal indentations made with an elastic cone indenter with an original inclined face angle β_o should be physically equivalent to indentations made with a rigid cone indenter with the decreased inclined face angle β_d.

A nominal quadratic *P-h* relationship for an elastic cone indenter can be depicted as follows:

$$P = k_{1n}h^2 \text{ for loading,} \tag{1}$$

where k_{1n} is the nominal indentation loading parameter. Here, h in Equation (1) is the nominal penetration depth because the decrease in β from β_o to β_d, due to the elastic deformation of a cone indenter, gives a decrease in real penetration depth. Thus, a physically meaningful *P-h* relationship can be written with a true indentation loading parameter k_1, which should be observed in a *P-h* loading curve using a rigid cone indenter with β_d as

$$P_{max} = k_1(h_{max} - \Delta h_d)^2, \tag{2}$$

where Δh_d is the decrease in h at the maximum penetration depth h_{max} due to the elastic deformation of a cone indenter (see Figure 1). The combination of Equations (1) and (2) leads to the equation:

$$k_1 = k_{1n}\left(1 - \frac{\Delta h_d}{h_{max}}\right)^{-2}. \tag{3}$$

This means that $\Delta h_d/h_{max}$ is a key parameter to estimating k_1 from the nominal k_{1n}. In other words, $\Delta h_d/h_{max}$ can be simulated as

$$\frac{\Delta h_d}{h_{max}} = 1 - \sqrt{\frac{k_{1n}}{k_1}}, \tag{3'}$$

where k_{1n} in Equation (3') is observed in a simulated P-h loading curve with an elastic cone indenter and k_1 in Equation (3') is evaluated with the mechanical properties inputted into the FEM model [9,10]. In the following paragraph, the effect of $\Delta h_d/h_{max}$ on a P-h curve is examined quantitatively through numerical analysis.

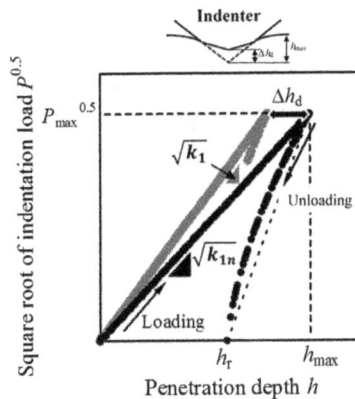

Figure 1. Schematic illustration of the effect of indenter elastic deformation on a P-h curve.

In addition to k_1, the relative residual penetration depth ζ, defined as h_r/h_{max}, where h_r is the residual penetration depth, characterizes a P-h curve and nominally decreased by the elastic deformation of a cone indenter to be ζ_n. A true ζ-value, which should be observed in a P-h curve using a rigid cone indenter with β_d, can also be evaluated with the mechanical properties inputted into the FEM model [9,10]. The numerical analysis revealed that the evaluated ζ can be correlated with the nominal ζ_n as a function of $\Delta h_d/h_{max}$

$$\zeta = \zeta_n\left\{1 - \left(\frac{\Delta h_d}{h_{max}}\right)^{0.85}\right\}^{-0.50}. \tag{4}$$

Figure 2 plots ζ estimated with Equation (4) and ζ_n against the true ζ evaluated with the mechanical properties inputted into the FEM model [9,10]. The results indicate the validity of using Equation (4) to estimate ζ from the nominal ζ_n and $\Delta h_d/h_{max}$. In addition, it is confirmed that ζ_n is smaller than ζ because of the overestimation of the penetration depth h due to the elastic deformation of the indenter. Moreover, a true indentation unloading parameter k_2 defined as $P_{max}/(h_{max} - h_r)^*$, which should be observed in an *P-h* unloading curve using a rigid cone indenter with β_d, can be estimated from a simulated *P-h* curve with an elastic cone indenter characterized by k_{1n} and ζ_n using Equations (3) and (4) as

$$k_2 = \frac{k_1}{(1-\zeta)^2} = k_{1n}\left(1 - \frac{\Delta h_d}{h_{max}}\right)^{-2}\left[1 - \zeta_n\left\{1 - \left(\frac{\Delta h_d}{h_{max}}\right)^{0.85}\right\}^{-0.50}\right]^{-2}. \tag{5}$$

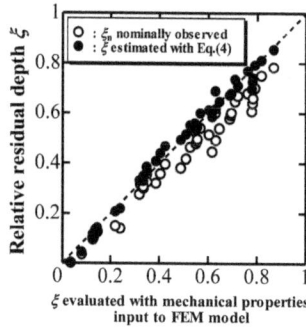

Figure 2. Relative residual depth ζ estimated with Equation (4) and ζ_n nominally observed plotted against ζ evaluated with mechanical properties inputted into the FEM model.

Figure 3 plots the estimated k_2 with Equation (5) as well as the nominal indentation unloading parameter k_{2n} determined from a simulated *P-h* curve with an elastic cone indenter against the true k_2 evaluated with the mechanical properties inputted into the FEM model [9,10]. Figure 3 indicates that k_2 can be estimated correctly by using Equation (5) with $\Delta h_d/h_{max}$, and that the nominal k_{2n} is quite far from k_2 owing to the overestimation of *h*.

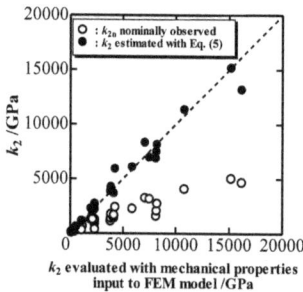

Figure 3. Indentation unloading parameter k_2 simulated and k_{2n} nominally observed plotted against k_2 evaluated with mechanical properties inputted into the FEM model.

The numerical analysis also revealed that $\Delta h_d/h_{max}$ is determined to be

$$\frac{\Delta h_d}{h_{max}} = 0.616\left\{\frac{k_{1n}}{E_i'\left(1 + \zeta_n^{1.5}\right)}\right\}^{0.84}, \tag{6}$$

where E_i' is defined as $E_i/(1 - v_i^*)$ and E_i and v_i are Young's modulus and Poisson's ratio of an elastic indenter, respectively. Figure 4 plots $\Delta h_d/h_{max}$ estimated with Equation (6) against $\Delta h_d/h_{max}$

evaluated with Equation (3′). Figure 3 indicates that $\Delta h_d/h_{max}$ can be estimated by using Equation (6) with a nominally observed P-h curve characterized by k_{1n} and ζ_n, and with the elastic properties of an elastic indenter characterized by E_i and ν_i.

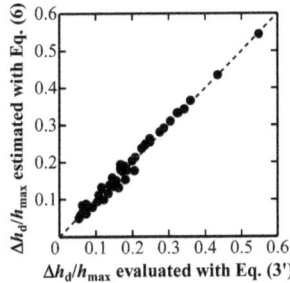

Figure 4. Degree of elastic deformation of conical indenter $\Delta h_d/h_{max}$ estimated with Equation (6) plotted against $\Delta h_d/h_{max}$ evaluated with Equation (3′).

In order to estimate mechanical properties from a P-h curve characterized with k_1, k_2 and ζ, we should know the inclined face angle β_d of the elastically deformed indenter. Numerical analysis revealed that β_d is given as a function of β_o, ζ_n and $\Delta h_d/h_{max}$

$$\frac{\tan \beta_d}{\tan \beta_o} = 1 - \left(1 - \zeta_n^{0.80}\right)^{0.83} \left(\frac{\Delta h_d}{h_{max}}\right)^{0.90}. \tag{7}$$

Figure 5 plots $\frac{\tan \beta_d}{\tan \beta_o}$ estimated with Equation (7) against $\frac{\tan \beta_d}{\tan \beta_o}$ observed in a simulated nanoindentation, and indicates the validity to estimate the inclined face angle β_d of the elastically deformed indenter with $\Delta h_d/h_{max}$.

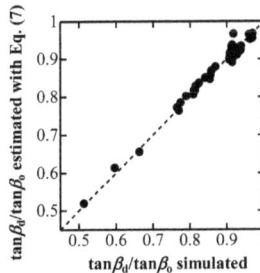

Figure 5. Degree of change in inclined face angle $\tan\beta_d/\tan\beta_o$ estimated with Equation (7) plotted against $\tan\beta_d/\tan\beta_o$ observed in simulated nanoindentation.

The representative indentation elastic modulus E^*, defined as $E^* = \dfrac{E}{1-\left(\nu-0.225\tan^{1.05}\beta_d\right)^2}$ in terms of β_d [9], can be estimated from the simulated P-h curve using Equations (3)–(7) if we know E_i and ν_i, whereas it is evaluated with E and ν inputted into the FEM model and with β_d observed in simulated nanoindentations. Figure 6 plots the estimated E^* (black circles) against the evaluated E^*. The white circles are E^*_n estimated with the nominal values of k_{2n} and ζ_n [9], which means that the elastic deformation of an indenter is not modified for the estimation of E^*. Figure 6 indicates the modification of the elastic deformation of an indenter can determine E^* correctly. The underestimation of E^* without the modification (the white circles in Figure 6) is caused by the overestimation of the elastic rebound during the unloading process because the extrinsic elastic deformation of the indenter is added to the intrinsic elastic deformation of the indented material.

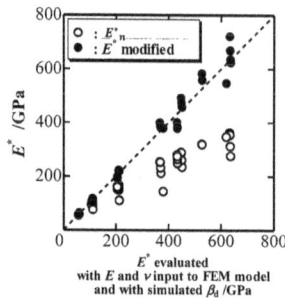

Figure 6. Representative indentation elastic modulus E^* estimated with simulated P-h curves plotted against E^* evaluated with mechanical properties inputted into the FEM model.

The representative indentation yield stress Y^*, defined as $Y^* = \dfrac{Y+0.25E_p\tan\beta_d}{1-\left(\nu-0.225\tan^{1.05}\beta_d\right)}$ in terms of β_d [10], can also be estimated using a simulated P-h curve and Equations (3)–(7) if we know E_i and ν_i. Moreover, it can be evaluated with Y, E_p and ν inputted into the FEM model and with β_d of the simulated indentation. Figure 7 plots the estimated Y^* (black circles) against the evaluated Y^*. Y^*_n estimated with E^*_n, ζ_n and β_o is plotted for comparison. This figure shows that the modification of the elastic deformation of an indenter more or less correctly estimates Y^* although the difference between the modified and unmodified Y^* is not so large with respect to the difference observed in E^* (see Figure 6). A relatively large difference in Y^* is typically found in the range of ζ less than 0.1, where plastic deformation is not dominant. The small difference observed in Figure 7 is attributed to the decrease in k_{1n} and ζ_n due to elastic deformation of an elastic indenter, where the former decreases Y^* nominally while the latter increases Y^* apparently.

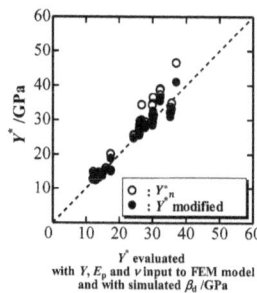

Figure 7. Representative indentation yield stress Y^* estimated with simulated P-h curves plotted against Y^* evaluated with mechanical properties inputted into the FEM model.

A previous study on the indentation hardness H_M found that it can be evaluated with the mechanical properties inputted into the FEM model and with the simulated β_d [9,10]. On the other hand, H_M can be estimated from a true P-h curve characterized with k_1, k_2 and ζ. Figure 8 plots the estimated H_M (black circles) against the evaluated H_M. The nominal H_{Mn} estimated with the nominal P-h curve is plotted as white circles in Figure 8, and a comparison reveals that the modification more or less correctly estimates H_M, although the difference between the estimated H_M and the nominal H_{Mn} is not so large with respect to the difference observed in E^* (see Figure 6). The difference is rather high in the large H_M region, where elastic deformation of the indenter is most severe. The small difference observed in Figure 8 owes to the decrease in k_{1n} and ζ_n due to elastic deformation of an elastic indenter, where the former decreases H_M nominally while the latter increases H_M apparently through the decrease of nominal contact depth.

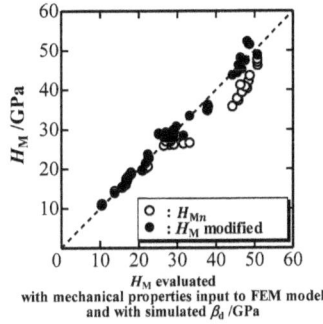

Figure 8. Indentation hardness H_M estimated with simulated *P-h* curves plotted against H_M evaluated with mechanical properties inputted into the FEM model and with simulated β_d.

We conducted nanoindentation experiments and reported E^*, Y^* and H_M for several materials [9,10]. These values were evaluated with modified *P-h* curves (see Figure 9) considering elastic deformation of a diamond indenter with Young's modulus and a Poisson's ratio of 1140 GPa and 0.07, respectively. Table 1 shows these mechanical properties as well as those evaluated with a nominal *P-h* curve made without considering any elastic deformation of the indenter. $\Delta h_d / h_{max}$ and β_d estimated with the numerical analysis developed in this study are shown in order to examine the degree of the elastic deformation of the indenter. $\Delta h_d / h_{max}$ and the change in β ($\Delta \beta = \beta_o - \beta_d$) are large for relatively hard materials (e.g., fused silica and alumina), which would cause a large elastic deformation of the indenter. Even in that case, the changes in Y^* and H_M due to the elastic deformation of the indenter are not so large. In contrast, the change in E^* is so large that it cannot be ignored. The underestimation of E^* without the modification is caused by the overestimation of the elastic deformation during the unloading process because the extrinsic elastic deformation of the indenter is added to the intrinsic elastic deformation of the indented material.

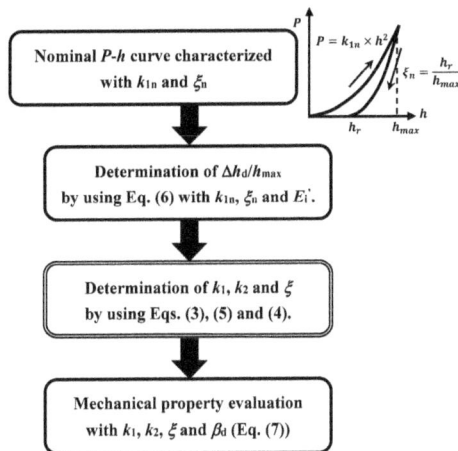

Figure 9. Flow chart of the procedure to evaluate mechanical properties.

Table 1. Effect of elastic deformation of diamond indenter on mechanical property evaluations.

Materials	$\Delta h_d/h_{max}$	β_d/β_0 (deg.)	k_2/k_{2n} (10^3 GPa)	ξ/ξ_n	E^* (GPa)	Y^* (MPa)	H_M (GPa)
						(Modified Value/Unmodified Value)	
Brass	0.020	19.6/19.7	7.65/4.78	0.930/0.913	102/81	584/597	1.26/1.29
Duralumin	0.020	19.6/19.7	4.37/3.03	0.909/0.893	77/64	605/618	1.31/1.33
Beryllium copper alloy	0.027	19.6/19.7	9.49/5.44	0.925/0.904	136/103	841/863	1.82/1.86
Fused silica	0.065	19.0/19.7	0.615/0.477	0.550/0.522	74/65	$5.21 \times 10^3 / 5.35 \times 10^3$	8.56/8.40
Alumina	0.158	18.3/19.7	4.90/2.25	0.690/0.614	340/225	$12.6 \times 10^3 / 12.9 \times 10^3$	24.3/22.8

According to Equation (6), the following equation can be derived

$$\frac{H_M}{E_i'} = \frac{\gamma^2}{g}\left(1 + \zeta_n^{1.5}\right)\left(\frac{\Delta h_d}{0.616 h_{max}}\right)^{1/0.84} \tag{8}$$

When indentation hardness is not affected much by the indenter elastic deformation, where γ is the surface profile parameter defined as $\gamma = h_{max}/h_c$, h_c is the contact depth, and g is the geometrical factor of a point-sharp indenter to be 24.5 for $\beta = 19.7°$. E_i' is required to be about 250 times larger than H_M for $\Delta h_d/h_{max}$ smaller than 0.05, where the effect of the indenter elastic deformation on a *P-h* curve may be ignored for indentations with Berkovich-type indenter.

4. Conclusions

The effect of the geometrical changes due to the elastic deformation of a point-sharp indenter was examined by conducting a numerical analysis of *P-h* curves simulated with FEM. The effect appears as a decrease in the inclined face angle β. The key parameter $\Delta h_d/h_{max}$, which can be utilized to derive the physically meaningful *P-h* curve and the decreased β, can be estimated with an equation derived by numerical analysis. The mechanical properties of indented materials, such as E^*, Y^* and H_M, can be estimated by using the *P-h* curve and β characterized by k_1, k_2 and ζ estimated with the key parameter $\Delta h_d/h_{max}$. The modification of a *P-h* curve and β with $\Delta h_d/h_{max}$ is most effective for the estimation of the accurate E^* with respect to Y^* and H_M.

Acknowledgments: Financial support for this research was provided in the form of a Grant-In-Aid for Scientific Research (C) (No. 18560652), Encouragement of Scientists (No. 13750622) and the Nippon Sheet Glass Foundation for Materials Science and Engineering.

Author Contributions: Takashi Akatsu conceived of the ideas for this research, performed numerical analysis on the simulated nanoindentations and wrote the paper. Shingo Numata and Yutaka Shinoda simulated nanoindentation. Fumihiro Wakai advised on the numerical analysis and reviewed the manuscript of the paper.

Conflicts of Interest: The authors declare no conflict of interest.

Appendix. Effect of the Tip Bluntness of a Point-Sharp Indenter on Nanoindentation Behavior

An indentation on a linearly elastic solid with a Young's modulus E and Poisson's ratio ν of 100 GPa and 0.499, respectively, was simulated using a rigid indenter with a truncated tip [8] in order to represent a tip in an extremely poor situation. The simulation was basically carried out in the same way described in the Section 2 of this paper. The inclined face angle β of the indenter was set to be 19.7°, which is the Vickers–Berkovich equivalent angle. The distance Δh_{tip} between ideally sharp and truncated tips (see Figure A1) was set to be 716 or 1790 nm with respect to the maximum penetration depth of 10 μm.

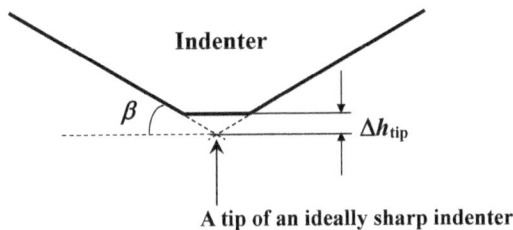

Figure A1. Schematic illustration of a truncated conical indenter tip.

The *P-h* curves shown in Figure A2 were obtained by simulating an indentation on a perfectly elastic body with a Poisson's ratio ν of ≈ 0.5 for a series of Δh_{tip}. We only examined a perfectly elastic body because the effect of bluntness would be most severe in this case. As shown in Figure A3, the curve deviates from that for an ideally sharp indenter ($\Delta h_{tip} = 0$) as Δh_{tip} increases and the amount of the shift in the h direction almost coincides with Δh_{tip}. The results in this figure indicate the effect of a blunt tip of a conical indenter on the P-h curve can be simply described as a change in the penetration depth with Δh_{tip}, especially in the large P and h region. The error in Young's modulus is evaluated as $\frac{|E_{P-h}-E_{inp}|}{E_{inp}}$ (expressed as a percentage), where E_{P-h} and E_{inp} are Young's modulus values obtained from the simulated P-h curve and inputted into an FEM model, respectively. Figure A4 plots this error as a function of penetration depth normalized by Δh_{tip}. It indicates that a reliable Young' modulus can be obtained with an error less than 10% from the P-h curve in the h region larger than $2\Delta h_{tip}$. In other words, P-h data for h-values shallower than $2\Delta h_{tip}$ should be omitted in order to obtain reliable mechanical properties. It may be possible to get more reliable P-h data in the shallow h region for other geometries, but the P-h data are actually obscure and depend on the unknown geometry of the indenter tip. We simulated a truncated tip of a conical indenter because it represents the worst case of a blunt point-sharp indenter; consequently, the P-h data obtained in the h region larger than $2\Delta h_{tip}$ guarantees accuracy in all cases. Figure A5, where h is plotted as a function of the square root of P, is an extrapolation of the linear relationship between h and P^{***} observed in the large P and h region to $P = 0$. It gives Δh_{tip} as an absolute value on the h-axis and is a way to estimate Δh_{tip} from the P-h curve. We have already reported an example of estimating Δh_{tip} from an experimental P-h curve [13]; Δh_{tip} is estimated to be about 50 nm for the Berkovich indenter that we actually used for the nanoindentation experiment.

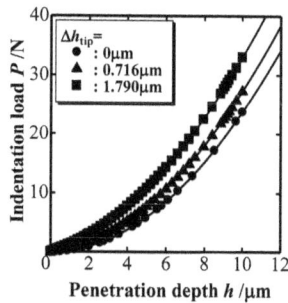

Figure A2. Indentation load P vs. penetration depth h curves obtained with truncated conical indenters.

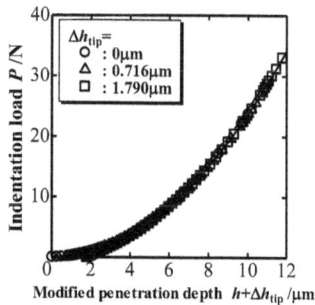

Figure A3. Indentation load P vs. modified penetration depth $h + \Delta h_{tip}$ curves obtained with truncated conical indenters.

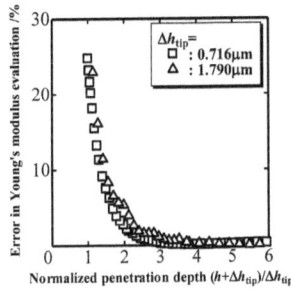

Figure A4. Error in Young's modulus evaluation with a truncated conical indenter as a function of normalized penetration depth $(h + \Delta h_{tip})/\Delta h_{tip}$.

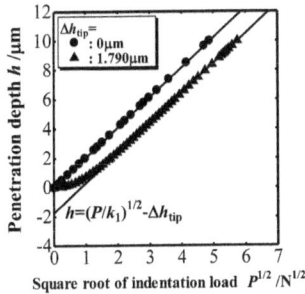

Figure A5. Penetration depth h vs. square root of indentation load P^{***} obtained with truncated indenters.

References

1. Mann, A.B.; Cammarata, R.C.; Nastasi, M.A. Nanoindentation: From angstroms to microns—Introduction. *J. Mater. Res.* **1999**, *14*, 2195. [CrossRef]
2. Sakai, M.; Shimizu, S.; Ishikawa, T. The indentation load-depth curve of ceramics. *J. Mater. Res.* **1999**, *14*, 1471–1484. [CrossRef]
3. Shih, C.W.; Yang, M.; Li, J.C.M. Effect of tip radius on nanoindentation. *J. Mater. Res.* **1991**, *6*, 2623–2628. [CrossRef]
4. Murakami, Y.; Tanaka, K.; Itokazu, M.; Shimamoto, A. Elastic analysis of triangular pyramidal indentation by the finite-element method and its application to nano-indentation measurement of glasses. *Philos. Mag. A* **1994**, *69*, 1131–1153. [CrossRef]
5. Yu, N.; Polycarpon, A.A.; Conry, T.F. Tip-radius effect in finite element modeling of sub-50 nm shallow nanoindentation. *Thin Solid Films* **2004**, *450*, 295–303. [CrossRef]
6. Oliver, W.C.; Pharr, G.M. An improved technique for determining hardness and elastic modulus using load and displacement sensing indentation experiments. *J. Mater. Res.* **1992**, *7*, 1564–1583. [CrossRef]
7. Knapp, J.A.; Follsteadt, D.M.; Myers, S.M.; Barbour, J.C.; Friedmann, T.A. Finite-element modeling of nanoindentation. *J. Appl. Phys.* **1999**, *85*, 1460–1474. [CrossRef]
8. Shimamoto, A.; Tanaka, K.; Akiyama, Y.; Yoshizaki, H. Nanoindentation of glass with a tip truncated Berkovich indenter. *Philos. Mag. A* **1996**, *74*, 1097–1105. [CrossRef]
9. Akatsu, T.; Numata, S.; Demura, T.; Shinoda, Y.; Wakai, F. Representative indentation elastic modulus evaluated by unloading of nanoindentation made with a point sharp indenter. *Mech. Mater.* **2015**, *83*, 66–71. [CrossRef]
10. Akatsu, T.; Numata, S.; Demura, T.; Shinoda, Y.; Wakai, F. Representative indentation yield stress evaluated by nanoindentation behavior made with a point sharp indenter. *Mech. Mater.* **2016**, *92*, 1–7. [CrossRef]

Materials **2017**, *10*, 270

11. Sakai, M. Energy principle of the indentation-induced inelastic surface deformation and hardness of brittle materials. *Acta Metall. Mater* **1993**, *41*, 1751–1758. [CrossRef]
12. Sakai, M. The Meyer hardness: A measure for plasticity? *J. Mater. Res.* **1999**, *14*, 3630–3639. [CrossRef]
13. Akatsu, T.; Numata, S.; Yoshida, M.; Shinoda, Y.; Wakai, F. Indentation size effect on the hardness of zirconia polycrystals. In *Fracture Mechanics of Ceramics (Volume 14 Active Materials, Nanoscale Materials, Composites, Glass and Fundamentals)*; Springer Science + Business Media Inc.: New York, NY, USA, 2005; pp. 13–20.

materials

MDPI

Communication

Micro-Mechanical Viscoelastic Properties of Crosslinked Hydrogels Using the Nano-Epsilon Dot Method

Giorgio Mattei [1,2,3], Ludovica Cacopardo [1,4] and Arti Ahluwalia [1,4,*]

1   Research Centre E. Piaggio, University of Pisa, Largo Lucio Lazzarino 1, 56122 Pisa, Italy;
    giorgio.mattei@centropiaggio.unipi.it (G.M.); ludovica.cacopardo@ing.unipi.it (L.C.)
2   Optics11 B.V., De Boelelaan 1081, 1081 HV Amsterdam, The Netherlands
3   Biophotonics & Medical Imaging and LaserLaB, VU University Amsterdam, De Boelelaan 1105,
    1081 HV Amsterdam, The Netherlands
4   Department of Information Engineering, University of Pisa, Via Girolamo Caruso 16, 56122 Pisa, Italy
*   Correspondence: arti.ahluwalia@unipi.it; Tel.: +39-05-0221-7050

Received: 6 July 2017; Accepted: 31 July 2017; Published: 2 August 2017

Abstract: Engineering materials that recapitulate pathophysiological mechanical properties of native tissues in vitro is of interest for the development of biomimetic organ models. To date, the majority of studies have focused on designing hydrogels for cell cultures which mimic native tissue stiffness or quasi-static elastic moduli through a variety of crosslinking strategies, while their viscoelastic (time-dependent) behavior has been largely ignored. To provide a more complete description of the biomechanical environment felt by cells, we focused on characterizing the micro-mechanical viscoelastic properties of crosslinked hydrogels at typical cell length scales. In particular, gelatin hydrogels crosslinked with different glutaraldehyde (GTA) concentrations were analyzed via nano-indentation tests using the nano-epsilon dot method. The experimental data were fitted to a Maxwell Standard Linear Solid model, showing that increasing GTA concentration results in increased instantaneous and equilibrium elastic moduli and in a higher characteristic relaxation time. Therefore, not only do gelatin hydrogels become stiffer with increasing crosslinker concentration (as reported in the literature), but there is also a concomitant change in their viscoelastic behavior towards a more elastic one. As the degree of crosslinking alters both the elastic and viscous behavior of hydrogels, caution should be taken when attributing cell response merely to substrate stiffness, as the two effects cannot be decoupled.

Keywords: nano-indentation; nano-epsilon dot method; strain rate; mechanical properties; viscoelastic models; soft materials; gelatin; glutaraldehyde

1. Introduction

In their native environment, cells are surrounded by the extracellular matrix (ECM), a complex network of glycosaminoglycans, adhesion proteins, and structural fibers, serving not only as a physical scaffold, but also providing biochemical and biomechanical cues that are critical for the regulation of cell adhesion, proliferation, differentiation, morphology, and gene expression [1]. Among them, the ECM's mechanical properties play a key role in directing cell fate and guiding pathophysiological cell behavior during tissue development, homeostasis, and disease [2–5]. Cells sense the mechanics of their surrounding environment (ECM) by gauging resistance to the traction forces they exert on it, and, in response, generate biochemical activity through a process known as mechano-transduction [6,7].

In the last few decades, several studies have focused on investigating the role of substrate elasticity (or stiffness) in cell mechano-transduction. A variety of biomaterials mimicking the native

Materials **2017**, *10*, 889

stiffness of different biological tissues have been proposed, particularly hydrogels (i.e., crosslinked three-dimensional (3D) networks of hydrophilic natural or synthetic polymers) [8]. Hydrogels have been widely used as cell culture substrates in mechano-transduction studies mostly because of their several advantages, such as high water content, biocompatibility, the availability of different crosslinking approaches, and high tunability, which allow recapitulation of the physicochemical and mechanical properties of native ECMs in vitro [9–13].

Studies on cell response to stiffness have significantly contributed to our understanding of cell mechano-transduction, designating substrate elasticity as a major determinant in the regulation of pathophysiological cell behavior and function [14,15]. For instance, stem cell commitment has been shown to depend on matrix elasticity [16], while tissue development, ageing, and disease progression are generally associated with tissue stiffening [17,18]. However, since native tissues [19,20] and hydrogels (e.g., gelatin [21,22], collagen [23], or fibrin [24]) typically exhibit viscoelastic behavior with stress-relaxation (i.e., a decrease in elastic modulus over time in response to a constant strain applied), focusing on stiffness only is generally an over-reductive way to describe their biomechanical properties. Moreover, culturing cells on primarily elastic substrates with constant (i.e., time-independent) elastic moduli is poorly representative of their native viscoelastic environment, where the resistance to the traction forces they exert is expected to relax over time due to flow and matrix remodeling [25].

There is thus a clear need to consider viscoelasticity when characterizing soft tissue biomechanics and developing biomaterials for cell culture and mechano-biology studies. To date, only a few studies have investigated the effect of substrate viscoelasticity on resultant cell behavior. Cameron and colleagues developed polyacrylamide gels with a shear loss modulus (G″, reflecting the viscous component of the material viscoelastic behavior) varying over two orders of magnitude (from 1 to 130 Pa) and a nearly constant shear storage modulus (G″ ~4.7 kPa, related to the elastic counterpart). In particular, increasing the substrate G″ led to increased spreading and proliferation of human mesenchymal stem cells (hMSCs), but decreased the size and maturity of their focal adhesions, possibly because of decreased cytoskeletal tension resulting from the dissipation of energy owing to inherent substrate creep. Another recent study from Chauduri et al. investigated cell spreading on either almost elastic (i.e., covalently crosslinked) or viscoelastic (i.e., ionically crosslinked) alginate gels [25]. Despite the current consensus that cell spreading and proliferation are suppressed on soft substrates, they reported that cells cultured on soft viscoelastic substrates behave differently than those cultured on elastic substrates with the same initial elastic modulus, increasing spreading and proliferation to a similar extent as that observed on the stiffer elastic substrate. Taken together, these results suggest that stress-relaxation can compensate for the effect of decreased stiffness and has a substantial impact on cell behavior and function.

In light of the above considerations, it is natural to start wondering what is the best method to derive "physiologically relevant" viscoelastic properties (i.e., those describing the biomechanical environment felt by cells in their native milieu) to develop better mechano-mimetic cell culture substrates for tissue engineering, in vitro models, and mechano-transduction studies. Indeed, different mechanical properties (i.e., parameter values used to describe a given material's mechanics) can be obtained when characterizing the same sample with different testing and analysis methods, likely leading to highly variable results that are difficult to interpret or not meaningfully comparable [18].

Among the available techniques, indentation testing with micron-sized probes is currently considered one of the most suitable methods for measuring a material's mechanical properties at typical cell length-scales [26]. It requires minimal sample preparation, and allows mechanical mapping at multiple locations (e.g., to characterize local gradients and heterogeneities), thus it is particularly suited for most soft tissues and biomaterials [27–29]. We suggested that an ideal testing method for deriving physiologically relevant mechanical properties should (i) not require initial force- or strain-triggers (unlike dynamic mechanical analysis or step response tests, such as creep and stress-relaxation); and (ii) involve quick measurements, in order to avoid sample pre-stress and minimize status alterations during testing, respectively. Moreover, mechanical properties should be derived in the physiological region of small deformations (e.g., the 0.01 ÷ 0.1 strain range, depending on the tissue of interest), and measurements should be performed at physiologically relevant strain rates/frequencies (e.g., a 0.001 ÷ 0.1 s^{-1} strain rate) [18,20,21]. In this context, we recently proposed the nano-epsilon dot method (nano-$\dot{\varepsilon}M$) to characterize the physiologically relevant micro-mechanical viscoelastic properties of soft tissues and (bio)materials through nano-indentation tests at different constant strain rates ($\dot{\varepsilon}$) [30]. Using data from the loading portion of the indentation curve and accurately identifying the initial point of contact, the nano-$\dot{\varepsilon}M$ allows for the derivation of "virgin" material viscoelastic properties (i.e., instantaneous and equilibrium elastic moduli as well as characteristic relaxation times) at typical cell length scales in the absence of pre-stress, unlike classical nano-indentation methods based on the analysis of the unloading curve (e.g., the Oliver–Pharr method [31,32]) or dynamic nano-indentation [33–35].

In this work, we used the nano-$\dot{\varepsilon}M$ to characterize the micro-mechanical viscoelastic properties of gelatin hydrogels. Gelatin is a low-cost, commercially available biomaterial derived from collagen, which is widely used as cell culture substrate mainly due to its inherent biocompatibility and bioactivity [36]. There are a variety of crosslinking strategies (e.g., chemical, enzymatic, and physical) available for improving its stability against enzymatic/hydrolytic degradation and tailoring its mechanical properties [37]. Glutaraldehyde (GTA) is one of the most widely used chemical crosslinking agents, particularly due to its highly efficient stabilization of collagenous materials through the reaction of free amino groups of lysine or the hydroxy-lysine amino acid residues of the polypeptide chains with its aldehyde groups [37,38]. Many studies have focused on characterizing the quasi-static elastic modulus (E) of GTA-crosslinked gelatin hydrogels, showing an increase in E with increasing GTA concentration [4,39]. However, as mentioned above, a single elastic modulus is an over-reductive way to describe the viscoelastic behavior of gelatin (and many other) hydrogels used in tissue engineering or cell culture applications. Therefore, the micro-mechanical viscoelastic properties of GTA-crosslinked gelatin hydrogels were characterized via nano-indentation tests, relating results to the crosslinker concentration.

2. Results

2.1. Apparent Elastic Moduli and Actual Sample Indentation Strain Rate

For all gelatin samples at different degrees of glutaraldehyde (GTA) crosslinking, experimental load-indentation (*P-h*) datasets collected at various constant theoretical strain rates ($\dot{\varepsilon}_t$) were converted into indentation stress-strain (σ_{ind}-ε_{ind}) according to the nano-$\dot{\varepsilon}M$ definitions [30], as outlined in Section 4.2. The linear viscoelastic region (LVR) was found to extend up to ε_{ind} = 0.05 for all samples and strain rates investigated (Figure 1).

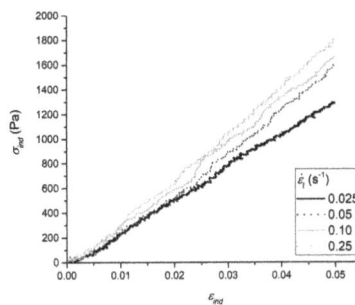

Figure 1. Examples of indentation stress-strain curves obtained testing 25 mM of GTA-crosslinked gelatin hydrogels. Sample viscoelasticity is reflected in the increase of apparent elastic modulus (i.e., stress versus strain slope) with increasing strain rate.

The actual sample indentation strain rates ($\dot{\varepsilon}_{ind}$) and strain rate-dependent "apparent" elastic moduli (E_{app}) obtained for the GTA-crosslinked gelatin hydrogels tested at different $\dot{\varepsilon}_t$ are summarized in Table 1.

Table 1. Actual indentation strain rates ($\dot{\varepsilon}_{ind}$) and apparent elastic moduli (E_{app}) obtained for GTA-crosslinked samples tested at different theoretical strain rates ($\dot{\varepsilon}_t$). Values are reported as mean $\pm$ standard error.

GTA (mM)	$\dot{\varepsilon}_t$ (s^{-1})	E_{app} (kPa)	$\dot{\varepsilon}_{ind}$ (s^{-1})
5	0.025	5.3 $\pm$ 0.3	0.021 $\pm$ 0.001
	0.05	9.3 $\pm$ 0.8	0.047 $\pm$ 0.001
	0.10	12.4 $\pm$ 0.6	0.070 $\pm$ 0.001
	0.25	17.3 $\pm$ 1.1	0.150 $\pm$ 0.001
25	0.025	27.5 $\pm$ 0.6	0.012 $\pm$ 0.001
	0.05	30.9 $\pm$ 2.1	0.024 $\pm$ 0.001
	0.10	35.2 $\pm$ 0.8	0.044 $\pm$ 0.001
	0.25	37.3 $\pm$ 0.9	0.124 $\pm$ 0.001
50	0.025	53.9 $\pm$ 0.8	0.008 $\pm$ 0.001
	0.05	57.8 $\pm$ 0.6	0.016 $\pm$ 0.001
	0.10	62.9 $\pm$ 0.2	0.031 $\pm$ 0.001
	0.25	65.3 $\pm$ 1.9	0.098 $\pm$ 0.001
100	0.025	76.7 $\pm$ 2.9	0.006 $\pm$ 0.001
	0.05	79.7 $\pm$ 1.3	0.013 $\pm$ 0.001
	0.10	83.0 $\pm$ 1.3	0.025 $\pm$ 0.001
	0.25	84.8 $\pm$ 1.1	0.067 $\pm$ 0.001

The apparent elastic modulus (E_{app}) was found to increase with both increasing GTA concentration and strain rate, as expected due to the higher molar ratio between GTA aldehydes and gelatin free amino groups involved in the hydrogel chemical crosslink and because of the rate-dependent behaviour exhibited by viscoelastic materials, respectively. Notably, the actual sample indentation strain rate ($\dot{\varepsilon}_{ind}$) was lower than the imposed theoretical indentation strain rate ($\dot{\varepsilon}_t$). The difference between $\dot{\varepsilon}_t$ and $\dot{\varepsilon}_{ind}$ increases with E_{app}, as outlined in Section 4.2. Briefly, for a given cantilever with stiffness k (constant in all experiments), the higher the sample E_{app}, the higher the cantilever deflection rate ($\dot{d}_c$). This results in a lower sample indentation rate ($\dot{h}$) with respect to the piezo z-displacement rate set by the user ($\dot{d}_p$), and consequently in a lower $\dot{\varepsilon}_{ind}$ with respect to the $\dot{\varepsilon}_t$.

2.2. Maxwell Standard Linear Solid (SLS) Lumped Viscoelastic Constants

The Maxwell SLS viscoelastic parameters estimated through the nano-$\dot{\varepsilon}M$ global fitting procedure (Section 4.3) are reported as a function of GTA concentration in Figure 2, where E_{inst} and E_{eq} represent the instantaneous (i.e., $E_0 + E_1$) and equilibrium (E_0) elastic moduli, respectively, while τ denotes the characteristic relaxation time calculated as η_1 / E_1.

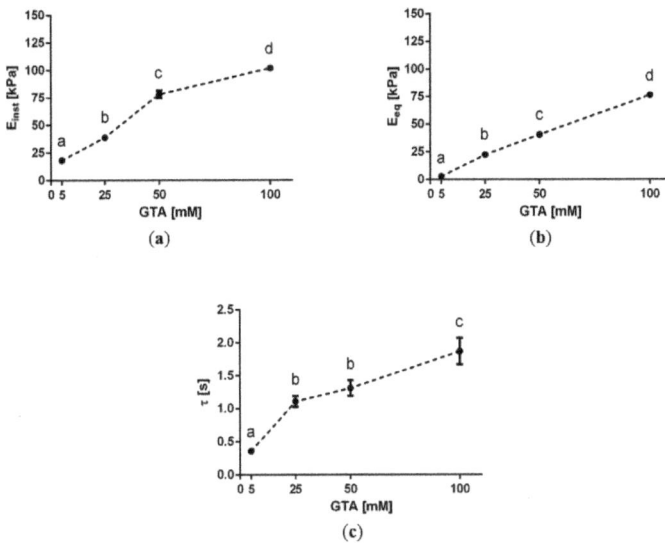

Figure 2. (**a**) Instantaneous (E_{inst}) and (**b**) equilibrium (E_{eq}) elastic moduli as well as (**c**) characteristic relaxation times (τ) as a function of glutaraldehyde (GTA) concentration obtained by globally fitting experimental nano-indentation stress-time data recorded at different constant strain rates to a Maxwell SLS lumped parameter model, as per the nano-$\dot{\varepsilon}M$. The error bars denote standard errors of estimation. Different letters indicate significant differences between samples (one-way ANOVA, $p < 0.05$), whereas the same letter means non-significant differences.

Both E_{inst} and E_{eq} significantly increased with GTA concentration ($p < 0.0001$). Moreover, a significant increase in τ. was also observed with increasing GTA concentration ($p < 0.001$), with the only exception between 25 and 50 mM GTA exhibiting statistically similar characteristic relaxation times ($p = 0.66$). Thus, the results obtained show that increasing the GTA concentration not only results in sample stiffening (reflected in the increased E_{inst} and E_{eq}), but also changes the viscoelastic behaviour from a more viscous towards a more elastic one, as indicated by the longer τ. This viscoelasticity shift is also reflected in the lower strain rate dependency of the E_{app} at higher GTA concentrations (Table 1).

3. Discussion

The results obtained in this study clearly show that characterizing sample stiffness only is generally an over-reductive way to describe the mechanical behaviour of GTA-crosslinked gelatin hydrogels. This consideration can be generalized to most soft biological tissues and hydrated biomaterials. We observed that gelatin hydrogels not only get stiffer with increasing GTA concentration, as demonstrated by increased instantaneous and equilibrium elastic moduli and as also expected from literature [4,39], but their viscoelastic behaviour concomitantly changes towards a more elastic one, as indicated by the longer relaxation time. The monotonic sample stiffening observed with increasing GTA concentration reflects an increase in the degree of crosslinking between gelatin free amino groups and GTA aldehydes. That the stiffness values do not reach a plateau suggests that the gelatin amines were not saturated [39].

Moreover, it is worth noting that the increase in elastic moduli with increasing GTA concentration obtained in this study using nano-indentation measurements (i.e., ~80 kPa in E_{inst} and ~70 kPa in E_{eq}, from 5 to 100 mM GTA) is significantly higher than the increase in bulk elastic modulus we have observed performing unconfined compression tests on gelatin samples prepared in the same manner (i.e., ~20 kPa, from 5 to 100 mM GTA) [4]. This could be due to a number of factors, including:

(i)   the scale-dependency of a sample's mechanical properties, i.e., surface micro-mechanical properties could be different from bulk volumetric ones [40];

(ii)   differences in testing and analysis methods, i.e., nano-indentation and unconfined compression techniques use different definitions of stress and strain, different models, etc., possibly affecting the mechanical properties obtained thereof [18]; and

(iii)   sample volumetric heterogeneity, i.e., GTA-crosslinking might be not uniform within the gelatin hydrogel volume due to the passive diffusion-reaction mechanism, which is established when submerging physically crosslinked gelatin hydrogels in GTA solution. This may lead to a highly crosslinked hydrogel shell and less crosslinked core, resulting in a lower increase of bulk mechanical properties with increasing GTA [41].

The above considerations should warn researchers that the mechanical properties of a material are likely dependent on the testing length scale as well as the experimental and analysis methods used to derive them [18].

In conclusion, since hydrogel crosslinking generally results in a concomitant alteration of both elastic and viscous mechanical behavior, caution should be taken when attributing cell response to stiffness in these materials as the two effects interact and cannot be decoupled.

To the best of our knowledge, this is the first report on the characterization of gelatin viscoelastic properties as a function of GTA concentration using nano-indentation, showing that crosslinking not only increases stiffness, but also gives rise to an increase in the relaxation time and hence a shift from viscoelastic towards a more elastic behaviour. This shift is also likely to occur in pathophysiological processes such as organ development and fibrosis, suggesting a need to consider viscoelastic properties when characterizing biological tissue mechanics or engineering biomaterials for cell cultures.

4. Materials and Methods

4.1. Sample Preparation

A 5% w/v gelatin solution was prepared by dissolving type A gelatin powder (G2500, Sigma-Aldrich, Milan, Italy) in $1\times$ phosphate buffered saline (PBS $1\times$, Sigma-Aldrich, Milan, Italy) at 50 °C under stirring for 2 h. Cylindrical gelatin samples with flat surfaces were obtained by casting the so-prepared solution in 8 mm height and 13 mm diameter molds, then leaving it to crosslink for 1 h at room temperature (RT). The physically gelled samples were removed from the molds and immersed for 48 h at 4 °C in glutaraldehyde (GTA; G5882, Sigma-Aldrich, Milan, Italy) crosslinking solutions prepared at different concentrations (i.e., 5, 25, 50, and 100 mM) in 40% v/v ethanol water solution, keeping a constant 5:1 volume ratio between the GTA solution and the gelatin samples to crosslink [4]. After chemical crosslinking, the samples were submerged in 0.5 M glycine solution (G7126, Sigma-Aldrich, Milan, Italy) for 2 h at RT to quench unreacted GTA, then copiously rinsed with deionized water. Finally, the samples were equilibrium swollen in PBS $1\times$ at RT to be in a stable and reproducible state for testing [4,18,20] (equilibrium weight was reached within 1 h, data not shown), and mechanically characterized as described in the following sections.

4.2. Nano-Indentation Measurements

Equilibrium swollen samples were glued onto the bottom of a Petri dish and tested at different constant indentation strain rates in PBS $1\times$ at RT according to the nano-$\dot{\varepsilon}M$ [30], using a displacement-controlled PIUMA Nanoindenter (Optics11 B.V., Amsterdam, The Netherlands). This instrument is based on a unique opto-mechanical ferrule-top cantilever force transducer operated by a z-axis piezoelectric motor, being very suited for nano-indentation measurements in liquids [42]. A probe having a 0.61 N/m cantilever stiffness (k) and a 70.5 μm radius (R) spherical tip was used in this study. The PIUMA Nanoindenter measures the indentation load as cantilever stiffness multiplied by its deflection resulting from pushing into the sample surface ($P = k \cdot d_c$), and calculates the actual indentation depth as the difference between the piezo z-axis displacement imparted to the probe and the resultant cantilever deflection ($h = d_p - d_c$). Thanks to the elastic nature of the cantilever, a constant piezo displacement rate results in a nearly constant "actual" sample indentation strain rate ($\dot{\varepsilon}_{ind}$) within the region of small deformation, as outlined in Mattei et al. [30]. Since only the piezo displacement rate ($\dot{d}_p$) can be set by the user, the $\dot{\varepsilon}_{ind}$ is generally lower than the "theoretical" strain rate ($\dot{\varepsilon}_t$) which would be obtained when using a cantilever with infinite stiffness (resulting in $d_c = 0$, $h = d_p$, and $\dot{h} = \dot{d}_p$), and depends on the cantilever-to-sample stiffness ratio. In particular, the higher the cantilever-to-sample stiffness ratio, the lower the cantilever deflection rate ($\dot{d}_c$), the closer the actual sample strain rate ($\dot{\varepsilon}_{ind}$) to the theoretical one ($\dot{\varepsilon}_t$). Notably, with an infinite stiffness cantilever no load (P) can be measured. In this study, the total piezo z-displacement rate ($\dot{d}_p$) was set to obtain $\dot{\varepsilon}_t = 0.025$, 0.05, 0.1, and 0.25 s^{-1} according to the following equation (Equation (1)) [30]:

$$\dot{d}_p = \frac{3}{4} \cdot \left(1 - v^2\right) \cdot R \cdot \dot{\varepsilon}_t \tag{1}$$

where v is Poisson's ratio, here assumed to be equal to 0.5 (i.e., incompressible material) for gelatin hydrogels [30,43,44]. Samples at different degrees of GTA-crosslinking were treated as mechanically isotropic and tested in triplicate ($n = 3$) performing $n = 10$ independent measurements per each strain rate investigated (i.e., a total of 40 tests per sample). Nano-indentation measurements were started out of sample contact and performed on different surface points (randomly selected) to avoid sample pre-stress and any eventual effect due to repeated testing cycles on the same spot, respectively.

4.3. Data Analyses and Viscoelastic Parameters Identification

Only data belonging to the loading portion of the load-indentation (P-h) curves measured at different $\dot{\varepsilon}_t$ were analyzed. The initial contact point was identified as the last one at which the load crosses the P-h abscissa towards monotonically increasing values [21,30]. Experimental P-h time data were offset to be zero in correspondence with this point. The load-indentation data were converted respectively into indentation stress (σ_{ind}) and strain (ε_{ind}) according to Equations (2) and (3) [30]:

$$\sigma_{ind} = \frac{P}{R \cdot \sqrt{hR}} \tag{2}$$

$$\varepsilon_{ind} = \frac{4}{3 \cdot (1 - v^2)} \cdot \frac{h}{R} \tag{3}$$

The linear viscoelastic region (LVR) was identified as the one in which σ_{ind} increases linearly with ε_{ind} ($R^2 > 0.99$), and strain-rate dependent "apparent" elastic moduli (E_{app}) were derived as the indentation stress-strain slope within the LVR. Then, the slope of the actual sample indentation strain rate ($\dot{\varepsilon}_{ind}$) was calculated as the slope of experimental strain (ε_{ind}) versus time (t) within the LVR, and used for the nano-$\dot{\varepsilon}M$ lumped viscoelastic parameters' identification [30]. In particular, the Maxwell Standard Linear Solid model (SLS) was chosen to represent the viscoelastic behavior of gelatin in this work [21,30,45,46]. The SLS model is the simplest form of the Generalized Maxwell lumped parameter model. It consists of a pure spring (E_0) assembled in parallel to a Maxwell arm (i.e., a spring E_1 in series with a dashpot η_1, defining a characteristic relaxation time $\tau_1 = \eta_1 / E_1$) [47], and exhibits the following stress-time response to a constant indentation strain rate input $\dot{\varepsilon}_{ind}$ (Equation (4)) [21,30]:

$$\sigma_{ind}(t) = \dot{\varepsilon}_{ind} \cdot \left(E_0 t + \eta_1 \left(1 - e^{-\frac{E_1}{\eta_1}t}\right)\right) \tag{4}$$

For each gelatin sample at a different GTA-crosslinking grade, experimental stress-time series within the LVR obtained at different indentation strain rates were globally fitted to Equation (4) for deriving the Maxwell SLS viscoelastic constants (i.e., E_0, E_1, and η_1). The global fitting procedure was implemented in OriginPro (OriginLab Corp., Northampton, MA, USA), performing chi-square minimization in a combined parameter space. In particular, for each set of stress-time data considered in the global fitting, the $\dot{\varepsilon}_{ind}$ value of the fitting equation (Equation (4)) was set to be equal to the actual one calculated from the experiments (i.e., ε_{ind} vs t slope), while the SLS viscoelastic constants to estimate were shared between datasets.

An annealing scheme based on multiplying and dividing each initial parameter guess by 10 while keeping the instantaneous modulus (i.e., $E_{inst} = E_0 + E_1$) at a constant value was adopted to obtain reliable and absolute SLS viscoelastic constant estimations, avoiding most of the local minima during the fitting procedure. Viscoelastic constants to estimate were constrained to be ≥ 0 to prevent the fitting procedure returning negative values.

4.4. Statistical Analyses

Results are reported as mean $\pm$ standard error (unless otherwise noted). Statistical differences between viscoelastic parameters of gelatin hydrogels at different GTA-crosslinking grades were tested using one-way ANOVA followed by Tukey's Multiple Comparison Test. Statistical analyses were performed in GraphPad Prism (GraphPad Software, San Diego, CA, USA), setting significance at $p < 0.05$.

Acknowledgments: This project has received funding from the European Union's Horizon 2020 research and innovation programme under the Marie Sklodowska-Curie grant agreement No. 705296 (ENDYVE). The authors are grateful to Antonio Jacopo Scardigno for his help in performing the experiments.

Author Contributions: G.M. and A.A. conceived and designed the experiments; G.M. and L.C. performed the experiments; G.M. and L.C. analyzed the data; A.A. contributed reagents/materials/analysis tools; G.M. and A.A. wrote the paper.

Conflicts of Interest: The authors declare no conflict of interest. G.M. is an employee of Optics11 B.V. The company had no role in the design of the study; in the collection, analyses, or interpretation of data; in the writing of the manuscript, and in the decision to publish the results.

References

1. Frantz, C.; Stewart, K.M.; Weaver, V.M. The extracellular matrix at a glance. *J. Cell Sci.* **2010**, *123*, 4195–4200. [CrossRef] [PubMed]
2. Mammoto, T.; Mammoto, A.; Ingber, D.E. Mechanobiology and developmental control. *Annu. Rev. Cell Dev. Biol.* **2013**, *29*, 27–61. [CrossRef] [PubMed]
3. Carver, W.; Goldsmith, E.C. Regulation of tissue fibrosis by the biomechanical environment. *Biomed. Res. Int.* **2013**, *2013*, 101979. [CrossRef] [PubMed]
4. Mattei, G.; Ferretti, C.; Tirella, A.; Ahluwalia, A.; Mattioli-Belmonte, M. Decoupling the role of stiffness from other hydroxyapatite signalling cues in periosteal derived stem cell differentiation. *Sci. Rep.* **2015**, *5*, 10778. [CrossRef] [PubMed]
5. Mattei, G.; Magliaro, C.; Giusti, S.; Ramachandran, S.D.; Heinz, S.; Braspenning, J.; Ahluwalia, A. On the adhesion-cohesion balance and oxygen consumption characteristics of liver organoids. *PLoS ONE* **2017**, *12*, e0173206. [CrossRef] [PubMed]
6. Ingber, D.E. Cellular mechanotransduction: Putting all the pieces together again. *FASEB J.* **2006**, *20*, 811–827. [CrossRef] [PubMed]
7. Vogel, V. Mechanotransduction involving multimodular proteins: Converting force into biochemical signals. *Annu. Rev. Biophys. Biomol. Struct.* **2006**, *35*, 459–488. [CrossRef] [PubMed]
8. De Volder, R.; Kong, H. Biomaterials for studies in cellular mechanotransduction. In *Mechanobiology of Cell-Cell and Cell-Matrix Interactions*; Springer: Boston, MA, USA, 2011; pp. 267–277.
9. Tibbitt, M.W.; Anseth, K.S. Hydrogels as extracellular matrix mimics for 3D cell culture. *Biotechnol. Bioeng.* **2009**, *103*, 655–663. [CrossRef] [PubMed]
10. Geckil, H.; Xu, F.; Zhang, X.; Moon, S.; Demirci, U. Engineering hydrogels as extracellular matrix mimics. *Nanomedicine* **2010**, *5*, 469–484. [CrossRef] [PubMed]
11. Chai, Q.; Jiao, Y.; Yu, X. Hydrogels for biomedical applications: Their characteristics and the mechanisms behind them. *Gels* **2017**, *3*, 6. [CrossRef]

12. Ahearne, M. Introduction to cell-hydrogel mechanosensing. *Interface Focus* **2014**, *4*, 20130038. [CrossRef] [PubMed]
13. Tirella, A.; La Marca, M.; Brace, L.-A.; Mattei, G.; Aylott, J.W.; Ahluwalia, A. Nano-in-Micro self-reporting hydrogel constructs. *J. Biomed. Nanotechnol.* **2015**, *11*, 1451–1460. [CrossRef] [PubMed]
14. Watt, F.M.; Huck, W.T.S. Role of the extracellular matrix in regulating stem cell fate. *Nat. Rev. Mol. Cell Biol.* **2013**, *14*, 467–473. [CrossRef] [PubMed]
15. Mason, B.N.; Califano, J.P.; Reinhart-King, C.A. Matrix stiffness: A regulator of cellular behavior and tissue formation. *Eng. Biomater. Regen. Med.* **2012**, *1*, 19–37.
16. Engler, A.J.; Sen, S.; Sweeney, H.L.; Discher, D.E. Matrix elasticity directs stem cell lineage specification. *Cell* **2006**, *126*, 677–689. [CrossRef] [PubMed]
17. Handorf, A.M.; Zhou, Y.; Halanski, M.A.; Li, W.-J. Tissue stiffness dictates development, homeostasis, and disease progression. *Organogenesis* **2015**, *11*, 1–15. [CrossRef] [PubMed]
18. Mattei, G.; Ahluwalia, A. Sample, testing and analysis variables affecting liver mechanical properties: A review. *Acta Biomater.* **2016**, *45*, 60–71. [CrossRef] [PubMed]
19. Fung, Y.C. *Biomechanics: Mechanical Properties of Living Tissues*, 2nd ed.; Springer: New York, NY, USA, 1993.
20. Mattei, G.; Tirella, A.; Gallone, G.; Ahluwalia, A. Viscoleastic characterisation of pig liver in unconfined compression. *J. Biomech.* **2013**, *47*, 2641–2646. [CrossRef] [PubMed]
21. Tirella, A.; Mattei, G.; Ahluwalia, A. Strain rate viscoelastic analysis of soft and highly hydrated biomaterials. *J. Biomed. Mater. Res. Part A* **2014**, *102*, 3352–3360. [CrossRef] [PubMed]
22. Martucci, J.F.; Ruseckaite, R.A.; Vázquez, A. Creep of glutaraldehyde-crosslinked gelatin films. *Mater. Sci. Eng. A* **2006**, *435–436*, 681–686. [CrossRef]
23. Knapp, D.M. Rheology of reconstituted type I collagen gel in confined compression. *J. Rheol.* **1997**, *41*, 971. [CrossRef]
24. Janmey, P.A. Rheology of Fibrin Clots. VI. Stress relaxation, creep, and differential dynamic modulus of fine clots in large shearing deformations. *J. Rheol.* **1983**, *27*, 135. [CrossRef]
25. Chaudhuri, O.; Gu, L.; Darnell, M.; Klumpers, D.; Bencherif, S.A.; Weaver, J.C.; Huebsch, N.; Mooney, D.J. Substrate stress relaxation regulates cell spreading. *Nat. Commun.* **2015**, *6*, 6364. [CrossRef] [PubMed]
26. Caliari, S.R.; Burdick, J.A. A practical guide to hydrogels for cell culture. *Nat. Methods* **2016**, *13*, 405–414. [CrossRef] [PubMed]
27. Oyen, M.L. Nanoindentation of biological and biomimetic materials. *Exp. Tech.* **2013**, *37*, 73–87. [CrossRef]
28. Oyen, M.L.; Cook, R.F. A practical guide for analysis of nanoindentation data. *J. Mech. Behav. Biomed. Mater.* **2009**, *2*, 396–407. [CrossRef] [PubMed]
29. Oyen, M.L. Mechanical characterisation of hydrogel materials. *Int. Mater. Rev.* **2013**, *59*, 44–59. [CrossRef]
30. Mattei, G.; Gruca, G.; Rijnveld, N.; Ahluwalia, A. The nano-epsilon dot method for strain rate viscoelastic characterisation of soft biomaterials by spherical nano-indentation. *J. Mech. Behav. Biomed. Mater.* **2015**, *50*, 150–159. [CrossRef] [PubMed]
31. Oliver, W.C.; Pharr, G.M. An improved technique for determining hardness and elastic modulus using load and displacement sensing indentation experiments. *J. Mater. Res.* **1992**, *7*, 1564–1583. [CrossRef]
32. Oliver, W.C.; Pharr, G.M. Measurement of hardness and elastic modulus by instrumented indentation: Advances in understanding and refinements to methodology. *J. Mater. Res.* **2004**, *19*, 3–20. [CrossRef]
33. Li, X.; Bhushan, B. A review of nanoindentation continuous stiffness measurement technique and its applications. *Mater. Charact.* **2002**, *48*, 11–36. [CrossRef]
34. Franke, O.; Göken, M.; Meyers, M.A.; Durst, K.; Hodge, A.M. Dynamic nanoindentation of articular porcine cartilage. *Mater. Sci. Eng. C* **2011**, *31*, 789–795. [CrossRef]
35. Hayes, S.A.; Goruppa, A.A.; Jones, F.R. Dynamic nanoindentation as a tool for the examination of polymeric materials. *J. Mater. Res.* **2011**, *19*, 3298–3306. [CrossRef]
36. Dawson, E.; Mapili, G.; Erickson, K.; Taqvi, S.; Roy, K. Biomaterials for stem cell differentiation. *Adv. Drug Deliv. Rev.* **2008**, *60*, 215–228. [CrossRef] [PubMed]
37. Rose, J.; Pacelli, S.; Haj, A.; Dua, H.; Hopkinson, A.; White, L.; Rose, F. Gelatin-Based materials in ocular tissue engineering. *Materials* **2014**, *7*, 3106–3135. [CrossRef]
38. Olde Damink, L.H.H.; Dijkstra, P.J.; Van Luyn, M.J.A.; Van Wachem, P.B.; Nieuwenhuis, P.; Feijen, J. Glutaraldehyde as a crosslinking agent for collagen-based biomaterials. *J. Mater. Sci. Mater. Med.* **1995**, *6*, 460–472. [CrossRef]

39. Bigi, A.; Cojazzi, G.; Panzavolta, S.; Rubini, K.; Roveri, N. Mechanical and thermal properties of gelatin films at different degrees of glutaraldehyde crosslinking. *Biomaterials* **2001**, *22*, 763–768. [CrossRef]

40. Kaufman, J.D.; Miller, G.J.; Morgan, E.F.; Klapperich, C.M. Time-dependent mechanical characterization of poly(2-hydroxyethyl methacrylate) hydrogels using nanoindentation and unconfined compression. *J. Mater. Res.* **2008**, *23*, 1472–1481. [CrossRef] [PubMed]

41. Orsi, G.; Fagnano, M.; De Maria, C.; Montemurro, F.; Vozzi, G. A new 3D concentration gradient maker and its application in building hydrogels with a 3D stiffness gradient. *J. Tissue Eng. Regen. Med.* **2017**, *11*, 256–264. [CrossRef] [PubMed]

42. Chavan, D.; Van De Watering, T.C.; Gruca, G.; Rector, J.H.; Heeck, K.; Slaman, M.; Iannuzzi, D. Ferrule-top nanoindenter: An optomechanical fiber sensor for nanoindentation. *Rev. Sci. Instrum.* **2012**, *83*. [CrossRef] [PubMed]

43. Zhang, X.; Qiang, B.; Greenleaf, J. Comparison of the surface wave method and the indentation method for measuring the elasticity of gelatin phantoms of different concentrations. *Ultrasonics* **2011**, *51*, 157–164. [CrossRef] [PubMed]

44. Czerner, M.; Fellay, L.S.; Suarez, M.P.; Frontini, P.M.; Fasce, L.A. Determination of elastic modulus of gelatin gels by indentation experiments. *Procedia Mater. Sci.* **2015**, *8*, 287–296. [CrossRef]

45. Clayton, E.H.; Okamoto, R.J.; Wilson, K.S.; Namani, R.; Bayly, P.V. Comparison of dynamic mechanical testing and mr elastography of biomaterials. *Appl. Imaging Tech. Mech. Mater. Struct.* **2012**, *4*, 143–150.

46. Pulieri, E.; Chiono, V.; Ciardelli, G.; Vozzi, G.; Ahluwalia, A.; Domenici, C.; Vozzi, F.; Giusti, P. Chitosan/gelatin blends for biomedical applications. *J. Biomed. Mater. Res. Part A* **2008**, *86*, 311–322. [CrossRef] [PubMed]

47. Roylance, D. Engineering viscoelasticity. *Dep. Mater. Sci. Eng. Inst. Technol. Camb. MA* **2001**, *2139*, 1–37.

MDPI AG

St. Alban-Anlage 66

4052 Basel, Switzerland

Tel. +41 61 683 77 34

Fax +41 61 302 89 18

http://www.mdpi.com

Materials Editorial Office

E-mail: materials@mdpi.com

http://www.mdpi.com/journal/materials

www.ingramcontent.com/pod-product-compliance
Lightning Source LLC
Chambersburg PA
CBHW051837210326
41597CB00033B/5685